# C AND THE 8051

## Volume II
## Building Efficient Applications

*Thomas W. Schultz*

*Associate Professor of*
*Electrical Engineering Technology*
*Purdue University*

Prentice Hall PTR
Upper Saddle River, New Jersey 07458
http://www.phptr.com

ISBN 0-13-521121-2
9 780135 211212
90000

Acquisitions editor: Bernard M. Goodwin
Cover designer: Design Source
Cover design director: Jerry Votta
Manufacturing manager: Alan Fischer
Marketing manager: Kaylie Smith
Compositor/Production services: Pine Tree Composition, Inc.


Prentice-Hall, Inc.
A Simon & Schuster Company
Upper Saddle River, New Jersey 07458

Prentice Hall books are widely used by corporations and government agencies for training, marketing, and resale.

The publisher offers discounts on this book when ordered in bulk quantities. For more information contact:

Corporate Sales Department
Phone: 800-382-3419
Fax: 201-236-7141
E-mail: corpsales@prenhall.com

Or write:

Prentice Hall PTR
Corp. Sales Dept.
One Lake Street
Upper Saddle River, New Jersey 07458



Printed in the United States of America
10 9 8 7 6 5 4 3

ISBN: 0-13-521121-2

Prentice-Hall International (UK) Limited, *London*
Prentice-Hall of Australia Pty. Limited, *Sydney*
Prentice-Hall Canada Inc., *Toronto*
Prentice-Hall Hispanoamericana, S.A., *Mexico*
Prentice-Hall of India Private Limited, *New Delhi*
Prentice-Hall of Japan, Inc., *Tokyo*
Simon & Schuster Asia Pte. Ltd., *Singapore*
Editora Prentice-Hall do Brasil, Ltda., *Rio de Janeiro*

# Contents

# Preface

Here I explain the general purpose and organization of this book. Specifically I define *efficiency* and identify the 8051 family of microcontrollers.

## EMBEDDED CONTROLLER APPLICATIONS

If you are developing embedded microcontroller projects of any sort, this is a book for you. The three objectives of this book are 1) to help you apply microcontrollers *efficiently,* 2) to help you develop the mindset of multitasking, and 3) to help you design software and hardware for embedded[1] applications. Over and over you will encounter an emphasis on efficiency and project development. While many examples rely on the *internal* peripheral devices of 8051 family devices, here you will also find applications for many commonly used *external* devices.

---

[1]I concentrate on the simpler *embedded* applications where the hardware is more likely switches and solenoids than disk drives and keyboards. An embedded *real-time* controller probably runs only one program for its entire life. It must be always ready to handle any number of inputs and outputs at the "same" time. A single controller can only do one thing at a time but it moves between the tasks so quickly that it appears to handle them simultaneously. While small microcontrollers do run more slowly, access much less memory, and seldom have elaborate math capability built in, they continue to be *much less expensive than the faster, bigger processors*! Instead of costing the hundreds of dollars of a Pentium-class processor, an 87C750, a complete computer in a 28-pin socket, costs perhaps four dollars plus the cost of a crystal. I do not want to detract from the importance of the high-power areas; rather, I illustrate the ideas and techniques that set the *embedded* controller field apart from the large-computer-system emphasis.

I hope you already have *Volume I (C and the 8051; Hardware, Modular Programming, and Multitasking).* Volume I is a better starting point if you are not familiar with microcontrollers or the 8051 family. There you will find more of the basics—this book instead emphasizes applications and project development.

## IS THIS BOOK FOR YOU?

If you are someone who learns best from examples rather than rules and lessons, you will enjoy this book.

***Engineers/developers:*** Engineer/system designer/developers producing small, embedded applications will find this book immediately applicable. You probably prefer to learn through practical hardware-oriented examples and you may be looking for new ways to replace existing hardware with a micro. This book describes a variety of new approaches you can store away for future projects.

***Students—classes:*** This book can *supplement* both beginning and advanced microcontroller classes. You might want to start with the general principles of microprocessors and computer architecture found in *Volume I.* After a quick overview of assembly language and C programming found there, you can return to this book as a resource for project ideas and electronic designs as well as a source of many additional programming examples.

***Students—design projects:*** If you already know micros well but are faced with an electronic design course,[2] this book can be a stand-alone guide to intelligently putting a micro to use in your project. Elegant, efficient design is far different from relying on raw computing power and speed. The application examples here can give you circuit ideas and schematics.

***Home Experimenter—self-taught:*** As an experimenter/hobbyist you want to see interesting applications running *right away*—not read about fine points of design. This book is not just a "designer's book." You needn't have advanced math and college training with microcontrollers as you would in some other engineering fields. You may already be familiar with the electronic devices on which many of the examples rely. You might want

---

[2]Many engineering and technology schools have added more emphasis on design because the accrediting bodies require it. Understanding the design and development process is important for many engineering jobs in industry—perhaps more important than the specific factual knowledge from other courses.

to start by getting one of the target boards described in *Volume I* and then go on to develop your own custom systems.

No matter which group you put yourself in, if you need a book of ideas along with enough specifics to get going without a lot of wasted effort, this is the book for you.

## PROGRAMMING

There is still a corner of the programming field where the efficient approaches developed with *small,* dedicated microcontrollers are best.[3] For example, you do not need an elaborate set of *print* functions in a system where the only outputs are a few LEDs and a motor! While in this volume I sometimes provide assembly language (machine instruction) program examples, most developers now focus on the C language. Along with the power of high-level languages like C, the elaborate software tools of today are a dramatic shift from the difficult programming techniques necessary for the first microprocessors.

## THE INCLUDED SOFTWARE

The CD included with *Volume I* included only Keil's demonstration tools—an assembler, a C compiler, and a linker. Those tools are included here as well, but in addition I have included *all* the code example source files of any appreciable size from *both* volumes, identified by the page on which they are found. With the free Keil tools (also available over the web) you can *compile* larger programs to check code size, but final code production is limited to 2K bytes.[4] While that is not a trivial program size, for large programs you will have to make an additional purchase. (Companies such as Keil still hope to make their living selling the software to industrial developers!) A full version is available for $1595.00. The compiler is among the best available for the 8051 family.[5] There *are* some less expensive compil-

---

[3]Like its predecessor, *C and the 8051: Programming for Multitasking*, and *Volume I* of this edition, the software examples in this book relate to embedded controllers rather than to general-purpose computers for data manipulation or graphics.

[4]I also discovered that the demo version forces the code to be located above 4000, making it useless for stand-alone one-chip systems—I believe they really want you to *buy* something from them!

[5]A significantly lower price can be negotiated for educational institutions—see Keil's Web page.

ers, but they lack features or behave much less efficiently with the (rather unusual) architecture of the 8051.

## THE AVAILABLE HARDWARE

Many of the examples in the book fit the microcontroller boards introduced in *Volume I.* The two boards used in many of those examples may be purchased in populated/tested form.[6] Most of the application sections begin with programs matched to the example board configurations. Unlike most magazine-article projects, with the software and information from this book you won't have to buy a pre-programmed EPROM for your projects—you'll be writing your own!

## WHAT THIS BOOK IS NOT

This book is not a book on computing.[7]

Unlike in *Volume I*, I don't emphasize assembly language programming here. In my opinion, with today's higher-level languages and optimizing compilers you should not stay with assembly language for any but the smallest projects. If you program for a living, save your time and reduce chances for human error by relying primarily on a higher-level language. Such benefits usually outweigh any potential assembly language efficiencies. With today's C compilers, assembly language programmers have to be quite skilled to even match the size and run-time speed of code produced by the knowledgeable C programmer.

This book is not a beginning electronics book. It *is* full of electronics design ideas. You need not be able to *invent* circuit designs, but to apply the circuit examples, you need at least a little knowledge of both digital and analog electronics. I supply many specific circuits along with advice on how

---

[6]I had hoped to include a blank microcontroller board in the back of every book, but economic realities reduced that to a schematic and parts list if you feel qualified to wire your own board. The original plan was to supply a copper pattern and offer bare boards so you could build exact replicas of the pre-built board. Surface mount parts in some of the designs made that inadvisable.

[7]While *Volume I* can be a tutorial on programming, it is still not the same as an introductory book covering computers and word processors. Likewise, these are not books on general programming.

to modify them, but you need to already have enough electronics background to understand the circuits as you read about them.

This book is not a code library. If you blindly copy code pieces from this book, you will be like a person walking in the dark along a narrow path with swamp on either side—the slightest misstep and you're stuck in the muck.

Finally, *this book is not perfect!*[8] I would love to say that all the code pieces have been tried and that all the hardware has been built, but that will probably not be true until I stop all the other things I do. View this book as a well of *ideas*—not a collection of guaranteed perfect examples to blindly copy.[9]

## HOW THIS BOOK CAME TO BE WRITTEN

For about twenty-three years, I have worked with dedicated microprocessor projects, and for about eighteen years I have been teaching in that area. About thirteen years ago, I began shaping a senior-level course in electrical engineering technology to cover real-time multitasking—first with Intel's RMX and eventually with their BITBUS and the 8044. I kept expecting to see a book appear on that processor family. Finally, in 1992 I revised class material to produce the first edition of this set, *C and the 8051; Programming for Multitasking*. It included PL/M as well as C and gave extensive coverage to real-time operating systems for the 8051.

Since PL/M dropped out of use and interest in commercial operating systems *for the 8051 family* failed to develop,[10] I revised the original material to become *Volume I* or *C and the 8051: Hardware, Modular Programming and Multitasking* that came out at the start of 1998. I had only intended to produce an updated version of the first edition, but as I gathered new ma-

---

[8]I make mistakes far more often than I would like, and the discoveries after the first printing of *Volume I* makes me wonder if I can do anything right! I am more of a broad-brush person than a detail person—if you find things that do not make sense, do not immediately assume the problem is with you.

[9]I might add that this book, *Volume II,* is not a coverage of the $I^2C$ bus or of development techniques and tools as I originally intended. First, I don't know enough yet to write a good book in those areas and second, I ran out of time, so I cut off after Chapter 7. Perhaps I'll be able to master that (and USB, I hope) enough to eventually have *Volume III*—I make no promises!

[10]Or to put it another way, those relatively few individuals who do use them do not need my help, and they could probably write a better book!

terial, it became evident that there was too much. To meet deadlines and produce a book more suitable for classroom use, I partitioned the material and released *Volume I.* There I added more on assembly language and cut back on the operating system information. That left all the "neat" cut-out stuff that I thought would be useful to advanced students and industrial designers lying on the floor. I have enhanced it and present it here as *Volume II.* This volume in particular has many more examples along with a *rationale for choices* in design. Students and readers of the first version of the book continued to cry for "examples to go to," and I hope this book will finally satisfy those requests.[11]

I am particularly grateful to my wife, Jill, who has consistently encouraged me. She has also unfailingly taken me to task over grammatical errors and helped me remove them. I am also grateful to Brian Witton and the folks with Keil Software for supplying the compiler and one of the training boards referred to throughout the book. Metalink loaned me an emulator to try, and it is the basis of some of the development tools materials.

If you would like to contact me by e-mail, my address is:

twschultz@tech.purdue.edu.

[11]I feel like some students just want to avoid independent thinking (or any thinking at all!), but the information explosion escalates every year, and who am I to fault a desire for efficiency just because it applies to learning! Perhaps the strong design emphasis of this new version will lure students and developers into further explorations on their own.

# Introduction

## WHAT IS AN "8051?"

Intel assigned the number 8051 to the single-chip microcontroller that followed their 8048—the first single-chip microcontroller on the market. The original version of the 8051 went on the market almost twenty years ago—prehistoric times for most electronic developments! Because of growing code size, many commercial applications no longer fit in the 4K bytes of on-chip code space in the original. The architecture and instruction set of the 8051 live on in what is now a large family of derivative microcontrollers having more code space, more timers, more interrupts, and more peripherals.[12] Going the other direction, Philips has developed derivatives that are smaller and less expensive. The wide base of development tools and experience with the 8051 family has made it the most common choice for middle-of-the-road projects needing embedded control.

[12]In the last few years, both Intel and Philips have introduced 16-bit microcontrollers specifically for upgrades from the 8051 family—either directly machine code compatible or assembly language compatible. Intel's 80C251 microcontroller retains code compatibility with the 8051 while adding additional features. This is the same strategy they used very successfully with the 8086, 286, 386, 486, and Pentium to capitalize on the installed software base. Philips has also introduced a 16-bit "upward compatible" 80C51XA which allows existing 8051 code to be translated into code for the new processor. They both acknowledge the huge installed base of 8051 devices and are trying to provide a smooth migration path for more demanding applications. If these controllers hang around another decade, perhaps I will get time to write another book!

Most programs presented here can run on *any* of the 8051 family. A few examples use features specific to only part of the family. Where small size and low cost are important, I love to focus on the 87C750 and 751. They add new meaning to the word efficient design! On a few occasions, I rely specifically on additional internal features of the 80C552. For example, some analog interfacing examples use on-chip AD or DA converters, but I supplement them with examples using traditional add-on devices. One section deals specifically with the $I^2C$ Bus devices. The basic ***8051 core*** remains the same—the differences relate to the built-in peripherals and most examples are applicable to the *entire* family.[13]

## MICROCONTROLLERS

The distinction between a micro*controller* and a general purpose computer or micro*processor* is not a sharp one. They all have a processor as part of the system to run the program—looking at each instruction to decide what to do, doing it, and going to the next instruction. To be useful, the processor must have memory and input/output (I/O) capability as well. If this is mostly on *one* chip, it is a micro*controller*. If, along with the *processor chip*,[14] you need several other chips for bus interface, memory, and I/O, the entire system is a *micro*computer.[15]

## OTHER MICROCONTROLLER FAMILIES

Today in terms of volume of units sold, most embedded applications rely on *8-bit* controllers—they give a high level of flexibility without high cost and the 8051 family is a leader in this area.[16] If you need *high-end* (high-speed)

---

[13]The exceptions to the "runs on any member" are the very small 750, 751, and 752, which, because of very small code space, do not support the long jumps and long calls. In C, it takes a special directive to the compiler to omit these unsupported instructions.

[14]A "chip" is another term for an integrated circuit (IC).

[15]A *mini*computer was a higher performance computer made up of even more, faster chips, but the distinction has faded as *micro*computers have increased in performance.

[16]I must add that the 68HC11 and the PIC (STAMP) processors are strong contenders in this market. Some of these processor cores are so common in low-to-medium performance applications that you can buy them as part of ASIC (Application Specific Integrated Circuit) design libraries much the same way as you could add a flip-flop or counter design to a custom integrated circuit.

capability, there seem to be new 16- and 32-bit controllers coming out yearly with no clear "winner" as of yet. For *low-end* (very high-volume, low speed) applications, there is still a place for 4-bit controllers such as National's COPS series. This is the extremely cost-conscious, high-volume area where the choice of a processor is decided by pennies of cost difference at the 100,000-piece level!

To try to cover all the different microcontrollers (68HC11, PIC, 8096, 68HC16, etc.) would make the examples very intimidating since the assembly languages differ and some hardware specifics change as well. *By including sections discussing the principles involved and providing flow charts before the examples, I try to keep the examples as broadly applicable as possible. The C programs, even allowing for the 8051 extensions, adapt easily to other microcontroller families if you have the corresponding C compiler.*

The add-on hardware also easily adapts to other processors, and most microcontrollers have similar internal timer and port features. Different commands may set up a timer, and interrupts may behave somewhat differently in another family, but the basic principles will remain the same. Some specific internal features may not be available in another controller, but you can often substitute equivalent functions.

## EMBEDDED MICROCONTROLLERS

What is an *embedded* microcontroller? Like a sliver *embedded* in your finger, an embedded microcontroller is not visible from the outside, but it has a significant effect. Embedded devices are the opposite of a PC—no one programs them after they arrive. In a microwave oven, for example, the embedded controller comes programmed "knowing" which output turns on the oven, which input comes from the door interlock, and so on. Day after day it runs the same program—the chip in the microwave will never control automobile ignition or play a video game because it is pre-programmed. You could program the same *type* of chip for those other jobs, but the embedded one in the microwave is not a general-purpose computer.

## EFFICIENT APPLICATIONS

I want to give you practical examples of microcontroller in use—*applications*. It is how you *use* a microcontroller that makes it valuable. When you know what to *do* with the devices or systems, you see the fruit of your understanding of programming and computer hardware. It is the responsibility of you, the designer, to produce the most user-friendly, reliable system for the least cost.

"Efficient use" is not always an intuitive or even obvious concept. With many years of experience in assisting customers and teaching students, I have strong personal opinions about what constitutes *efficient* use of microcontrollers.

*An efficient application of a microcontroller is a design with the minimum of external hardware that still allows the software to keep up with all of its tasks.*

*Efficient software is easy-to-understand software made up of subroutines (functions) with tables, interpolation, and simplified calculations with the smallest possible variables.*

These ideas are expanded in detail in the next chapter.

> **EFFICIENCY TIPS:** You will occasionally find highlighted sections like this discussing tips for efficiency—not just in code design, but also in hardware-software tradeoffs, and in hardware design.

## ORGANIZATION OF THIS BOOK

***Efficiency and Planning.*** Chapter 1 illustrates a general approach to *project planning* by taking a hypothetical project from start to finish. In addition to suggestions about software development, it gives a survey of typical hardware development steps for embedded systems and introduces tools such as simulators and emulators. It also discusses the easily overlooked system integration phases of a project!

***Multitasking and RTOS.*** If you are going to write *efficient* embedded applications code, you must develop a *new way of thinking* where your controller does multiple things "at the same time." Probably any programming courses you have taken concentrated on single, data-processing-related tasks with no real-time requirements at all. Most books and courses about programming stop there. Sadly, even high-powered operating systems for the PC manage to reduce multitasking to (slow) switching between application programs. Few programming books show multitasking for small, fast real-time control projects.

The *concepts* of Chapters 2 to 3 were introduced in *Volume I* with interrupts, timers, and a scheduler. Here I pick up by walking you through the development of a pair of operating systems of increasing complexity. If the "do it yourself" approach of Chapter 3 is intimidating, the material in Chapter 2 on commercial operating systems can prepare you to ask the right questions of such suppliers.

***Input/Output, and Algorithms.*** Chapters 4 to 6 constitute the bulk of the book and cover the use of just about every sort of peripheral electronic thing I can imagine as well as ways to do clever things in software. That is not to say that I have given you everything that is possible or that I'm *that* clever, but rather that here is an organized file of common devices and techniques for you to *consider*. The focus is on the principles but there are usually simple examples provided as well. You can branch off with variations of your own, and might well think of something new.

***Examples.*** Chapter 7 is my opportunity to illustrate how both hardware and software tradeoffs can combine for efficient, low-cost systems. Specifically, I walk you through the development of a moderate-sized zone heat control system, and through a tiny photographic light meter, show part of the spectrum of 8051-family applications.

# 1

# Efficiency and Planning

## EFFICIENCY

***Efficient*** is not always an intuitive or even obvious concept. With many years' experience in assisting customers and teaching students, I present my own opinions about what constitutes *efficiency*. The alternative to more efficient is more powerful—bigger faster—go the longer way but go faster to arrive at the same time. I cringe, but that seems to be what drives the PC market these days.

### An Efficient Microcontroller

> An efficient application of a microcontroller is a design with the minimum of external hardware that still allows the software to keep up with all of its tasks.

***Software over hardware.*** You do as much as possible in software to avoid hardware. By transferring debouncing and calibration functions to software, you reduce construction costs and simplify hardware troubleshooting.

***Minimize I/O.*** Your designs use as few port pins as possible. With the 8051 family, the least expensive devices have the 28-pin packages and only 19 I/O pins. The packages above 40 pins increase the socket cost and the package cost.

***Rely on hardware or other processors only when too busy.*** If there is only one processor and it is quite busy, you might add more hardware, but otherwise you always choose the more economical and thus more efficient route, doing as much as possible to software.[1]

This is the perspective I have throughout this book when I talk about *efficient*.

## Efficient Electronic *Hardware*

These are things you consider in designing efficient hardware:

***Fewest voltages.*** You use as few power supply voltages as possible. The logic supply (usually 5 V, but recently going as low as 3.3 V) can often handle a limited range of analog input. A microphone signal never gets to 5 V peak to peak, nor do most sensor signals get that large. There is no law that you *have* to do all analog processing with ±15 V—often modern A-D converters work over a 2.5 or 5 V range so it is unnecessary to amplify signals.

***Weigh relative currents.*** For projects with multiple voltages, you look carefully at the current requirements for each voltage. By choosing a transformer whose secondary winding is sized for the main (high current) voltage, you efficiently generate *one* voltage and then use small switching regulator(s) to step that voltage up for other low-current devices.[2] Particularly if you are running off batteries, you will choose to use the simplest, cheapest, most readily available batteries and use a switching regulator to make the actual needed voltage. If you make 5 V by boosting up the 3 V from two 1.5 V alkaline cells, a good switcher will continue to give a steady 5 V out until the battery voltage has almost collapsed. Without a switching regulator, you either waste power when the battery voltage is high or you must design circuitry that works over a wide voltage range.

***Conserve power by smart choices.*** You use as little current as you safely can by using the largest pull-up resistors and by designing so the high

---

[1]I am here assuming designs for volume production. It might be argued that for a one-of-a-kind project, the software development costs could outweigh the cost of additional hardware. Certainly for low-volume products the economics can favor already-developed commercial hardware.

[2]For example, it is common to have a system with high current for the 5 or 3.3 V and needing only a very little 12 or 15 V for linear circuits. The worst you could do is generate the 12 or 15 V and then use a *linear* regulator to get the high-current 5 or 3.3 V—you would waste lots of power and have to heat sink both the primary regulator and the low voltage one.

current drains are the shorter duration ones. Since activating relays takes lots of power, you pick the normally open (NO) or normally closed (NC) contact that requires activating the relay the smaller part of the time.[3] You do not design so you drive a relay 99% of the time to get a closed contact when you could use an NC configuration and drive it open only 1% of the time. Along the same line, if you are driving a solenoid to, say, clamp off a hose most of the time, you consider adding a spring and have the solenoid pull *against* the spring to open the hose.

***Shut down when not needed.*** When you have sensors that consume significant power and supply data that is needed only infrequently, you add circuitry to shut them off when not in use. For example, a robot navigating by use of LEDs or a camera needs position information only a few times a second and no information at all when it stops. This becomes an obsession when you are designing for battery operation!

***Availability.*** You choose the sort of devices readily available from local electronics supply houses—preferably with ready substitutes. You know that the exotic new devices save board real estate, but you assess how difficult they may be to obtain quickly, and you recognize that they may go out of production due to lack of demand. You prefer second-sourced devices for products that should have a long life and you try to "go with the flow" in parts choices. You only risk new devices when the benefits far outweigh the risks, and even then you mentally review your fallback options.

***Stability and calibration.*** You are conscious of temperature stability, but you favor self-calibration over precision, stable analog devices. You make all the accuracy of your design trace back to a single precision voltage reference or a precise crystal oscillator.[4] Where you have a microcontroller that can do its own calibration in software, you get the necessary accuracy by putting the cost into a single stable, precision device.

***Develop to the eventual form.*** You design your packaging from the start and keep it in mind all through the circuit design process. Knowing the shape that your customer prefers, you choose components accordingly. If the final package has to be thin, you hunt for a different configuration of

[3]Of course, you balance this by fail-safe considerations so, for example, a furnace thermostat opens the relay when the battery fails—you don't want the furnace to stay on if the battery fails.

[4]Analog devices are inherently sensitive to inaccuracy and drift—there is no thresholding to allow restoration of signals that pick up noise. The more stable an analog device, the more expensive it usually is. If your system requires a lot of calibration by setting pots, there is a good chance they will be maladjusted as well.

power transformer, as well as using axial rather than stand-up capacitors. The form also governs your choice of buttons and displays. Your prototype board layouts anticipate the final configuration so you can catch noise and interaction problems sooner.

***Fewest boards.*** You put as much circuitry as possible on a single board or else, from the start, consider where you will have interconnects to make good partitioning. In production, it is generally cheaper to have more on one board, since handling costs and interconnect hardware costs are less. It looks nice to have a rack of small boards in a bus, but, except where you *need* modularity for ease of repair, it is a luxury—particularly for consumer products.

## Efficient *Software*

> Efficient software is easy-to-understand software made up of subroutines (functions) with tables, interpolation, and simplified calculations using the smallest possible type of variables.

The result of efficient software is code that is faster and smaller without unnecessary library functions.[5] Here are some immediate suggestions:

***Easy to understand.*** You use enough comments—meaningful comments that don't just echo the information directly available from the instructions—so the next programmer (or you, several months later) can easily pick up what you did, how it fits the bigger picture, and why you did it in the given way. You choose variable names so their purposes are apparent.[6]

***Functions, functions, and more functions.*** You use functions (subroutines) rather than straight-line programming. That way your code can be re-used. Your functions roll up program functions into clearly delineated

[5]Contrast this sort of software with straight-line programming where code must be repeated over and over. I see it often in my students' programs where they send messages out to displays a character at a time with all the handshaking repeated with each character—usually with the ASCI codes rather than using the function of the single or double quotes in C. It is particularly bad when the software is developed under time pressure. There is usually no time to plan ahead, but lots of time to type endless repetitive code lines. It is much like the saying, "No time to do it right, but plenty of time to do it over."

[6]I also hate the fifty-character-sentence-type names used by Microsoft in their examples. I suspect that they were made so verbose only after the development work was done. Efficiency relates to how much typing you have to do!

pieces with names that make the function clear to the reader. For example, you separate the process of getting a number out to your display from the calculations that produced the number. This structured programming is the opposite of software where a long string of instructions involves criss-crossed jumps to unpredictable places up and down the instruction list.[7]

***Only necessary precision.*** You seriously evaluate how much precision you really *need.* You use the smallest possible (*unsigned char* byte or *unsigned integer*) variables that will get you the necessary resolution. Bytes before integers; integers before long (double precision); integers before floating-point; unsigned before signed. With the 8051 family, you can even have bits over bytes if there are not too many needed (128 maximum). The worst thing with the 8051 family is to use floating-point math—there is no co-processor and the compiler will have to call for a large library of functions that take a lot of instructions cycles to execute. Of course there are some cases where floating-point is unavoidable, but be sure that is really the case.

***Tables over calculations.*** Where possible you use tables and interpolation rather than complex calculations. Pre-simplified calculations make your code faster and may avoid the large libraries for math operations.[8] Look-up tables are covered in much more detail in a part of Chapter 6 on Algorithms (page 324).

***Real-time interrupt.*** You rely on a timer-driven interrupt to free up your software from managing scanning and putting out regular signals. Your software is free from moment-by-moment concerns because they are handled through the interrupt (see the chapter on the scheduler in *Volume I*). Instead the main program manages variables and data processing as well as the high-level decisions.

***Multitasking.*** Your overall approach is multitasking oriented. Chapters 2 and 3 discuss this, but in a nutshell, your software is able to flit from one job to another as needed. Your software is never stuck in the middle of a

---

[7]I have heard of "the 60-line rule" that says that any piece of code must be short enough to fit on one page. If not, break it into several subroutines so the user can comprehend a given function without having to look over several pages.

[8]Once you have brought in a particular calculation anywhere, say a floating-point square root, then you might as well use it over and over—it takes no extra code space after the first time it is used. On the other hand, just because you have a floating point add, you haven't given up the space for the floating-point multiply if you don't multiply—the library has separate modules for various functions and a good linker only brings in the modules for the operations actually used.

long calculation that makes the processor unavailable for more urgent input from a user or external hardware.

## PROJECT PLANNING

If you're like most developers, you want to start right in. Please let me first walk you through the planning *process* as I understand it. Project planning is perhaps less dramatic than learning a new language or a new processor, but in the long run it is more important. You can really learn planning only by experience—reading about it is not enough. If you grit your teeth and seriously set out to "learn to plan," it can look formidable. I encourage you, instead, to *skim* the rest of this chapter for suggestions on *how to get started* and then go on to later chapters where I describe the long, convoluted process of finding, say, the best sensor for a hot water system or a robot navigation system. Then take a quick look at Chapter 7 where planning is illustrated in very specific detail. Good planning is partly orderly development, but it also results from continuously replenishing your stock of new ideas waiting to be tried!

### Plan First?

Unlike most school assignments, real projects seldom come fully defined. Your *customer* expects the project will be quite easy.

*It is very much like what is currently being done.*

*There are so many new devices it should be easier than ever.*

*Software? I'll get my brother-in-law to do it—he's a whiz at it.*

Or you have discovered an application just waiting for a microcontroller, but the current approach is working.

*Just let me redo it. My new version will surely be less expensive, more flexible, and more reliable!*

Be careful—understand the current approach thoroughly before you make any promises. There are *years* of experience to show up shortcomings with the present approach, and *no* experience with your proposed alternative.

For the planning example of this chapter, picture a small company wanting to enter the controller business.[9] They have many sales contacts

---

[9]I have obscured an actual application just enough so I can talk honestly without reflecting on the companies involved. If you want to know, the one thing I can say is that it is not actually an envelope handler!

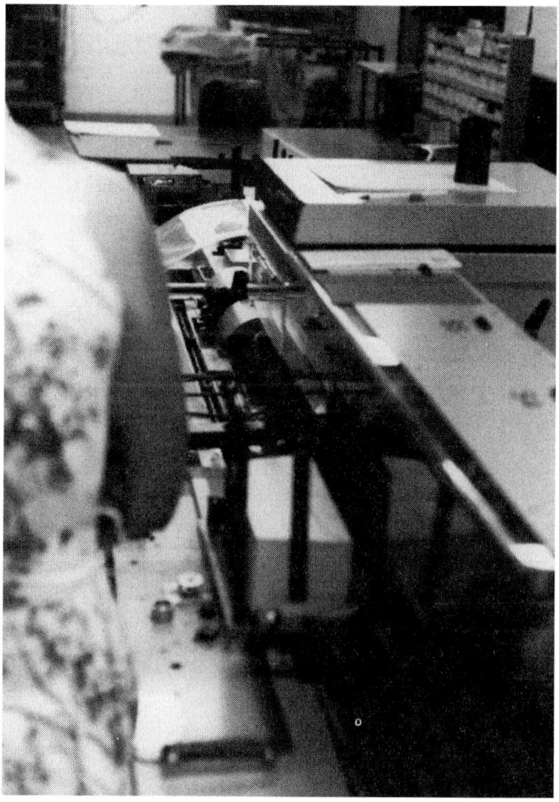

**Envelope Handler Controller**

with end users. Focus in on one manufacturer of mail-handling systems whose present controller is expensive, unreliable, and unsuitable for new products.

Envelope handling systems for mass mailings look like that shown above. As envelopes come out of a box, they are addressed by a printer, picked up by a handler, and moved down a conveyor. Then the flap is opened, a letter is inserted, and the flap is sealed. Imagine the system works fairly well, but sometimes noise causes the controller to reset, requiring the operator to manually clear and restart the system. A separate company supplies the custom controller. This supplier allegedly ignores the manufacturer's complaints and refuses to invest more development effort. Here is an opportunity to start supplying controllers!

Various management texts and classes formalize an entire decision-making/problem-solving process. The outline below shows one version of the process.

## THE DECISION-MAKING PROCESS[10]

***Identify the Problem.*** The problem must be separated from its symptoms and properly identified (a critical step). Then you should be able to:

***State the Goal of Your Work.*** The goal defines the specific result you expect to accomplish through your decision-making work. The next step should be to:

***Gather Information.*** This step may also occur at other places in the process. The questions *Who?, What?, When?, Where?, How?,* and *Why?* should be useful. The answers are separated into columns of *What is?* and *What is not?* The next step is a:

***Component Analysis.*** This is used to identify the major components of the problem/situation or system. The key question here is "*What major factors can be changed?*" Then you should:

***Generate Possible Solutions.*** These will achieve the goal. The question here is *"How can each component be changed?"* When this creative, divergent task is completed, it is appropriate to specify the:

***Constraints and Assumptions.*** Constraints limit what can be done. Assumptions help simplify the problem. The next step is to make a:

***Choice.*** A choice is made by applying the constraints and any other criteria that are relevant to the possible solutions and carefully examining the positive and negative consequences of different options. A table is sometimes used to help organize this work and the solutions may be ranked from best to poorest. The best solution is selected for further work. This is also a critical step in the decision-making process because a great deal of detailed work will follow. This work begins with an:

***Analysis of the chosen solution.*** An excellent way to perform this step is to identify all the components that must be considered and ask as many questions as you can about the choice using the words *Who, What, When, Where, How* and *Why.* After all the components have been identified, it is time to begin the:

***Synthesis.*** In step 1, the analysis questions are answered; in step 2, a detailed solution is devised. When this work is completed, an:

---

[10]Reprinted courtesy of Center for Guided Design (Charles E. Wales), 137 Engineering. West Virginia University, Morgantown, WV 26506-6101.

***Evaluation*** is performed to see if the plan satisfies the goal and meets any criteria that are appropriate. Here again the positive and negative consequences should be examined. If the evaluation shows that the solution is not acceptable, it may be necessary to return to the possible solution step, make another choice, and proceed again through the analysis, synthesis, and evaluation steps. If the work is acceptable, a plan to evaluate the implemented solution should be devised. Then:
***Recommendations*** may be made and a:
***Report*** prepared. If the work is approved, you will implement the plan and prepare to:
***Check the Results.***

The critical first step, best done in a group, is to ***identify the problem.*** Do not develop a detailed solution to the *wrong* problem. ***Define*** what you want to accomplish. You *think* you *know* what needs doing and immediately start off to do it. Much later, you will finally slow down enough to notice that you are developing a great solution to the wrong problem. Far better to *first* sit down with the *customer* to determine what is (and is not) desired. Often the customer may be unsure what is difficult or easy to accomplish. Negotiate *at the start* so a difficult feature appears only if actually necessary. You can also suggest alternative features not even considered. Reach a solid agreement at the start to avoid ***creeping features*** where, as the project progresses, the customer asks you to include this or modify the goal to do that.[11]

For this example the goal was to help a small company develop a product to meet the controller needs of the envelope-handler manufacturer better than the present controller. It was to also provide an entry into the custom controller market.

Not knowing enough to proceed, I took an ***information-gathering*** visit to the manufacturer.

[11]If you put functionality first, *creeping features have crept in.* Features are nice, but they are useless if they do not enhance the usefulness of the product to the regular user. Serious advance planning should narrow down what is important to the customer as opposed to what the customer thinks may be "nice" but to which the customer does not attach a high value. For example, you might be tempted to replace a few printed instructions on the enclosure by an electronic display used only once.

The handler was impressive to watch with envelopes zipping along and being picked up to have letters inserted. The people in the plant demonstrated the set-up process for a newly built machine—it was quite complicated. There were a host of individually labeled membrane pushbuttons having multiple functions that varied depending on the current mode of the system. It was particularly illogical during setup, but during normal operation there were single, specifically labeled buttons for each function. It was a very *custom* controller. It had a plastic box with foil-wrapped circuit boards—I expect it was a belated shielding effort. The triac drivers (for pneumatic valves) and their circuit traces were close to the microcomputer chips, so inductive spikes were a real possibility. Opto-isolators did provide some degree of protection. A power-line filter at the input was obviously part of the effort to reduce the random failures due to noise.

Having gathered information, your next step is to determine answers to ***what, when, where, why, how*** that go with journalism and also help define a problem. *What is* the problem? *What is not* the problem? *When is* it a problem? *When is it not* a problem?

In the envelope controller, the basic functioning of the controller was not the problem. Many end users had utilized envelope systems for years, and they were still buying them. The isolation techniques were good. The price to the original equipment manufacturer (OEM) must have been acceptable—actually it was much less than that of a commercial controller from a company like Allen Bradley. The custom user buttons were an absolute essential because it gave the OEM a prominent name on the system. All of these features were *not* a problem.

What was a problem was noise susceptibility, which caused occasional failures. The lack of friendliness for setup and the lack of flexibility, while considered a problem by the manufacturer, could easily be corrected in a new design.

So at this point, I digress to discuss user interfaces.

## User Interfaces

Any design you do should focus on the *actual user*, not the bystanders who may admire it or be impressed by the colors of the display or the lights and buttons. In a nutshell, *do not put aesthetics first*. "*It probably won a prize'*" is no compliment to a design if it implies that you have made it showy for the sake of display. Beauty is nice, but if it does not make the product more efficient with fewer user errors, you are going in the wrong direction.

Plan how the user will interact with the product *before* the design details are set. Planning what the user wants should take place *before* you set the details of the actual design.

***Research first.*** How do customers use your sort of product? Be sure you know what is already on the market before you decide to do things differently. There may be good reasons to change, but do not make a design in ignorance of the competition. I once encountered an LCD display module that had a significantly different pin-out than all similar modules on the market. The product manager told me that the engineer was no longer working there and they were changing the layout. Rather than locking *in* the product's customers, the unusual arrangement locked the module *out* of many designs.

When you have a user interface planned, test it by *playing dumb*. Find the least initiated potential users and have them go through installation or use of the product—ideally with a mock-up or a hand-built prototype before you finalize the design. *You* are the worst possible person to use your design because you know it inside out and are probably very technically oriented.

***Make it easy to use.*** If I do not know your system but I know what it does, I should be able to use it with no help or instruction! People lose instruction manuals and hate the delay of lengthy tutorials. Take note of the increasing use of on-screen tutorials and cue cards with modern software. Help files only bring up information if the user knows enough to make the request. Even better than good instructions, however, is a design that is inherently intuitive. For example, make your power switch be a toggle where up is on—push-push power switches aren't as good. Do any non-critical settings with slide potentiometers—not digital keypads. Up-down or clockwise-counterclockwise should agree with common expectations. Arrange important controls where they are easier to reach; arrange menus so you reach the most frequently used choices first.

***Make it friendly.*** Let the user be sloppy. Design software so the user can enter a number or skip on to another entry. Let users make entries in the order they choose. One money-management program forces the user to enter the date, then the expenditure and finally the amount. Any other way the user must move around with the mouse or skip back with the shift-tab key.[12] Do not design the interface to require completeness in all entries. If necessary, have default values to supply if the user makes no entry.

***Visibility.*** What you need to know is what you see. Avoid complicating the user interface with information that is unimportant at the moment. Like aircraft cockpits, put the most urgent controls near the center of atten-

[12]Many older programs insisted on a valid entry before going on. In Windows® that corresponds to a "modal dialog" where the software stops until the user responds to the message or supplies the input.

tion and the only occasionally used ones behind a cover, out of sight, or off to the side.

***Feedback.*** Keep the user informed. When there will be a lag between the user input and some visible result, at least supply a *beep* to acknowledge the entry. If the process is long, have a light flash occasionally or give a message on the display that changes frequently.

***Mapping.*** Mapping is relative to *familiar, natural, intuitive,* or *conventional.* If you have done your research on existing products, you know what is familiar to your class of users. Take advantage of that because that is where operating systems like Windows® are making PC software design easier. There are now "Windows® feel" features that are recommended to be included in all outside software. Clicking and dragging are almost second nature to PC-literate users, so the amount of learning for a new program can be much less if it builds on the familiar skills and behaves in a similar way to other programs. It is a bit like saying that if you want to write a book, use the common alphabet rather than inventing a new set of symbols. Someone else has already taught your readers how to recognize the alphabet—take advantage of that!

## Component Analysis

The step of ***component analysis*** consists of determining which major factors you can change. With the envelope handler, because there is a large installed base of envelope sorters, the manufacturer wants no radical changes, but *claims* to be open to other proposals. Because the existing controller supplier is not at all open to licensing or changing their design, clearly a new design is in order for this start-up company.

Next we ***generate possible solutions***. The control application, although it looks impressive, actually involves less than two dozen control outputs and a few opto-sensor inputs to determine if the envelopes have become hung up. Speed is not at all a problem. Because the entire process is effectively just a series of timers, I considered using the 8051 family microcontroller because I have experience with it. Obviously other controllers could be used. The base of my previous experience suggests that the 8051 family is about as inexpensive as any similar microcontroller, is multiple-sourced, is in wide usage, and is likely to remain on the market a long time. It has built-in timers and interrupts, which should reduce the chip count, and it is expandable to 64 K bytes of code space. What I am really doing is short-cutting the process of analyzing *possible solutions* in the choice of a microcontroller!

In embedded controller applications you want to determine what aspects of the project have the ***most uncertainty***. To be sure you can succeed, give *those* parts the most attention *early* in the project. Perhaps no one is familiar with a planned keyboard. Possibly a math algorithm may run too slowly. Whatever is most uncertain, the focus of the early activity should be there.

### Hardware/software Tradeoffs

The decision process outlines the steps for ***constraints and assumptions*** and the ***choice***. A critical part of this for embedded control projects is the hardware/software tradeoffs. In general, if software can do it, the cost will be less. Once you have developed software, you can use it over and over at no added cost, but you have to purchase and assemble hardware for each unit. If the system is fairly busy, however, the choice of more software gets into processor-speed and memory-space questions.

One of the first steps in getting specific is the decision of where to be ***hardware or software intensive***. Will all the switches be debounced with flip/flops? Or will you write software to do a periodic scan of the switches and decide the state has changed only after two successive readings agree? The former is easier for the programmer, but the latter does not involve the *repeated* cost of the flip-flops and increased board real estate. The software-intensive approach may involve more program memory, but memory has gotten cheaper—EPROM smaller than 8 K bytes is rare and may be no less expensive than much larger ones. Besides, unused space in an EPROM does not bring a refund from the chip supplier!

Other tradeoffs relate to the ***user interface*** to the unit. Since LCD modules have become so inexpensive, the use of large arrays of custom buttons has been declining. A menu approach is much more flexible because you can send any message you choose to the display with a software routine. It is then easy to change the user language for international markets. Some displays even have versions supporting a Japanese character set. Software-defined buttons could allow flexibility—put the buttons just below a two-line display, and the lower line can show the current function and name of each button. Incidentally, such an approach makes the hardware much more universal because you can adapt it to other customers with only a software change.

In the envelope-handler example, the decision to use an 8051 microcontroller probably comes prematurely.[13] The justification is that the start-up company has *no* experience with any controller chip. Because all their projects are likely to be relatively small ones with simple on/off control, it seems unlikely they will soon have projects needing regular computers or

[13]No one is *totally* objective!

high-performance controllers. They can also do the two or three other applications they are considering with an 8-bit controller. If they were to get into image processing, digital filtering, etc., they would have to hire an entire new department rather than upgrade their present people! Because the 8051 has built-in timers and UART, it can reduce the chip count from the earlier envelope controller design. Cost is not critical, but there is no good reason to purposely choose a higher performance approach.

The question of noise immunity is a more difficult one because no one knows precisely how the disturbances enter. Because the routing of the solenoid driver printed circuit traces is close to the microcomputer, we assume that electromagnetic radiation is the problem rather than power-line transmitted noise. We choose two approaches. First, put the unit in a steel box to improve the magnetic shielding, and arrange the package to have a middle wall separating the low-level signals from the AC side. Ideally the opto-isolators would be right at the wall between the two halves. Much more experimentation is necessary to really determine the cause, and the time schedule does not allow that luxury.

The second prong of the noise "war" is the careful use of cold and warm start routines. Some sort of watchdog timer arrangement will recognize when the processor is behaving strangely and will force a warm restart. Ideally the software can then recover from a random disturbance. Usually noise corrupts an instruction as it is being read into the processor and causes the processor instruction fetches to get out of step. Once the instruction count is off by one, the next "instruction" may really have been the second byte of the previous instruction. From there it is a sort of roulette game as to whether the program sequence will recover or go jumping off to places unknown. Assume the program has been periodically saving the state of the process (what is happening). When the program fails to reset, the watchdog timer, the timer can cause a jump to a restart location (warm start). The program then resumes from the last stored state. There is particular safety where you store the state in an EEPROM because the device's very slowness to take new values means a lost program cannot quickly corrupt the data.

Another common hardware/software tradeoff relates to ***analog/digital interfacing***. It is straightforward to simply connect up an A-D chip to a port or use an 8051 family member with a built-in converter. An alternative is to bring the analog voltage into a V-F (voltage to frequency) converter. Because you can measure a frequency or period using the internal clock/timer, you can digitize the voltage with only a single input pin and a small V-F chip. Because the amount of time spent counting determines the accuracy of the reading, you can realize the 12-bit accuracy of the V-F or you can obtain a 6-bit reading quickly. The V-F converter is quite noise-immune because the *average* of the incoming voltage (technically the integral) determines the frequency—noise pulses just make small changes in the time the next pulse of the V-F comes along.[14]

Finally, a common tradeoff relates to the question of ***keeping track of time*** with an external clock/calendar chip. By using an external chip, it is easy to keep track of the time of day, month, year, leap year, and so on, but the cost is an extra chip and crystal as well as possible address decoder and buffer. In projects where the low standby power of a separate chip is unimportant, to still use separate clock/calendar chips to avoid the relatively minor programming required to track time using the internal 8051 timers seems a poor choice. If the application already involves a real-time interrupt for other timing functions, then the calendar function is truly a minor software item.

In the envelope handler example, the calendar features are not significant and the internal timer of the 8051 is more than sufficient.

### Hardware Planning

All of these hardware decisions rely on a good knowledge of electronic devices on the market. Such knowledge comes through experience, but you can find some information in the application sections of this book.

Having defined the envelope handler *problem,* it seems that the project needs buttons for user input, ports and opto-isolated triacs for solenoid drivers, and some sort of display. A timer function is needed to keep things in sequence, and a flexible data storage structure is necessary to sort out the user set-up inputs into sequence commands. For the envelope handler, once the overall hardware is determined and that design job is proceeding, serious software planning can start.

### Software Planning

There is an approach to programming called ***top-down,*** which relates to the idea of understandability and the use of routines. Essentially it is an

---

[14]Incidentally, the terms "analog" and "digital" date back to the early days of electronics. An analog computer was a collection of operational amplifiers, integrators and differentiators that used the voltage to represent the analogue of physical parameters such as the position of a car above a road with springs, mass and shock absorbers represented by capacitors, resistors and inductors. The importance of the specific voltage value is the key to analog electronics. Digital computers (probably related to fingers as digits) ignore the exact value and represents a "1" as, say, any voltage above 2.4 V and a "0" as any voltage below 1.0.

approach that starts with the broad overview (the "top") and then moves ("down") to the details. Using such an approach, you *first* write the main program (the "main" function in C). In doing so, you make liberal use of routines that you haven't written yet, and assume that these routines will perform in specified ways. For example, if you need a printing device, call your (at that moment non-existent) *printf* routine with a pointer to the desired message. If you need an A-D reading, you assume some yet-to-be-written routine will exist to return the reading when called.

Once you've written the main program, the "down" of "top-down" consists of writing the first level of routines. These in turn may use other routines, which you will write later. Finally, you write the detailed drivers—the most basic routines.

Such an approach looks good on paper, but you will find most hardware-oriented projects have details requiring practical experience. At the beginning, it may be better to use a ***bottom-up*** approach to software development, which starts with the most basic drivers. *First,* write a routine to initialize the display and show a single character. *Then* write a canned message program. *Then* write routines to accept varying messages. *Then* write fixed and scrolling message routines. *Then* perhaps write messages with constant text and variable values intermixed. *Finally,* go on with the rest of the project!

The best solution may be an ***ends-in*** approach that maintains an overall perspective while developing the most basic details. That avoids the "*Where do I go from here?*" sense that comes from the bottom-up approach as well as the "*You mean the hardware can't do it?*" problem of the top-down approach. From any direction, there is no substitute for experience in this phase of project development.

## Rules for Task Partitioning

In planning software, *divide the overall job into tasks.* The term **task** simply means *the thing to be done* and corresponds to the various routines that go with modular programming. Sending a message to a display or getting an interpolated value from a look-up table is an example of an individual *task.* If several programmers do the work, task partitioning may even up the programming load while letting individuals work where they can be most productive. In partitioning tasks, it is easiest to go bottom-up and first set up ***driver tasks*** for the individual hardware components. If you need a joystick, you might make an interrupt task to interface the joystick. The task could set a flag bit when it sees changes, or a separate background routine could return the most recent readings. A display task could get the actual information or could get brief codes that refer to a library of canned messages. The partitioning may even involve a display driving task called only by a code-to-message conversion task. At some point, you will have to resolve the question of how general purpose to make the driver tasks.

Once you've defined the driver tasks, then plan the ***"thinking" tasks***. If there is significant math processing to do, make the math a background task. If you need serial communication, then write a single task to interface the "outside world." The general idea is to break the overall job into logical, isolated pieces.

Finally, there may be ***management tasks*** that define the sequence of operations or determine when user input is calling for some other mode of operation. These tasks are isolated from the I/O hardware and interact with pre-processed information such as flags or simple bytes of information.

When defining tasks, it may be appropriate to establish priorities. The section on operating systems goes into a little of this. In all the planning, it is important to include testing and system integration so that the pieces come together smoothly.

For the envelope sorter, it is decided that the LCD display interface will be two separate tasks—one task will be the initialization of the display and a second task will take the pointer to a message and convert it to sequence of LCD drive commands and data. A third task will scan the user-input keys. At least initially, there are a large number of keys, so scanning will be broken up into pieces to distribute the load among the interrupt cycles. If one-quarter of the keys are scanned every 10 msec, then the time to the next scan of the same key will be 40 msec. That is long enough for a key to stop bouncing without missing any button pushes or appearing to have delay. Several tasks are defined to get the different user commands onto the display and take the resulting entries into a data table. Another task will take the key scan results and convert them into coded values. Then one more task will convert the user set-up entries into actual sequential operation times, because the user will set the duration of pneumatic activations, but the actual times when they are turned on and off have to be measured from the start of the cycle. Then the main cycle-timing (scheduler) task is set up. It will run off the timer interrupt every 0.01 sec because that is the smallest significant time increment. Every 0.01 sec the scheduler checks to see if any outputs are due to change (and changes them) as well as doing one-quarter of the button scan.

## Real-time Data Flow Diagrams

When a job involves multiple tasks, the tasks are often interrelated and it is useful to show this in a "block diagram" manner. In its simplest form, the conventional task diagram includes rounded rectangles for the tasks, with arrows showing the information that passes between tasks, and rectangles for the actual hardware interfaced by the tasks. The diagram below shows the planned tasks for the envelope handler.

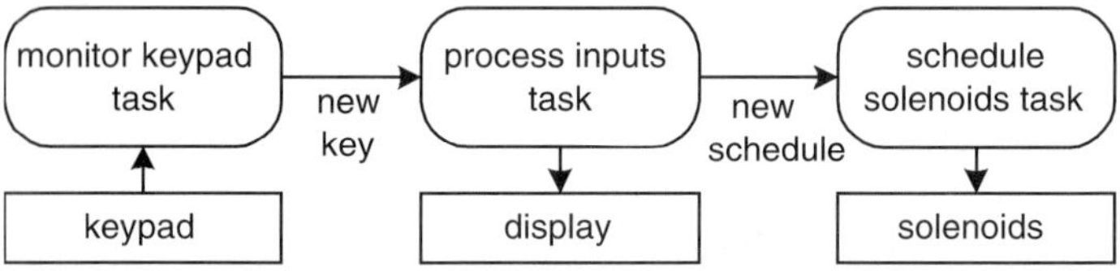


**Envelope Handler Task Diagram**

## Defining Task Interfaces

Once you've outlined and named the tasks, it is important to define *carefully* the ***interfaces between tasks***. Where several people are writing program modules, it is particularly important that they agree about the uses of flags or variables. *What does each flag mean? How long should a task wait for a flag? Would flags work to signal successful data transfer? Will things go right during development even if you haven't written some tasks yet?*

The task interfaces for the envelope handler example involve sharing data with the display task as well as determining if the button state has changed. The user-interface (background) tasks and the solenoid-output (interrupt-driven) task exchange sequence times. When the user changes a value, as background time becomes available, the values are sequentially updated. If the updating catches up with the current operation, the cycle parameters immediately change for the rest of the current cycle and then into later cycles. The only danger with updating during operation is that the change in "on" and "off" times might cause the processor to miss an "off" in the current cycle. This relates to the idea of *protected regions or resource management,* which is also part of multitasking. In this case, the problem either can be avoided by requiring the operation to be stopped before new parameters can take effect, or else be avoided by using two lists and setting up the new one while the system continues to use the old one.

### How Busy Is the Processor?

There are excellent software tools that can evaluate software time in loops (such as DS51 described in *Volume I*). When using multitasking, if you suspect processor overloading, devise a temporary background task to replace the regular background task. The new task could simply count free time. The faster the *free-time* count increases (perhaps you could even run a stepper motor off the overflow to make it more graphic!), the more free time there is. If the counting stops, then there is *no* free time, and there is obviously system ***overload***.

If you have too many things to do on a continuing basis, the processor will never be able to catch up. You can send major math algorithms to a math chip, or you can get a faster controller. It is also possible to go to several processors with a serial communication scheme between them. In the end, the real test of the system is "*Is it fast enough?*" Where people are the sources of inputs to the controller or the users of outputs, "fast enough" is a subjective question, but it still needs to enter into the planning.

In the project involving the envelopes, there does not appear to be any problem with overload because the running of the system has higher priority than the display update or user interface processing. The scheduler interval of 10 msec should be slow enough to easily get through the pre-processed list of start/stop times. It might have trouble if the entire user-input-to-actual times conversion must finish during one interrupt. The same concern is behind the division of the scan into four pieces, but it is probably unnecessary. Probably the only measures of *fast enough* for the example are 1) the speed of the display update when the user pushes buttons and, 2) the requirement that all the outputs change on time. The speed of detecting an envelope jam is probably not critical because the system only needs to respond in the approximately 1-sec index time.

### System Integration

Planning the way in which you will gradually develop and smoothly fit together the pieces is an often-neglected part of planning. *How will you test the individual driver tasks when everything else is incomplete? How can you test a supervisory task when none of the worker tasks are ready yet? How can you test the basic interaction before adding all the bells and whistles?* These are all questions of system integration.

Probably the best way to start is to ***test the hardware*** before testing any software. Write a short test program to read and write to the ports. See which logic level turns on the LEDs, which pins come out where on the cables, what sequence is necessary to initialize the LCD, etc. Despite your presumably carefully executed hardware design and schematic layout, you need to be sure the hardware works the way you supposed it would.

Having done that, ***start small***. Write a *simple* program to read in from the keypad and store in a variable location that you can send out to a parallel port. Make a task to put a *simple* message on the LCD. Get the speech chip to say a short phrase over and over.

In the example project, the integration steps involve the following:

1. Get the breadboard to power up without any chips, and check the presence of correct voltages at the proper pins.
2. Install the processor (with power off while installing!), and check for a clock signal or ALE.
3. Burn and try a simple program that reads a port and echoes to another port.
4. Write and debug a program to send a fixed message to the display.
5. Write a program that echoes the key pushes to the display.
6. Get the real-time interrupt routine going, and have it send a count to the display.
7. Develop the sequencing routines, and test the triac (solenoid driver) outputs (with neon bulbs).
8. Work out the conversion routines for the user input/output sequence.
9. Take the finished unit to the manufacturer to test on an actual machine.
10. Debug, revise, revise, package, etc.

### Testing and Maintaining?!

One of the easiest planning steps to overlook is the planning of testing and maintenance. In the hardware design stage, it is easy to forget that troubleshooters need a place to connect to ground and power for using a simple logic probe or scope. If a logic analyzer is available, have a row of pins accessible for connecting to the various busses. Certainly, at least the processor and EPROM should be socketed. The mechanical design of the packaging should include ways to easily extend out the boards for access while the controller is running. Designers seem to seldom make circuit layouts with any logical order to help in following the traces, so a good component layout diagram and schematic are essential. This could include silk-screened component layout on the board and separate diagrams of component placement and orientation. There are much more elaborate testability schemes including ones involving a special mode on the buffer chips or additional "glue logic" devices, but this is beyond the scope of this book. It bears watching for serious high-volume designs, however.

### Self-test Routines

A well-planned project will include self-test and self-calibrate features. The software could start by going through its memory space looking for locations that are bad. Evaluate the entire EPROM space for a checksum to determine if some location is corrupted. But, beyond the usual computing-hardware sorts of tests, build in special test modes so that, say, the D-A converter outputs a staircase waveform, the pushbuttons can individually show on the display when pushed, each of the outputs activates directly by a push of a button, and so on. Such test routines would make it easier to diagnose the bad components (given that the processor works). The degree of effort placed here will depend first on the expected production quantities. If the first system is to be the only one, it may be easier to troubleshoot it in the lab. Secondly, the anticipated repair philosophy governs effort—if it is a throwaway, why bother. Finally, the location of the repair shop and the skill-level of the individuals charged with maintenance are influence the self-test development effort.

### Closure on Decision-making

The process above, in an embedded-control setting, includes the remaining steps of the decision-making sequence—***analysis, synthesis, evaluation,*** *and* ***recommendations***.

For the envelope-handler, it was done over several meetings and involved comparing the existing controller with the proposed design. A prototype was developed and shown to the customer for feedback. At the time of this writing, some seven years after the project, the customer has used the new controller and shipped many envelope-handler systems with the new controller installed.[15]

[15]In the for-what-its-worth category, the LCD display was replaced with a similar module using a vacuum fluorescent (blue-green) technology, several iterations of the custom membrane keyboard were required, and the software has grown to push the 64 K limit on code due to a high degree of custom set-up features added (creeping features). It would be a prime candidate a multitasking operating system if there were time and resources to start over.

# 2

# Real-Time Multitasking

If you are doing a large project where you work with other programmers, or if you need a formalized approach, this chapter introduces the ***real-time operating system (RTOS)***. Such systems are not magic—they are merely more elaborate versions of the sort of programs I develop in the next chapter. They use the same processor and resources as your software, but they bring order and standardization, which is especially valuable when someone else comes along later to modify your code. You've most likely used operating systems such as DOS or Windows® on a PC to simplify disk access or graphical functions. Now consider if a real-time operating system might work for your microcontroller.

I have adopted a top-down approach—*this* chapter introduces the high-level concepts of real-time multitasking and shows the use of the system calls of several commercial RTOS while the *next* chapter takes you one level down and develops both a scheduler and two relatively simple operating systems from scratch. You will see how an operating systems works at the code level.[1]

With multitasking, your microcontroller can seem to do many things at the same time—it will never seem unavailable and will never be hung up waiting when there is something it should do.

---

[1]Incidentally, if you understand the code-level details of the next chapter, you are in an excellent position to understand the functioning of these and other operating systems (such as Windows®). You will be in a much better place to understand any event-driven programming such as that on a PC using Windows and languages like Visual BASIC or Visual C. Perhaps you think of multitasking as an "advanced" topic. Here I show you how to develop a real-time application and in the next chapter, I show you how to develop an operating system of your own.

## BEYOND SINGLE-PROGRAM THINKING

Traditional programming involves *one* program flow branching to different functions as determined by decision structures. You are probably used to programming where only one thing happens at a time. If you need an A-D reading, for example, you issue the *start_convert* signal, *wait* until the *end_of_convert* signal is true, and then collect the finished reading. With such an approach, while the program is waiting for the A-D converter to finish, the processor cannot do anything else.

When applications grow more involved and time-critical though, the single program approach becomes awkward. If you have several external hardware devices that need to be served at the same time, the challenge is to make sure the software satisfies each device. To write tasks for multitasking you need to developed a *new way of thinking* about program flow. Never again will you let your software become stuck if something is not ready yet.

Before I describe that way of thinking, I need to define some of the terms you might encounter.

## TERMINOLOGY

For the terms that follow, you will find that no one definition is universally accepted.[2] For example, some people view multitasking as best accomplished without a real-time clock. There are different definitions of real-time. Recognize that you are seeing *my* definitions—not those selected by a body of experts.

### Real-time

In this book, ***real-time*** will refer to any system that responds to inputs and supplies outputs *fast enough* to meet hardware or user requirements. For example:

A keyboard entry system is real-time if it gives the user some feedback quickly enough to feel confident that the system noticed. If the system puts out a *beep* within 100 msec, the user will feel confident that the system recognized the input *right away*. So, a key scanning routine that repeated every 100 msec would probably meet the requirement of *fast enough.*

---

[2]In fact, there is no clear agreement as to whether the term should be realtime (one word), real-time (hyphenated), or real time (two words). I use the hyphenated form because my spell checker seems to like it.

Humans cannot assimilate new digital display information more quickly than perhaps five times a second, so updating a display of rapidly changing numeric values only a few times a second is *fast enough.*

A moving stepper motor needs new step pulses fairly consistently—here millisecond precision is *fast enough.*

Serial ports usually have only a one-character buffer, so at 9600 baud, if every *incoming* character is picked up within about 1 msec (10 bits @ 9600 bits/sec = 960 char/sec ≈1 msec) that is *fast enough. Outgoing* (asynchronous) characters can be sent whenever the processor is not busy so, depending on the application, sending characters within a minute could be *fast enough.*

The same considerations would apply to collecting data with an A-D converter. Depending on how rapidly the incoming voltage is changing, retrieve the reading immediately or let it sit until you have time to process it.

*How rapidly should you adjust a flow valve for a process?*

*How quickly should you supply a new speed setting to a motor?*

Both of these are questions relating to what is *fast enough.*

At the other end of the spectrum from real-time systems are traditional ones that do data processing for large organizations—updating insurance records, inventory lists, and so on. They take a batch of input data from a file, and store the results in another file or send them to a printer some time later. Whatever the data and whatever the processing, there is a significant delay from the request for processing to the obtaining of the results. If the delay were less, presumably it would be more useful or make the user more effective. If a data processing job takes 20 sec, perhaps it would be more useful to have it done in 1 sec. For a job normally finished 100 msec from when the operator pushed the last key, it is doubtful that speeding it up to 1 msec could even be perceptible to a human. So to a human user, overnight processing is not *real-time*, 20-sec processing is not *real-time*, but 1 sec or less is probably *real-time*.

Some areas where there is a push for real-time systems include speech recognition, image processing, and digital signal processing. Imagine a dynamic display of the spectral energy distribution of a speech signal as a microphone picks it up rather than later from processing a recording! Consider the possibilities of issuing canceling signals to make an airplane invisible to radar, or an acoustic system to cancel background noise. How about a recognition system that can identify objects as they travel down a conveyor line to direct a robotic system?

Much of the *high-tech* focus of the microcontroller field involves very compute-intensive applications. This book only discusses much less demanding parts of the real-time application world. Some of the high-tech applications are impossible in real-time with a single 8051, but the difference between the large, fast applications and an 8051 job is more one of scale than of technique.

> **REAL-TIME:** There is no need to believe that real-time means fast. A good design only does things as fast as necessary to satisfy the system needs.

### Job

For a dedicated microcontroller, the ***job*** is the total mission for the device and its associated hardware. A job might be to control a microwave oven. It consists of multiple *tasks* (hence the term, *multi*tasking) such as scanning the user touch-pad, updating the displays, tracking the time, monitoring the safety interlock devices, and driving the magnetron.

### Task

A ***task*** is *one* thing to be done by a controller. Tasks again represent this *new way of thinking*. You logically divide a job and find it easier to effectively use the microcontroller. Imagine a human analogy: the overall mission or job could be:

- *Drive out the dictator.*

The tasks might be:

- *Drive this supply truck to the front lines.*
- *Sit in this trench until ordered to advance.*

For a microcontroller, tasks might be:

- *Wait for and process input from a keyboard.*
- *Keep a motor running at a set speed.*

A task can be a single function consisting of a few lines of code, or it can be an entire set of interrelated functions with nested calls. The task divisions are arbitrary, but it is logical to relate them to functional parts of the overall mission.

## Event

This is something that comes from outside a task—an *external* interrupt, the expiration of a waiting period marked by a hardware timer, or an *internal* signal from another task to the task in question. An event might cause the system to change between tasks. Events are the things that drive the operating system software.[3] Such an operating system is said to be ***event-driven***. When there are no events, there is no activity. Sometimes it is said that at such times the *idle task* is running.

## Priority

This refers to something being more important or urgent than something else—which task gets done first. Priority is a common idea—we expect to yield to fire trucks and ambulances because they have urgent missions. Having different priorities in a system implies that some activities are more urgent or important than other activities. The same can apply to multitasking.

Priority can reflect duration instead of importance—you might give a very short task higher priority in the same way that a kindly person at a grocery store checkout with a full cart allows a person with only a jug of milk to go ahead.

Assume several tasks are currently ready for processor time. I will say that their status is *READY*. Should a given task *move to the head of the line (queue)* if other tasks have been waiting longer to run? Sending a pulse to a stepper motor probably is more urgent than computing an average of several data samples—stepper pulses should come at regular times, whereas you can do computation whenever time is available.[4]

## Setting Priority

In any real-time system, it is important to establish priorities. *How* to establish them is a good question, but start by considering,

*How long can this task wait?*

---

[3]This is particularly true of Visual BASIC where virtually all the software runs in response to timer events or mouse-click events on the Windows display.

[4]Likewise, with a multistage data collecting and processing job, if there is a chance you might overwrite old readings with new ones before the old ones are fully processed, then first-level processing of incoming readings needs higher priority than later stages of computing.

For example, take a project that includes the following tasks:

- Read a character from a serial port before the next character arrives (typically within 1 msec);
- Recognize a person pushing a button (this could wait for 100 msec before you begin to sense a delay); and
- Maintain a real-time clock (ticking, say, at a 10-msec rate), by restarting the timer and counting up by one before the next tick expires.

Now to set priorities:

This clock interrupt ought to be recognized *at once,* because any delay will slow down the clock. You have to balance the urgency of the interrupt with the amount of processing time it takes. Whether writing your own, or with an RTOS, keep interrupt service routines as short as possible so the system can remain sensitive to *other* interrupts; set flags (or use RTOS signaling) to request that any time-consuming and less urgent processing be done in the background tasks.

For the case above, if the timer must be software reset (say in mode 1), it would be logical to make the real-time clock a high-priority interrupt, but write a very quick service routine. The serial input could be low-priority, and the button scan could be a non-interrupt background task.

Then again, for the case above, you might wish to make the serial reception high-priority because it takes only a few microseconds. Make the real-time clock low priority but longer running by including the button scan. Your decision will depend on the impact of occasionally missing a character or occasionally slipping the clock a few microseconds. Only the specific application can determine the answer to such questions!

Most RTOS will mask task interrupts when equal or higher priority tasks are running. Put seldom-used tasks that run quickly at a high priority or write them as Interrupt Service Routines (ISRs) without context switching at all. One candidate would be a stepper drive task, which uses only a few microseconds every few msec, yet if it is held up too long, causes the motor to appear to "stutter."

**CHOOSING PRIORITIES:** A high-priority task can be a non-urgent one that is very quick to run. In choosing priorities, weigh not only the importance of tasks, but also the impact of a task being blocked for a long time.

## Preemption

Sometimes a task should not just move to the head of the line, but even *kick out* the task that is currently running. If this happens, the new task has ***preempted*** the running task.

The concept of preemption is not difficult, but the implementation in software can be challenging. You have to save enough information to be sure the preempted task can resume with all its temporary results intact. For a human analogy, imagine what would happen if the checkout person had already started handling the cart full of groceries when the person with the jug of milk arrived. To preempt the current batch would require saving the current total, moving groceries aside, handling the jug of milk, and then somehow resuming the previous checkout. Obviously grocery checkouts are not designed to be preemptive!

## Operating Systems, Kernels, Executives, and Schedulers

All 4 terms are used in this book. A ***scheduler*** is a simple form of system where there are provisions for tasks to run at regular times—the schedule. It can be an operating system or it can be a way of using some of the microcontroller hardware with programs entirely of your own writing. A simple scheduler is developed in the next chapter.

An ***operating system*** is the software interface between your software and the hardware on which your programs run. Common operating systems are DOS (disk operating system) and Windows®. The operating system handles the details of reading and writing to storage devices, keyboards, mice, and displays. Some handle the connection to ports and various other hardware devices. The operating system may manage timing and the switching between jobs or tasks (see the next section).

A ***kernel*** is either a very basic operating system or the most primitive part of one. If you make a distinction, then I believe an ***executive*** is the stand-alone version while a kernel works in conjunction with higher-level operating system software. Thinking in terms of PCs, the kernel might be the part that handles loading and unloading programs, initializing the various devices, setting up the mode of the processor, and allocating memory space for various functions. The rest of the operating system might include higher-level functions such as file management, virus checking, display card drivers, and power management.

## USX

United States Software Corporation markets a multitasking operating system it names *SUPERTASK!*, but which I will designate *USX*. It is a considerably more complex system with thirty-seven system calls (DCX51 has only ten). It runs in a round-robin time-slice mode for equal priority tasks, so it can preempt tasks with other equal-priority tasks when a time-slice expires. In a sense, the operating system is a scheduler with interrupt features. There can be up to 255 tasks (configurable) with 255 different priorities, a whole host of mailboxes and semaphores, as well as resource control. The operating system is basically a block of C source code, which *you* compile and link in with your own application code. You supply the C compiler and may write tasks in either C or assembly.

USX is configurable. You can omit timing features, event and resource controls, or even omit message capability (no mailboxes). You can limit the number of active tasks and vary the number of mailboxes, events, and resources. These configuration choices can reduce the size of the system code as well as RAM usage. The code size is said to be about 10 K bytes for a fairly full system. The literature does not make clear *how* context switching is managed for the 8051.[13]

## CMX

The CMX Company markets a multitasking operating system that includes a version for the 8051. Like USX, it is a more general software product written in C and ported to the 8051. It has forty-seven system calls and is configurable much like USX. Context switching for the 8051 varies depending on the particular C compiler used. You may want to check if it supports your particular C compiler, since the variation in compilers makes it difficult to have a generic operating system while handling non-reentrant functions.

In addition to message, resource, and event functions, CMX has functions to manage flags (bits within a single byte) and an unusual set of func-

[13]I wondered if the system copies the *entire* on-chip RAM to off-chip space whenever it switches tasks. The time for a cooperative context switch on an 8051 is said to be 345 μsec @ 16 MHz (460 μsec @ 12 MHz). Given the following assembly instruction sequence, it takes about five instruction cycles to swap off-chip with on-chip RAM, so the stated time is about enough to move 90 bytes. *MOVX A, @DPTR* (1); *XCH A, @R0* (1); *MOVX @DPTR, A* (1); *INC DPTR* (1); *INC R0* (1). The only other data on times are for events (1212 μsec @ 16 MHz) and messages (1815 μsec). This explains why the system favors large tasks and encourages direct user-written interrupt handlers. As long as an interrupt can avoid such a lengthy context switch, you save a lot of time.

tions to manage lists. It makes a distinction between a task and an interrupt; presumably you can write an interrupt to quickly modify the appropriate event flags and signal the tasks. Unlike DCX51, events are strictly task-defined and task-managed. An interrupt handler is the only way to set an interrupt-related event. Probably you would do that without context switching where possible. Both CMX and USX make a point of specifying a special procedure for calling a system function *from* an interrupt routine.

### Byte-BOS

Byte-BOS Integrated Systems offers an operating system that seems to straddle the large-small gap. It comes in versions specific to the particular C compiler being used. Like the previous two RTOS, it is primarily a C program, but it comes with part of the kernel coded in assembly. It too is one of a family of RTOS for various processors.

Byte-BOS combines non-preemptive and preemptive scheduling. The former allows tasks to run as long as they desire and then schedules the highest priority "ready" task when the running task puts itself to sleep. The task queue is a linked-list so there is no inherent limit on the number of tasks. An *interrupt service routine* (ISR) invokes the preemptive scheduler. The ISR overrides the currently running task and can have the scheduler suspend the interrupted task and execute higher priority tasks. Byte-BOS can emulate the simpler preemptive RTOS by adding a user-written ISR to cause rescheduling on every *timer0* interrupt. The literature indicates that any function that places a task in the *READY* state will force rescheduling, so most system calls are preemptive.

The documentation lists forty-two system calls including timer, event (signal), message, resource, and interrupt management. Its complexity seems on a par with the two other big systems and the combination of round robin and preemptive scheduling is fairly complicated at the start. Byte-BOS is unique in having an entire set of calls related to managing of serial communication via a UART.

### Observations on Two Groups of RTOS

The two *small* RTOS, DCX51 and RTX51/tiny, differ slightly but they both represent the very small, possibly restricting approach, which is probably superior for small, fast systems tailored specifically to the 8051. (I draw your attention to the two small ones, RTXS and RTXB, I develop in the next chapter as well.)

USX, CMX, and Byte-BOS operating systems are significantly larger and more flexible than the first two. I suspect they were written in C for some other processor and *then* adapted to the 8051's unusual architecture. You can write

ANSI-standard C code that will port over to other processor families. In preserving some of the features of the operating system from the other processors, I suspect efficiency and speed may suffer. In return you get a much wider range of system calls and a higher degree of flexibility that may outweigh the drawbacks. You can choose to set up events, resources, and memory pools, but *you* have to provide the interface to the specific hardware (including the standard 8051 features relating to interrupts, priorities, and timers). They help you by giving you sample files or configuration examples to copy, but you certainly will have a good number of details to handle before getting down to writing tasks.

Make your choice guided by the size and complexity of your applications as well as your prospects of future code migration to other processor families. I have no figures on purchase costs but, to my knowledge, none of them charge a per-copy license fee for applications that ship using portions of their code.

## SYSTEM CALLS FOR COMMERCIAL RTOS

The next table shows most of the system calls for the 8051 commercial operating systems. Several actually precede the "meat" of the call with a several letter code that makes it clear that it is a system call (this is clearer in the examples that follow). Studying these can give you a better picture of what operating systems can do.

| | DCX[14] | (RTX)tiny[15] | USX[16] | CMX[17] | ByteBOS[18] |
|---|---|---|---|---|---|
| TASK MANAGEMENT (sometimes a distinction exists between delete and suspend) | createtask<br>deletetask | createtask<br>deletask<br>(getrunning-taskid) | runtsk<br>klltsk<br>pritsk<br>slttsk<br>scdtsk | tcretrmvt<br>pri<br>tendtwat<br>m twaketwakf | createtask<br>runtask<br>sleep<br>pause<br>deactivatetask<br>endtask<br>halttask<br>resumetask<br>denyreschedule<br>forcereschedule<br>wakeup<br>wakeupall<br>wakeuphipri<br>wakeuppri |

*(cont.)*

[14] DCX calls are all preceded by RQ as in rqwait.

[15] RTX calls are preceded by os_ and words are separated by underline as well. Calls in parentheses are not in RTXtiny.

[16] USX calls are all six characters long but have no special prefix.

[17] CMX calls all begin with cx but have no separation characters.

[18] BvteBOS calls all begin with bb_ and words are separated by underlines.

| | DCX[14] | (RTX)tiny[15] | USX[16] | CMX[17] | ByteBOS[18] |
|---|---|---|---|---|---|
| TIME (some of these calls involve periodic timers while others provide simple delay or timeout) | wait<br>setinterval | wait<br>(setslice) | dlytsk<br>wketsk<br>setclk<br>getclk<br>period | tpcre<br>tpstt<br>tpstp<br>tpread | delay<br>wait<br>timeoutoccurred<br>disabletimeout<br>enablesampling<br>enabletimeout<br>acktimeout<br>updatetimers |
| EVENTS, FLAGS, AND SEMAPHORES (Some are simple shared flags while others are counters. There is a significant variation here.) | | sendsignal<br>clearsignal<br>wait | setevt<br>clrevt<br>incevt<br>decevt<br>chkevt<br>wteset<br>wteclr<br>setgrp | ecre<br>esig<br>ewatm<br>eread<br>fgsig<br>fgwatm<br>fgread | enableevent<br>ackevent<br>sleepforevent<br>sleepforallevents<br>eventoccurred<br>waitforevent<br>waitforallevents<br>disableevent<br>eventenabled<br>postevent<br>posteventall<br>posteventhipri<br>posteventpri |
| RESOURCES (a resource is managed as a simple flag) | | | chkres<br>getres<br>reqres<br>relres | rscre<br>rsget<br>rsrsv<br>rsrel | initresource<br>captureresource<br>capturesourcenb<br>releaseresource<br>releaseresource-hipri |
| MESSAGES (all quite similar except for timeout and handshake features) | sendmes-sage<br>wait | (sendmessage)<br>wait | sndmsg<br>rcvmsg<br>chkmsg | mssend<br>mssdst<br>mssenw<br>msswst<br>msget<br>mswatm<br>mswake | putmessage<br>putmessagenb<br>enablemessage<br>messageoccurred<br>ackmessage<br>getmessage<br>disablemessage<br>messageenabled<br>sleepformessage<br>waitformessage |
| MEMORY BLOCKS (a few are on-chip but most are off-chip) | allocate<br>deallocate | (createpool)<br>(getblock)<br>(freeblock) | rqmem<br>chkmem<br>relmem | bfcre<br>bfget<br>bfrel | |
| QUEUES, LISTS UART, MISC. | getfunc-tionids | | | lcre<br>lrst<br>latt<br>latb<br>lrft<br>lrfb | (24 UART calls will not be listed here) |

| | DCX[14] | (RTX)tiny[15] | USX[16] | CMX[17] | ByteBOS[18] |
|---|---|---|---|---|---|
| INTERRUPTS (a marked difference in power here) | disableinterrupt enableinterrupt | isrsendsignal (attachinterrupt) (detachinterrupt) (disableisr) (enableisr) (issrssndmessage) (isrrecvmessage) | usxcmd_c | intin intex | enterisr exitisr |

## EXAMPLES WITH COMMERCIAL RTOS

An outline of the examples that follow may be helpful. First there are two examples repeated for each operating system just introduced. If you are comparing various RTOS, look these over. Following those examples, there are three additional, more complex examples using only USX. A *detailed* study of individual operating systems with extensive examples would require another book!

In the examples, the focus is the software—the system calls and how you use them to get things done. Since later chapters go into electronic hardware, I omit details of such here.

## EXAMPLE: SOLENOID CYCLER

Many real-time applications depend on knowing the time. For example, the envelope handler of Chapter 1 should operate solenoids at specific times to synchronize the conveyors, the envelope stuffer, the sheet folders, etc. Likewise, a system may need to measure the time between two inputs to determine motor speed. In RTOS, a hardware counter that counts the system (crystal-stable) clock usually manages time. This is entirely different from timing by software loops even though both ultimately rely on the system clock. Other things are going on while the hardware count—a software timer would not allow anything else to happen.

In its simplest form, timing can come directly from a real-time interrupt. In the ***scheduler*** of the next chapter, the timer overflows every 1 msec, causing a regular interrupt. That is all you need to produce a scheduler. By counting the interrupts, you can determine at which tick each scheduled task should be called. As long as the task is done before the next tick, all the timed interrupts will be caught. Any background task (the *main* function) would work out of a separate context, but the normal two register banks would be sufficient.

This example calculates the incremental time delays to the next state change from the table of absolute times; *nowtime*. The key benefit of an operating system is replacing the real-time interrupt with a system call. The example hardware would connect to two port pins with the assumption that a logic 1 at the port activates the given solenoid.

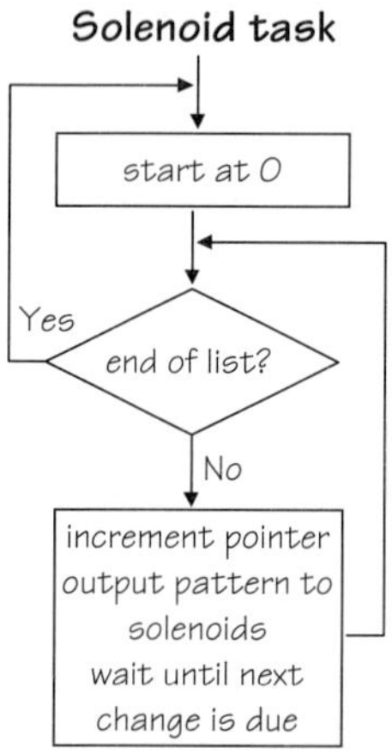


**Solenoid Cycler Task Flowcharts**

With the DCX and RTX calls, two parameters involved. One is the *timeout* count. If no other event comes within that time, the task will move to the *READY* state (if something else is using the processor) or *RUNNING* state (in the much more likely case that nothing else is going on) to toggle the next LED. The word "timeout" (actually a zero for DCX or an 8 with RTX) specifies that the task is to ignore any other possible events. Thus, by default, *wait* will timeout. The time delays do not tie up the processor and the solenoids could go on operating seemingly quite independently of whatever else the processor might be doing.[19]

[19] In addition to the system calls, the use of DCX requires each task to have a block of identification information, called the *Initial Task Descriptor* (ITD), which identifies to the operating system such things as where the task code (subroutine/procedure/function) begins, the amount of stack space to reserve, which interrupts to unmask, and what priority to assign to the task. It is a quirk of the languages that the ITD must be written in assembly code to get the linker to supply the correct address *differences*. The details should not be of concern here, but the DCX examples show the ITD at the top of the file—actually it is a separate module because it cannot be in-line assembled.

The USX, CMX, and Byte-BOS examples are quite similar. The one dramatic difference between operating systems is the size of system tick. DCX and RTXtiny all support 1-msec times. The other three carry enough overhead that they don't recommend ticks at less than 10 msec. Perhaps by careful tailoring they can run faster, but the context-switching load is significant.

**Solenoid Cycler—DCX**

```
[20]#include <dcx1.h>
uchar i;
void solenoid(void){
[21] struct code{uint abstime;uchar pattern;} next[]=
               {{0,0x00}, {2000,0x01}, {2100,0x03},
               {4000,0x02}, {4500,0xff}};
 forever{
  for (i=0;next[i].pattern!=0xff;i++){
   P1=next[i].pattern;
[22] rqwait(next[i+1].abstime-next[i].abstime,timeout);
  }
 }
}


$INCLUDE(RMX51A.MAC)
EXTRN CODE (solenoid)
CSEG AT 0FFF0H
ITD1
[23]%ITD(ENTRY1,10,81H,101B,2,0,ITD1+1)
```

---

[20]This include file (not shown) has the definitions of the system calls (like *rqwait*) as well as shorthand defines for *uchar, uint,* and *forever*. The custom-written interface between the C compiler and the DCX firmware is be prototyped here.

[21]Values in an initialization statement may be missed where the C compiler does some initialization in a separate code piece that links with the *main* initialization. Where added tasks are linked in with existing the initialization code may not be reached where there is no *main* function. As an automatic variable, the array is filled by instructions in the function rather than a separate initialization.

[22]*rqwait* is the system call. Embedding the subtraction may obscure the fact that two parameters are passed. The first parameter is the timeout delay value in "system ticks" (normally 1-msec units with DCX). DCX allows only byte values for timeout for efficiency and speed. The system call as shown here in C actually uses an assembly module that interfaces the C code to the DCX firmware. Otherwise parameters are "passed" by pre-loading *B* and *ACC*—efficient but less "friendly."

[23]This is the initial task descriptor (ITD) which comes at the head of the task. It assembles to 11 bytes (plus the *ljmp*) and has to be done in assembly to work well with multiple tasks. The

```
ENTRY1:LJMP solenoid
 END
```

```
                    Solenoid Cycler—RTXtiny
 #include <RTXtiny.h>
 uchar i,status;
[24]void solenoid(void) _task_ 0 {
 struct code {uint abstime;uchar pattern;}next[]=
              {{0,0x00}, {2000,0x01}, {2100,0x03},
              {4000,0x02}, {4500,0xff}};
 for(;;){
  for (i=0;next[i].pattern!=0xff;i++){
   P1=next[i].pattern;
[25]   status=os_wait(K_TMO, next[i+1].abstime-
            next[i].abstime);
  }
 }
}
```

```
                    Solenoid Cycler—USX
#include <usxcfg.h>
#include <usx.h>
uint status;
uchar xdata free_memory[MEM_ALLOCATION];
uchar solslot;
uchar i;
```

% is the ASM51 directive to include a macro—in this case it puts together a block of constants using the 7 parameters within the parentheses. The macro involved makes it easier to enter the parameters that make up the ITD. It was originally done in assembly because the PL/M compiler would not handle the subtraction of two addresses which are both unknown at compile time. Breaking down the parameters: The 10 indicates the fixed amount of stack space to be reserved for this task (all the on-chip "real" stack. 101B indicates that this task will use register-bank 1. The 2 indicates that this task has a fixed priority of 2 (4 = highest; 1 = lowest). Unlike some larger systems, priority is not dynamically alterable. The 0 indicates that no external hardware or timer interrupts are used.

[24]*RTXtiny* initializes the task numbered zero automatically. Other tasks need to be created in task 0.

[25]*os_wait* is the system call. Embedding the subtraction obscures the fact that two parameters are passed. The first parameter (a byte) is the mask which identifies the possible event(s) to awaken the task. The second parameter is the delay value (here system ticks assumed to come at a 10-msec rate).

```
void solenoid(void){
 struct code {uint abstime;uchar pattern;}next[]=
              {{0,0x00}, {200,0x01}, {210,0x03},
              {400,0x02}, {450,0xff}};
 for(;;){
  for (i=0;next[i].pattern!=0xff;i++){
   P1=next[i].pattern;
26 dlytsk(solslot,1, next[i+1].abstime-
             next[i].abstime);
  }
 }
}
void main(void{
 MEM_DEF *memptr;
27 usxini();
 usrclk_init();
 memptr=(MEM_DEF *)free_memory;
 memptr->mem_size=MEM_ALLOCATION- sizeof(struct
              mem_def);
 memptr->task_slt=SYSTEM;
 memptr->mem_nxt=NULL;
 relmem((usxmem_t *)(++memptr));
28 solslot=runtsk(5,&pulse,20);
29 begusx();
}
```

**Solenoid Cycler—CMX**

```
#include <cmxfunc.h>
uchar solenoidslot;
void solenoid(void){
```

[26]The delay has three parameters. The first is the slot of the particular task to be delayed (here the given task, arbitrarily in slot 5). The second parameter is the type of delay—the 1 indicates system ticks while 2 and 3 indicate units of seconds or minutes respectively. The third term is the number of units of delay—here the number of ticks until the next solenoid state change. It is assumed that the clock is initialized for 10-msec units.

[27]All these details are involved with initialization of the operating system—where it can keep its variables, tables, etc.

[28]This actually puts the solenoid task on record with the RTOS. The 5 is the priority with zero lowest and 255 the highest priority. The second parameter is a pointer to the task. The final parameter is the stack size—here kept fairly small. The returned parameter is the slot number assigned to the task.

[29]This actually starts the operating system. There should never be a return.

```
 struct code {uint abstime;uchar pattern;}next[]=
                 {{0,0x00}, {200,0x01}, {210,0x03},
                 {400,0x02}, {450,0xff}};
 uchar i;
 for(;;){
  for (i=0;next[i].pattern!=0xff;i++){
   P1=next[i].pattern;
[30]   cxtwatm(next[i+1].abstime-next[i].abstime);
  }
 }
}
void main(void){
 cmx_init();
[31]cxtcre(4,&solenoidslot,0,solenoid,20);
[32]cxttrig(solenoidslot);
[33]cmx_go();
}
```

---

**Solenoid Cycler—ByteBOS**

```
#include<bbdefs.h>
BB_TCB_TYPE solenoid_tcb;
uchar xdata xstack[256];
uchar xdata context[64];
uchar data istack[20];
uchar i;
void solenoid_task(void){
 struct code {uint abstime;uchar pattern;}next[]=
                 {{0,0x00}, {200,0x01}, {210,0x03},
                 {400,0x02}, {450,0xff}};
 for(;;){
  for(i=0;next[i].pattern!=0xff;i++){
   P1=next[i].pattern;
```

[30]*cxtwatm* is the system call. It actually returns an unsigned status character, but since no testing is being done, it is omitted from the system call. It would indicate either the timeout or else that some other task, timed procedure, or interrupt caused the task to resume—not the intention here. Embedding the subtraction to get the timeout may obscure the fact that only one parameter is passed—the timeout delay value in system ticks. The value is an unsigned integer.

[31]This creates the task of priority 4 (0 is highest, 255 is lowest) with zero message envelopes (it receives no messages) and having a 20-byte stack space. The stack size is not likely to need to be larger since no library functions are used.

[32]This puts the task in the *READY* state so it can run when the operating system starts.

[33]This starts the operating system. There is no return.

```
34    bb_delay(next[i+1].abstime-next[i].abstime);
    }
  }
}
void main(void{
35  bb_init_bos();
36  bb_cteate_task(&solenoid_tcb,context,
                  istack,xstack,0,0,1);
37  bb_run_task(&solenoid_task,solenoid);
    bb_start_bos();
}
```

## EXAMPLE: PULSE GENERATOR

This second example brings about a regular, variable pulse width signal, which can be used to drive a motor or as a low-cost D-A converter.[38,39] The output would go out a single port pin.

Although the values of width and frequency are fixed in this example, they would normally be changed from outside by some other task. The pulse could quite easily be created by two timeouts in succession where the first was for the pulse width and the second was for the off time! A simple subtraction should handle it. To show the flexibility of the systems, the timeout

[34]This call delays with no sensitivity to other events. It is equivalent to a simple timeout in some other systems. The event-sensitive call, *wait,* is used in the next example.

[35]This initializes the operating system.

[36]The *create_task* call requires all sorts of details from the user. Specifically you must define a *task control block* (tcb), a *context storage block* (array), both internal and external *stack blocks,* a *message buffer* and *buffer length* (here set to 0 since we use no messages) and the priority of the task (here zero which is highest priority).

[37]This places the task in the *READY* state (but does not start until the operating system is started with the *bb_start_bos* call).

[38]Incidentally, some of the 8051 relatives have additional pulse-width modulated (PWM) timers, which can provide such signals without regular software involvement.

[39]Depending on the repetition rate and the DC motor inductance, it is possible to vary the speed of the motor by varying the duty cycle of the drive. Such a signal can control speed with very little hardware—sometimes one transistor is enough—and the use of only one port pin. For slowly-changing D-A outputs, the mere addition of a resistor and capacitor can deliver an analog voltage output by integrating the variable-width waveform. You will have to trade ripple against maximum rate of change, but many applications have no need of the fast performance of a normal D-A.

and the periodic interval event are combined where possible to allow the pulse *frequency* to be maintained by one value (the interval value) while the pulse *duration* is maintained by another (the timeout value). This is certainly a more complex way for some of the RTOS. No attempt is made to assess the impact of the different approaches in a more heavily loaded system where preemption could interrupt tasks and throw off timing. Specifically, does the pulse recover lost time later if the task can't run right away, or does the system slip all the following pulses?

Software for a periodic event is built-in for ***DCX*** (*setinterval*), ***USX*** (*period*), and ***Byte-BOS*** (*enable_sampling*).

***RTX*** and ***CMX*** are quite different. For RTX you write a separate task that produces a signal for the task in question. For CMX you write a timed procedure (not a task which is an endless loop) which starts and runs to completion each time the time runs out (or just once, if you set a parameter differently). Incidentally, CMX requires you to write the system tick also. These require more user code, but are no problem where many tasks are allowed to exist at the same time.

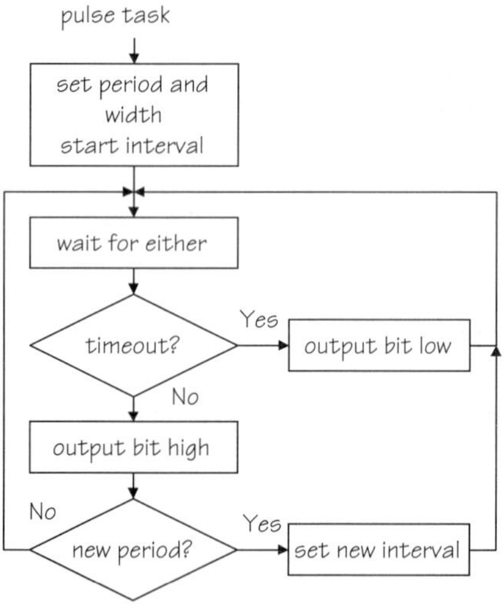


**Pulse Generator Flowchart**

**Pulse Generator—DCX**

```
$INCLUDE(RMX51A.MAC)
EXTRN CODE (solenoid)
CSEG AT 0FFF0H
ITD1:
```

```
%ITD(ENTRY1,10,81H,101B,2,0,ITD1+1)
ENTRY1:LJMP pulse
END

#include <dcx1.h>
void pulse(void){
 uchar width=50; /*20% duty cycle*/
 uchar freq=250;
 uchar oldfreq=250; /*4/second*/
 rqsetinterval(freq);
 forever{
[40] rqwait(width,timeout|interval);
[41] P1=(status)?0:1;
  if (freq!=oldfreq){ /*changed*/
    rqsetinterval(freq);
    oldfreq=freq;
   }
  }
 }
```

---

**Pulse Generator—RTXtiny**

```
#include <RTXtiny.h>
uchar i,status;
uchar freq,width;

[42]void interval(void) _task_ 0{
 freq=250;
 width=50;
[43]os_create(1);
 for(;;);
[44] os_wait(K_TMO,freq);
```

---

[40]This is the wait for either a timeout or an interval. The interval comes every 250 msec and signals the start of the next pulse. The timeout (*freq*) comes when the pulse is done. The timeout may come several times before the next pulse starts, but zero is still zero no matter how many times it is sent.

[41]This line sets the port high if status is greater than zero (here an interval) or low if status is zero (a timeout).

[42]This is the definition of the first task.

[43]The first task to start must start up at least one of the other tasks.

[44]This is a simple wait for a timeout at *freq* system ticks. It is used to make an interval for the pulse task because no resources exist to do it in a system call.

```
[45]    os_send_signal(1);
    }
  }
[46]void pulse(void) _task_ 1 {
    for(;;){
[47]    status=os_wait(K_TMO | K_SIG,width);
[48]    P1=(status==K_TMO)?1:0;
    }
  }
```

---

**Pulse Generator—USX**

```
#include <usxcfg.h>
#include <usx.h>
#define period_event 1
uchar width=5; /*20% duty cycle*/
uchar freq=25;
uchar oldfreq=25; /*4/second*/
uchar pulse_slot; uint status;
uchar xdata free_memory[MEM_ALLOCATION];

void pulse(void){
[49]period(period_event,freq);
 for(;;){
[50] status=wteset(period_event,width);
   P1=(status==period_event)?1:0;
```

[45]This causes a signal (event) for task 1 (the pulse task). The actual effect is the setting of a bit in the signal byte for task 1.

[46]The pulse task waits actually drives the lsb of P1 high for "width" system ticks and low the rest of the "freq" ticks that make up the repetition period. Technically *freq* should be called something like "period" since it is the inverse of the frequency.

[47]The status byte which is returned is used here to determine whether the task timed out or received the interval signal from the other task.

[48]The *() ? :* operator condenses the *if() ;else ;* operations. If the test is true then it timed out and the port is set low because the width time has expired. Otherwise it is set high to start a new pulse.

[49]This system function sets an event, here arbitrarily chosen to be event #1 (described by a *define* to make more readable code) to be set every 10 system ticks. The assumption is that the USX system is configured with a system tick time of 10 mSec.

[50]At this point the task will wait for either a timeout number of system ticks *(width)* or for the period event (the period of the waveform being generated by the task).

```
51 if(status==period_event)decevt(periodevnt);
   if(freq!=oldfreq){ /*changed*/
    period(period_event,freq);
    oldfreq=freq;
   }
  }
 }
 void main(void{
                  /*same initialization as previous
                  example*/
  pulse_slot=runtsk(5,&pulse,20);
  begusx();
 }
```

---

**Pulse Generator–CMX**

```
52#include <cmxfunc.h>
53uchar slot,events,timer;
  uchar width=5; /*20% duty cycle*/
  uchar freq=25;
  uchar oldfreq=25; /*4/second*/
54void interval(void){
   status=cxesig(events,1,1); /*interval event*/
  }
  void pulse(void){
    for(;;){
```

---

[51]If a period event caused the task to resume, in USX it is necessary to clear the event. Actually *period_event* is a 1-byte *counter* which is incremented every *freq* system ticks and this line of code subtracts one from the count to cancel the increment done by the period function. By counting rather than using a single bit flag, it is possible to get several events behind in processing and still not lose any of them.

[52]This header file prototypes all the system functions.

[53]These are the various variables used by the operating system.

[54]With CMX, the interval causes a procedure to run. In order to get an *event* for the task, you must write a procedure to produce the event, shown here. CMX behaves differently for periodic timed tasks. The event group is a byte that can hold up to eight event bits. The task can be awakened by any combination of events, but all the events have to be user defined and must be managed by the user-written tasks. In this case, a periodic-procedure system call is used to make the regular interval event and only one event is used in the group. Notice that the periodic procedure is not in a *forever* loop; apparently the operating system picks up when it finishes.

It would be possible to produce the same result without the interval, perhaps, by simply using a simple wait for the pulse high time and a second wait for the difference between *freq* and *width*.

```
[55] status=cxewatm(events,1,1,1,width);
[56] if (status){ /*interval event set*/
      P1=1;
[57]  status=cxesig(events,0,1);
                /*clear interval event*/
     }
     else P1=0; /*timeout=pulse width*/
     if(freq!=oldfreq){/*change interval*/
[58]  cxtpstt(timer,freq);
      oldfreq=freq;
     }
   }
 }
 void main(void){
  cmx_init();
[59]cxtcre(4,&slot,0,pulse,20);
  cxttrig(slot);
[60]cxtpcre(&timer,&interval,freq);
  cxtpstt(timer,freq);
  cmx_go();
 }
```

**Pulse Generator—ByteBOS**

```
#include <bbdefs.h>
#define interval 0x0001
BB_TCB_TYPE pulse_tcb;
BB_TCB_TYPE interval_tcb;
uchar xdata pxstack[256];
uchar xdata ixstack[256];
```

[55]This is an *event wait* with a timeout of “width” time slices. The task will wait for the first bit of the event or will timeout in *width* ticks. The status will be the key to deciding which happened.

[56]A zero indicates a timeout so here anything else has to be the occurrence of an interval event.

[57]Once an event has occurred, it must be manually cleared, as shown here.

[58]The interval timer must be reset if the frequency value is changed. Here it would have to come from some other task changing the value. Ultimately it would better be done with a message.

[59]This line creates the *pulse* task—puts it on record with the operating system.

[60]This pair of system calls first sets up the periodic interval timer within the system and then starts it. It even has control of how long to wait before the *first* time. The interval is the frequency for the waveform.

```
uchar xdata pcontext[64];
uchar xdata icontext[64];
uchar data pistack[20];
uchar data iistack[20];

uchar width=5; /*20% duty cycle*/
uchar freq=25;
uchar oldfreq=25; /*4/second*/

void interval_task(void){
[61]bb_enable_sampling(freq);
 forever{
[62]  bb_sleep();
[63]  if(bb_timeout_occurred()){
    bb_ack_timeout();
[64]    bb_post_event(&pulse_tcb,interval);
[65]    if (freq!=oldfreq){ /*changed*/
      bb_sample(freq);
      oldfreq=freq;
     }
    }
   }
 }
void pulse_task(void){
```

[61]Contrary to the first impression, ByteBOS doesn't allow a task to have both a sampling interval *and* a timeout running at the same time (*Sampling* is the term used for a periodic event to cause the task to wake up). The solution, as with RTX and CMX, is to make a *separate* task which manages the interval and then signals the pulse task. Getting the regular time could have been done with *delay* here, but it was decided to illustrate the variety of "low-level" calls in the operating system. The *enable_sampling* causes a periodic event for *this* task.

[62]The *sleep* here is the generic one that returns for any possible event. Several other versions exist where the task wakes up only if *specific* events occur.

[63]The status test has the effect the name implies. Since no other event is possible for this task, this may be unnecessary.

[64]This call will signal the pulse task that it can start a new pulse. The event number was defined with the word "interval" to give more clarity; in this case it was arbitrarily defined as event #1.

[65]Notice that *freq* would never change unless it was shared with some other task. The messages of the next chapter would be a better way to handle this; a message could come from an input task tied to a keypad.

```
[66]bb_enable_event(interval);
 forever{
[67]  status=bb_wait_for_event(interval,width);
[68]  P1= (status)?1:0;
 }
}
void main(void{
 bb_init_bos();
 bb_cteate_task(&pulse_tcb,pcontext, pistack,
                pxstack,0,0,1);
 bb_cteate_task(&interval_tcb,icontext, iistack,
                ixstack,0,0,0);
 bb_run_task(pulse_task,&pulse_tcb);
 bb_run_task(interval_task,&interval_tcb);
 bb_start_bos();
}
```

## COMMUNICATION AND SYNCHRONIZATION

The earlier examples involve separate, unrelated tasks, which begin at set times. To produce a flexible, adaptive system while still maintaining separate tasks, it is necessary to have communication *between* tasks. Earlier in this chapter, I discussed some of the communication mechanisms of RTOS. In the various RTOS, there are significant differences between mechanisms for signaling, messages, pools, resource, and protection.

### Signaling

***DCX*** has no signaling mechanism as such, but empty messages can serve a signaling purpose.

***RTXtiny*** is also limited in signaling. Each task has only one signal bit. Each task wishing to send a signal may send the signal to one specific task

[66]This task has to be set up to accept events. Here it is actually only allowing one event with the mask of 0x0001.

[67]All the other programming was to get this line to work! The interval is the event for which the task waits and the width value is the timeout delay. In this system, it might be simpler to solve this problem by computing an off-delay by subtracting *width* from *freq* and using two simple *delay* calls.

[68]The zero status indicates the task did time out and the on time of the pulse should end (a zero out port *P1*). Otherwise, since the periodic event called *sampling* is the only other event for this task, it is time to start a new pulse.

at a time. Clearly this is quite simple and must be fast to execute, but the complex options of the following systems are not going to be easily reproduced.

***USX*** calls semaphores *events*—their *event* holds a number. They are created at configuration time and belong to no particular task. Any task(s) can set, clear, decrement, or increment them and can wait for a zero or non-zero condition to 216 come to any particular flag. Event *groups* are more like binary semaphores—16-bit sets of flags that are either set or cleared. A task can be *ASLEEP* waiting for a particular combination of any or all of those 16 bits before proceeding. As with the next two systems, the options become rather mind-boggling!

***CMX*** has two categories of events that need to be distinguished. *Flags* relate to a specific task while *events* are separate signals that can relate to all the tasks. Both *flags* and *event groups* are 8-bits wide. Each task automatically has a single flag byte while *event groups* must be created. Like USX, waiting for a flag or an event involves a mask and a condition test so you can wait for only specific bits being high or low while ignoring all the rest. There are a huge number of system calls because, for most calls, there are waits with *and* without timeouts both for flags *and* events. There are also calls that can wake another task that has suspended itself—a directed signal with no conditions. The options are so many that you will probably go through life using only some of the commands without noticing the lack of the others. There are differences between calls that, much like the choice of variable type, can affect efficiency, but the job can be done several ways. Very careful planning is suggested before using event groups with multiple tasks.

***Byte-BOS*** has provision for user-defined *events* (signaling). There are at least sixteen event bits and events can be selectively *posted* to one task or all tasks. Each task automatically gets just one event holder (integer). Likewise, one task can wait for any of a number of event bits to be set or it can wait for all of a specific set of event bits. It will take some time to really use the options to their fullest, but the number of calls will seem larger than with some other systems because the various options within the same function are again broken out as separate calls.

## Messaging

***DCX*** automatically gives just one mailbox per task and the messages go directly to tasks. Except for off-board messages, the only information that moves is the pointer to the start of the message. A 7-byte header carries information about the destination (address) and the source (return address) of the message as well as the length. The sending task must consider the RAM space as unavailable once the message is sent; otherwise it could corrupt the data before the receiving task has used it. The messages on-board can be in on-chip or off-chip RAM and the latter can be of essentially any length.

***RTXtiny*** has no message mechanism and would have to alert a task to a message by use of signaling.

***RTX*** apparently has eight mailboxes total with fairly short messages, but the details were not available.

***USX*** passes a pointer to the message, which can be any arbitrary array or structure. The message pointer is sent to a mailbox, of which any number can be configured at startup. There is no specific tie between a task and a mailbox. Since only the pointer to the message moves, the sending task can only know if the actual message data space is free for re-use by a return signal from the receiving task or by use of memory buffers, which can be requested and released. Essentially, like DCX, the user has to ensure the *envelope* is free for a new message. USX adds a unique feature where the messages to a mailbox are assigned priorities and the first-in-first-out aspect can be overridden to have a high-priority message picked up first by the task checking the mail.

***CMX*** is the only system that actually transferrs the *contents* of the message when it is sent. The default length (envelope size) is 12 bytes but messages can be as small as 1 byte if so initialized. All envelopes are configured the same size. For efficiency, the system might best be configured for 2- or 3-byte messages which serve as pointers to the actual messages. There are several variations of sending and receiving depending on whether the sender should go to sleep, wait for a reply, wake up a sleeping destination task (have the mailman ring the doorbell?), etc. Likewise, the receiving task can sit and wait until a message arrives, wait for a specified time, or just check the mailbox with delay if no message has arrived. The large variety of choices allows the system to perform actions that would otherwise require combining several simpler calls.

***Byte-BOS*** has what it calls *low-level* and *high-level* functions. The distinction is that low-level calls do not suspend the task—either the message has already arrived or the call returns with the indication that no message was waiting. High-level calls can wait for the message to come (or timeout). The high-level call *sleep_for_message* is a *wait_for_message* without provision for a timeout.

## Pools, Resources, and So On

***DCX*** has a memory pool manager of on-chip RAM.

***USX*** has a memory pool manager for off-chip RAM management and a resource manager. Incidentally, the next example illustrates resource management in accessing the LCD display *(getres* and *relres)*.

***CMX*** has a memory pool manager for off-chip RAM management, a resource manager, and a unique list manager.

***Byte-BOS*** has a resource manager and a unique UART buffering system.

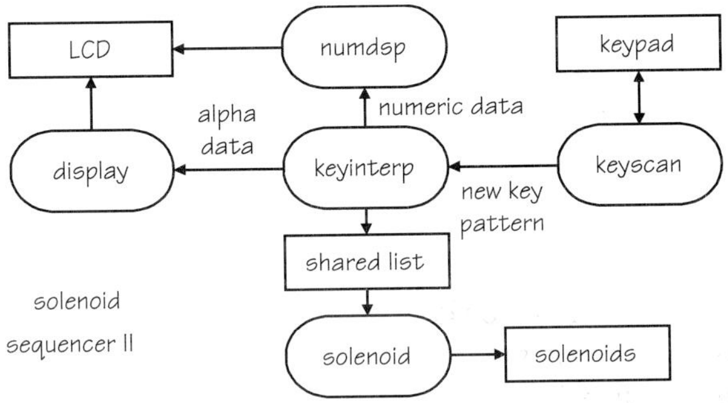


**Solenoid Sequencer II—Task Diagram**

## EXAMPLE: SOLENOID CYCLER II

By adding an LCD display task, adding a control task, and revising the keyscan task to the first solenoid cycler, a user programmable sequencer is produced. The task block diagram will be particularly helpful in visualizing the interaction.

The lead-in to *display* task initializes the LCD. The details are not of particular concern here; it is just a succession of writes to the LCD module which happens to be a standard $2 \times 40$ module.

The *keyscan* task handles the reading of a user keypad. The keypad is a $4 \times 4$ matrix. The *keyscan* task takes care of the scanning at a 40-msec cycle time to eliminate bounce. When it is recognized that a button has changed state, the value is forwarded to the *keyinterp* task.

The *keyinterp* task updates the solenoid sequence list. The list is shared with the *solenoid* task, which operates much like an earlier example. The information is also sent to the *display* task so the user can see the result of pushing a button.

To simplify the example, only a bare minimum of user inputs and displays is used—user-friendly features should be added for a real system.[69]

---

[69]For example, the solenoid sequence is not reorganized if times are put in out of order. A three-key sequence is demanded even if the first digits are zero, and there is no backing up from a wrong entry.

The task deciphers the *keyscan* input to assemble a decimal number or to use three of the six alpha keys to carry out pre-defined functions. The absolute time in decimal is entered numerically. The ***A*** key means go on to the next time in the sequence. The ***B*** key increments the two states of the two solenoids (four possibilities) with "on" and "off" showing under the respective labels on the display. The ***F*** key toggles the start/stop of the actual solenoid sequencing. Only when stopped is the user allowed to change the sequence table. Presumably custom key labels would be put on the alpha keys of the keypad. This could be refined and made more friendly, but the goal here is to illustrate the multitasking.

When the *display* task starts up it initializes the LCD display.

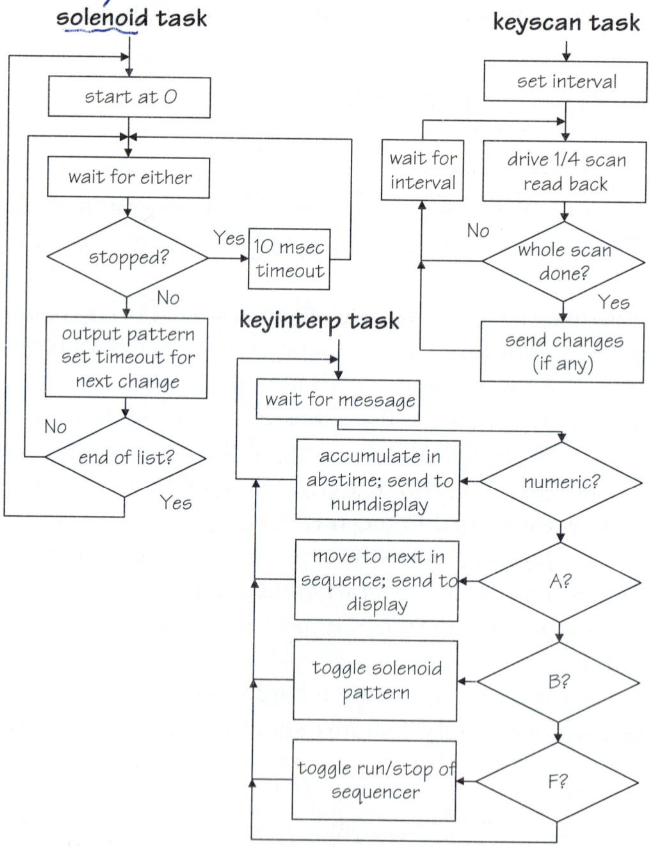


**Solenoid Sequencer II—Flowcharts (Part 1)**

All alphanumeric LCD modules seem to be based on the same controller chip, which makes set-up sequences identical. The LCD programming sequence used is better understood from the data sheets; here just accept it as a given.[70]

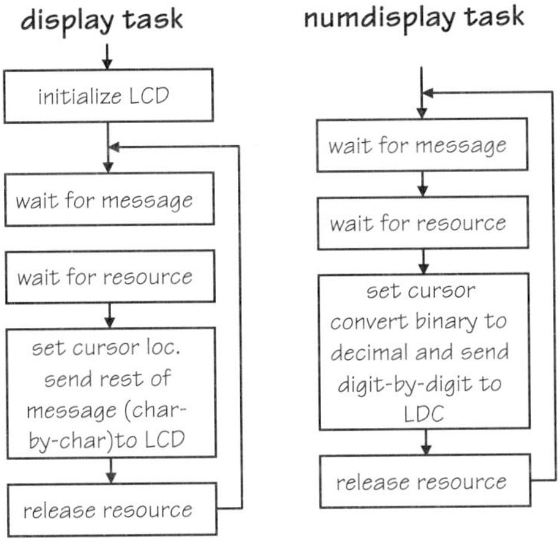


**Solenoid—Flowcharts (Part 2)**

To avoid conflict with the *numdisplay* task, resource management is used where the multitasking system could otherwise allow both tasks to access the LCD at the same time.

The *numdisplay* task illustrates converting decimal to ASCII.

This is the first true *multi*tasking example in the book. The task block diagram is an excellent way to visualize the interactions. The example employs protected resource management so the LCD is not addressed by two tasks at once. As the examples grow in complexity you begin to see the advantage of multitasking as well as the reasons to make "generic" driver tasks that can be reused.

---

[70]The modules have a 450 nsec access time, which makes interface directly to a 12 MHz 8051 somewhat marginal, but it seems that hundreds of projects have gotten by because it is a worst-case specification. If you don't want to drop back to 10 MHz but want to stay strictly within specifications, it is possible to I/O drive the display by generating a write pulse through toggling a port bit in software. Well-written software will isolate the interface in a driver routine so, if the hardware changes, there is only one place needing the software change.

**Solenoid sequencer II—USX**

```
#include <usxcfg.h>
#include <usx.h>
#include<reg51.h>
#include<absacc.h>
#define forever for(;;)
[71]#define uchar unsigned char
#define uint unsigned
#define tenmsecs 1
#define keymbx 1
#define dspmbx 2
#define nummbx 3
#define lcddat XBYTE[0xc000]
#define lcdcmd XBYTE[0x8000]
#define pcmd XBYTE[0x2003]
#define PORTA XBYTE[0x2000]
bit goflg;
struct {uint abstime;uchar pattern;} next [20];
uchar entry;
uint status,t;
uchar slotini,slotdsp,slotnum,slotsol,slotks,
      slotki;
uchar xdata free_memory[MEM_ALLOCATION];

void solenoid(void){
 uchar code scanpat[]={0x10,0x20,0x40,0x80};
               /*solenoid drive portion*/
 uint nowtime;
 period(tenmsecs,1);
 forever{
[72]  wteset(tenmsecs,0);
  decevt(tenmsecs);
  if (goflg){ /*freeze if stopped*/
   if (++nowtime==next[i].abstime){
    if (next[i].pattern==0xff)i=nowtime=0;
    PORTA=next[i++].pattern;
   }
  }
```

[71]Labels help make more sense out of system calls.

[72]This models the non-RTOS version with a regular 10-msec event and a counter. The approach could be rather to do a simple wait for the computed time until the next state change for the solenoid. Depending on the overhead for context switching, the latter might be more efficient for only the solenoids, but by combining the *keyscan* in the same task, the one event does double duty.

```
[73] P1=scanpat[j]; /*added keyscan portion—1/4 scan
                 each time*/
     newsc=(newsc<<4) | (P1 & 0xf);
     if(++j==4){
      j=0;
      scan=oldsc=newsc;
[74]  if (newsc!=oldsc) sndmsg(keymbx,&scan,5);
      newsc=0;
     }
    }
   }

   void display (void){
    struct {uchar loc;uchar ch[12];}*dmsg;
    uchar i;
[75] lcdcmd = 0x30; for(t=0;t<400;t++);
    lcdcmd = 0x30; for(t=0;t<400;t++);
    lcdcmd = 0x30; for(t=0;t<400;t++);
    lcdcmd = 0x38; for(t=0;t<400;t++);
    lcdcmd = 0x0e; for(t=0;t<20;t++);
    lcdcmd = 0x01; for(t=0;t<20;t++);
    lcdcmd = 0x06; for(t=0;t<20;t++);
    forever{
[76]  dmsg=rcvmsg(dspmbx,0);
[77]  getres(lcd,0);
     lcdcmd = 0x80 | dmsg->loc; /*start location*/
     for(t=0;t<20;t++);
     for (i=0;dmsg->ch[i]!=0;i++){
      lcddat = dmsg->ch[i];
      for(t=0;t<20;t++);
     }
```

[73]The scan is done in four separate parts at a 10-msec rate to even up the loading on the individual events.

[74]The message holding the changed scan result is sent to the *keyinterp* task. It is given an arbitrary priority as a message.

[75]This is the setup of the LCD module. The details are derived from the data sheets and are of no particular interest here.

[76]The task waits *forever* (0) for a message in its mailbox. The returned result is the pointer to the message—the message probably sits where the sending task put it and should not be overwritten by the sending task before the receiver is finished with it. The part ahead of the forever is the one-time initialization of the LCD.

[77]This resource manager claims the exclusive use of the LCD. There is nothing inherent to stop another task from use of it except good programming practice.

```
[78] relres(lcd);
  }
 }

void numdisplay(void){
 struct {uchar loc;uint dat;}*nmsg;
 uint number;
 forever{
  nmsg=rcvmsg(mummbx.0);
  getres(lcd,0);
  lcdcmd = 0x80 | nmsg->loc;
  for(t=0;t<20;t++);
  number=nmsg->dat;
  uint m = 10000;
  bit ldgzero = 0;
  do{
   if ((((number/m)>0) || (m == 1)) || ((m == 100)
              && (dpt == 1))) ldgzero = 1;
   if (ldgzero == 1) lcddat = number/m+'0';
   else lcddat = ' ';
   for(t=0;t<20;t++);
   number %= m;
   if (m == 100) {
    lcddat = '.';
    for(t=0;t<20;t++);
   }
  }
  while((m /= 10)>0);
  relres(lcd);
 }
}

 /*following 2 functions are not tasks—go with
              keyinterp*/
void solenoidupdate(void){
 if (next[entry].pattern & 1) sndmsg(dspmbx,&(92,
              "on"),5);
 else sndmsg(dspmbx,&(92,"off"),5);
 if (next[entry].pattern & 2) sndmsg(dspmbx,&(81,
              "on"),5);
 else sndmsg(dspmbx,&(81,"off"),5);
}
```

[78] This releases the LCD for some other task's use.

```
void row2update(void){
 struct {uchar loc,uint numb} envelope;
 envelope.loc=65;
 envelope.numb=entry;
 sndmsg(nummbx,&envelope,5);
 numdsp(71,next[entry].abstime);
 envelope.loc=71;
 envelope.numb=next[entry].abstime;
 sndmsg(nummbx,&envelope,5);
 solenoidupdate();
}

void keyinterp (void){
 uint *msg;
 uchar x;
 uint scan;
 bit newnumbflg;
 sndmsg(dspmbx,&(0,"entry"),5);
 sndmsg(dspmbx,&(8,"time"),5);
 sndmsg(dspmbx,&(15,"solenoid 1"),5);
 sndmsg(dspmbx,&(26,"solenoid 2"),5);
 forever{
79 msg=rcvmsg(keymbx,0);
  scan=msg; /*new key condition to process*/
  if (goflg==0){ /*update only if stopped*/
   if (scan==0x0000);/*0->key released*/
80 else if (scan<0x0200){ /*0-9 key*/
   if (newnumbflg==0) {
           next[entry].abstime=0;
           newnumbflg=1;
   }
   for(x=0;scan & 1==0;scan>>=1)x++;
   next[entry].abstime= 10*next[entry].abstime+x;
   sendnumb(71,next[entry].abstime);
  }
  else if (scan==0x0400){ /*"a" key*/
```

[79]The task waits *forever* (0) for a message from the keypad.

[80]The first ten bits represent the 0 to 9 keys. Multiplying the previous entries by ten carries out the numeric conversion since each additional key indicates that the previous keys were one place over from the previous knowledge.

```
    if (entry<18) entry++;
    newnumbflg=0;
    row2update();
   }
   else if (scan==0x0800){ /*"b" key*/
    newnumbflg=0;
    next[entry].pattern&=0x3;
    solenoidupdate();
   }
   else if (scan==0x1000); /*"c" key*/
   else if (scan==0x2000); /*"d" key*/
   else if (scan==0x4000){ /*"e" key*/
    if (entry) entry—;
    newnumbflg=0;
    row2update();
   }
  }
81 if (scan==0x8000) goflg=~goflg;/*"f" key*/
   }
  }

void main(void{
 MEM_DEF *memptr;
 usxini();
 usrclk_init();
 memptr=(MEM_DEF *)free_memory;
 memptr->mem_size=MEM_ALLOCATION-sizeof(struct
              mem_def);
 memptr->task_slt=SYSTEM;
 memptr->mem_nxt=NULL;
 relmem((usxmem_t *)(++memptr));
 pcmd=0x80; /*initialize solenoid I/O*/
 nowtime=newsc=goflg=0;
 for(i=0;i<20;i++) next[i].abstime=
              next[i].pattern=0;
 next[0].pattern=0xff;
 slotdsp=runtsk(5,&display,20);
 slotnum=runtsk(5,&numdisplay,20);
 slotsol=runtsk(10,&solenoid,20);
```

[81]The “F” key must be interpreted outside the *goflg* test; otherwise the flag would block ever changing the flag back.

```
    slotks=runtsk(5,&keyscan,20);
    slotki=runtsk(5,&keyinterp,20);
    begusx();
}
```

## INTERRUPTS

### Built-in Interrupt Events

From an RTOS-user standpoint, to ***DCX*** the interrupts are simply other events, with the details handled in the operating system. Hardware interrupts are permanently assigned to specific tasks by the initial task descriptor (ITD) file. Tasks created later can take over the specific interrupt because it "belongs" to the last-encountered task claiming it. For ***RTX51*** (*not* RTX-tiny) the same mechanism is in place, but the interrupt is assigned or reassigned to a specific task by *rtx_attach_interrupt*.

### Interrupt Handlers

For the other RTOS, due to the context switching concerns, interrupts are a major issue. Since the RTOS is just some code which happens to be running, if your interrupt is unmasked and takes place, the processor does the usual saving and hands control to your interrupt task. You do not have to give control back to the RTOS! You could trash stacks, delay time-slices or timeouts, and generally foul up the system! At the same time, you could get something done quite quickly without all the delays inherent in the RTOS. As a result, all the systems admit you *can* write interrupt handlers. As long as you don't unmask anything and don't use the operating system calls within the handler, you have essentially preempted the operating system. If you take too long in your handler, of course, the regular system tick will be preempted so timeouts and time intervals will get stretched. Assuming your interrupts are carefully written to be fast and short—if you need to do lots of processing, set a flag to be picked up in the main program or in a task—you can get quick urgent things done efficiently.

### Interrupts with the Operating System

Signaling low-priority tasks to carry out the work leads to the another use of interrupts. If you want to interact with the RTOS from your interrupt handler, you get into context switching. Some RTOS make provision for this.

***RTX51*** allows interrupts to send and receive messages as well as assigning interrupts directly to specific tasks.[82]

***USX*** has an indirect call, *USXCMD_C*, which can activate any of the normal system calls (probably by adding the necessary context switching).

***CMX*** has *cxint_in* and *cxint_ex* which inform the RTOS to save and restore the previous context; from there all the system calls can be used in the normal manner. Incidentally, CMX expects the user to write the interrupt to cause the system tick rather than hiding it in other code.

***Byte-BOS*** lets the interrupt service routines post events (signaling), wake-up tasks, and invoke the task scheduler. If the normal system calls are to be used, the *enter_isr* and *exit_isr* calls perform context switching.

In most RTOS, task interrupts will be masked when equal or higher priority tasks are running, so the use of different priorities may become unavoidable. Tasks that run quickly and are not used again for long times can be put at a high priority or they can be written as ISRs without involving the operating system and context switching at all. A candidate for this would be a stepper drive task, which takes only a few microseconds every few milliseconds, yet if it gets held up too long, the motor appears to stutter.

### Example: Jam Detector

The envelope system of Chapter 1 could easily include hardware interrupts. Assuming there are a number of sensors along the path of the system, the microcontroller would have to either poll the sensors or use some sort of *wired-OR* to bring the sensor signals into a single hardware interrupt.[83] For this example, only the particular sensor causing the interrupt (from subsequent polling) will be sent to another task (for further processing). Each version of the example assumes the resources of the previous examples, but to save space, the tasks will not be repeated here.

---

[82]RTXtiny avoids the question by only allowing the sending of a *signal* to a task from the interrupt (probably just setting a bit).

[83]A possible system for bringing in multiple photo-interruptor signals as interrupts could use a 74148 priority encoder chip. Alternately, the simple open-collector signals from the photo-interrupters could provide a wired-OR directly, but then it would be impossible to separate out the source to determine which input caused the interrupt.

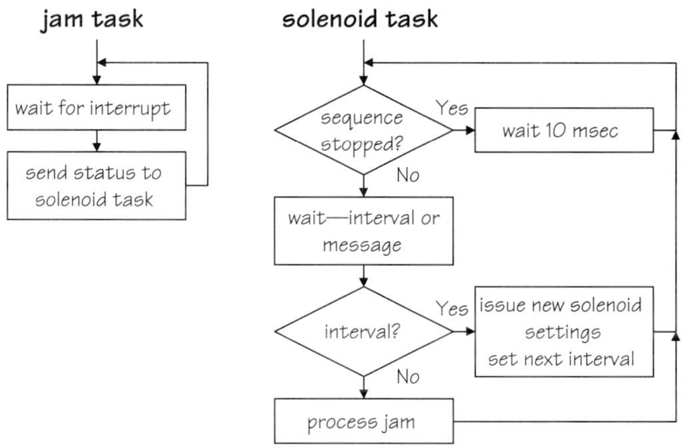


**Jam Detector—Flowchart**

---

**Envelope detector—USX**

```
#include <reg51.h>
#include <absacc.h>
#include <usxcfg.h>
#include <usx.h>
#define forever for(;;)
#define uchar unsigned char
#define uint unsigned
#define solmbx 4
#define solenoids XBYTE[0X2000]
#define portcmd XBYTE[0X2003]
uchar slotini,slotsol;
uchar xdata free_memory[MEM_ALLOCATION];

[84]id jam (void) interrupt 2 using 2 {
  uchar envelopestatus;
  envelopestatus=P1;
[85] usxcmd_c(SNDMSG,solmbx,&envelopestatus,100);
 }
```

[84]Notice that *jam* is not an endless loop—it is an interrupt rather than a task to USX.

[85]System calls can be invoked only indirectly as shown. It probably leads to a context switch.

```
void solenoid(void) {
 uchar *envelopestatus;
 forever{
  for (i=0;next[i].pattern!=0xff;i++){
   if (goflg){ /*freeze count if stopped*/
    solenoids=next[i].pattern;
    dlytsk(slotsol, next[i+1].abstime-
             next[i].abstime);
86   if (chkmsg(solmbx)){
       envelopestatus=rcvmsg(solmbx,0);
       while (P1>0){ /*until jam is cleared
       sound alarm,shut off motors */
    }
   }
  }
  else {
   dlytsk(slotsol,1);
   i--;
  }
 }
}

void main(void{
 portcmd=0x80; /*set up 8255*/
 MEM_DEF *memptr;
 usxini();
 usrclk_init();
 memptr=(MEM_DEF *)free_memory;
 memptr->mem_size= MEM_ALLOCATION-sizeof(struct
             mem_def);
 memptr->task_slt=SYSTEM;
 memptr->mem_nxt=NULL;
 relmem((usxmem_t *)(++memptr));
 IP=0x04; /*set IE1 high priority*/
 IE=0x86; /*enable IE1 and IT0*/
 /*other task initialization here*/
 slotjam=runtsk(10,&jam,20);
 solslot=runtsk(200,&initialize,20);
 begusx();
}
```

[86]It is not possible to produce a delay with the possibility of a message all in one call. Here the checking for a message comes only at the end of each step in the sequence.

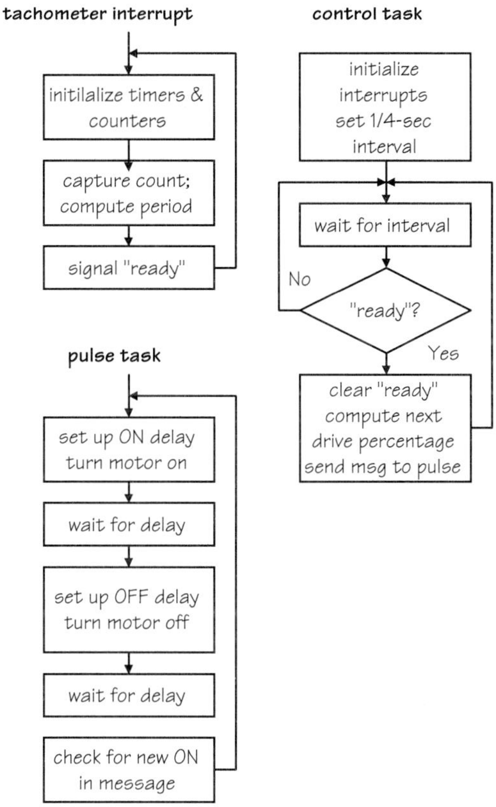


**Speed Control—Flowchart**

## EXAMPLE: MOTOR SPEED CONTROL SYSTEM

For a completely digital motor speed control system it is interesting to combine a PWM drive to a DC motor with an optointerrupter monitoring the shaft turns. Although some systems have hardware that puts out multiple pulses per turn and then count pulses for a fixed time, it is also possible to determine rpm where the hardware provides only one pulse per revolution by measuring the time between pulses (one revolution). Interrupt latency might be of concern because the goal would be to read the time as quickly as possible after the turn is completed.

In the example code, the rpm-measuring task determines the time for a revolution by reading timer 1 after an interrupt and forwarding the difference from the last time to a control task, which does a simple up-one-or-down-one control algorithm. It might someday provide proportional-integral-derivative (PID) algorithms but such calculations are not the focus here. The control task sends a new *width* to the *pulse* task adapted from an earlier example. A number of tradeoffs can be considered, but they will be discussed in *Algorithms* (Chapter 6).

**Speed control—USX**

```
#include <reg51.h>
#include <usxcfg.h>
#include <usx.h>
#define forever for(;;)
#define uchar unsigned char
#define uint unsigned
#define ready 1
#define ctrlmbx 4
uchar slotctrl,slotpulse;
uchar xdata free_memory[MEM_ALLOCATION];
uint target;

[87]void tachometer(void) interupt 3 using 2 {
  union split{uint word; struct{uchar hi;uchar
             low;}bytes}};
  union split newcount;
  uint oldcount,period;
  newcount.bytes.hi=TH1;
  newcount.bytes.lo=TH0;
  period=newcount.word-oldcount;
[88] usxcmd_c(setevt,ready);
  oldcount=newcount.word;
}

void control (void) {
  uint period;
  IP=0x04; /*set IE1 high priority*/
[89] TMOD=0x11; /*T1 in endless 16-bit overflow mode*/
  TCON=0x54; /*T0,T1 running, IE1 edge triggered*/
  TH1=0; TL1=0; /*set initial count (unnecessary)*/
  IE=0x8c; /*enable IE1 and IT1*/
```

[87]This is an interrupt rather than a task. It cannot be a *forever* loop. It is not clear whether a separate register bank is needed.

[88]An interrupt cannot receive an event signal. So it has to *send* a signal to the other task saying that a new reading is available.

[89]The initialization of timer 1 and the external interrupt for the tachometer can be done here or in the initialization.

```
  period(ctrlintvl,25);
  forever{
90  wteset(ctrlintvl,0); /*update every 1/4 Sec*/
    decevt(ctrlintvl);
91  if (chkevt(ready)){
     clrevt(ready);
      if (period < target){
       if (percent<100) ++percent;
      }
      else if (percent>0) —percent;
      sndmsg(pulsembx,&percent,5);
     }
    }
  }

  void pulse(void){
    for(;;){
92      P1=1;
        dlytsk(slotpls,1,percent);
        P1=0;
        dlytsk(slotpls,1,100-percent);
    }
}

  void main(void{
   MEM_DEF *memptr;
   usxini();
   usrclk_init();
   memptr=(MEM_DEF *)free_memory;
   memptr->mem_size= MEM_ALLOCATION-sizeof(struct
                mem_def);
   memptr->task_slt=SYSTEM;
   memptr->mem_nxt=NULL;
   relmem((usxmem_t *)(++memptr));
   /*other task initialization here*/
   slotctrl=runtsk(10,&pulse,20);
```

[90]To keep the period between updates of the control parameter constant, an interval of 25 × 10 msec is used.

[91]The *ready* event comes from the tachometer every time it gets a new reading. If a new reading has not arrived the control update is not done. Note that here the event is set and cleared rather than incremented and decremented. If multiple new readings have arrived in the 250 msec, it is only the latest one that is desired.

[92]This is simplified before because there is no real need for an interval.

```
    slotctrl=runtsk(5,&control,20);
    begusx();
}
```

## BENEFITS OF ANY RTOS

1. When you break jobs into tasks, it becomes ***easier to grasp the overall picture*** without becoming lost in the details. In addition to formalizing the process, the operating system can make the interrelation between tasks more comprehensible to someone else. With a formal relation between tasks, there will be less chance of overlooking or forgetting details. When the inevitable revisions come, there will be less start-over-from-scratch work.
2. A second benefit is the ***simplicity of handling real-time inputs***. Compared to the single monolithic program approach, multitasking can make it easy to respond to real-time demands. If you need to scan the keys every 40 msec, design a task to sleep just that long, no matter what else is happening. If the display *ought* to be updated every 100 msec, because it is not mandatory, arrange it to have a lower priority. The operating system does not perform magic. There is never more than 100 percent of the processor's time available, but an operating system makes it possible to clearly define the different activities and the degree of urgency (priority) of the activities. One still faces the *"What if this needs to happen when that is going on?"* questions, which are part of any real-time system. Compared with single monolithic programs, however, the solutions are more obvious to someone else, and are not buried in the inter-relation of large code modules.
3. Third, ***inter-task communication (signals and messages) provides a solid method of controlling execution order and timing***. In other words, it is possible to move from one activity to another in a way determined by the events that have happened. Unlike with a scheduler you would write yourself, you do not lock in the sequence. At one time, you can send a message from one task to start a second task running, but at some other time you can send the message to a third task instead.
4. Fourth, ***tasks can be independent*** so some of the interaction details of single-program solutions are avoided. They can be *small,* making them easier to manage.

## BENEFITS OF A COMMERCIAL RTOS

*All* real-time operating systems aid in *formalizing* the relationship between activities so you can simplify software development. Having briefly described commercial *real-time operating systems* (RTOS), it is time to compare them with traditional monolithic programs and the "write-your-own" approaches of the next chapter. What are the reasons to go to a *commercial* RTOS?

1. It is likely that commercial systems have been fully debugged and tested. Someone has an ongoing commitment to improving and fixing them. Commercial RTOS probably have a stability that is particularly important for projects in which there are *several* programmers working over a long period of time and where frequent revisions are likely.[93]
2. The larger 8051 RTOS are portable to other processors. If you develop C code that has to migrate up to faster machines with different architecture, the operating system calls can move with the rest of the code.
3. You are building on a much broader knowledge base. My do-it-yourself systems (see Chapter 3) are not based on any large-system experience, nor do I have any formal course work in real-time systems.
4. There is standardization to using a commercial RTOS. By its very nature, the software is not customized to one application, so when it is moved to a new project, you do not have to remove the special features built in for the first job.

On the other hand, there are some benefits to the build-your-own approaches of the next chapter:

1. There are no license fees to pay or any initial purchase costs.
2. They may run more efficiently; they can be quite small.
3. Given enough programming effort, they can have many of the same features as commercial operating system.
4. All the code is visible and you control it. You can fully understand the program. You are free to use and modify any of the source code incorporated in any application you develop. There is no risk of the company going out of business and leaving you without support.[94]

---

[93]That probably describes most microcomputer projects!

[94]On the other hand, there is the very real risk for your *company* that you may leave and no one else can support your obscure programming!

## COSTS OF RTOS

What does multitasking *cost?* I have mentioned some of the costs already.

1. The ***hardware must have a specific timer and interrupt structure***. Versions of some RTOS exist for a variety of processors but there still must be certain hardware in those systems and you must configure the RTOS to match the specific hardware.
2. In addition to the hardware requirement, ***multitasking costs time***. The *overhead* of all the operating system activity uses up computer time. Periodic interrupts to update tables cost time. Sorting through to determine which task should run takes time. If the task relationships are very simple and inflexible, it might be better to write a monolithic program. Of course, if the processor load is too heavy, it may mean that a different approach is in order—a faster processor, multiple processors to handle different pieces, or specialized devices such as math coprocessors.
3. A third very real cost is the ***time it takes to learn the system***. The application programmers must get up the *learning curve* far enough to become productive. Again, for a single, very small project, multitasking probably does not make sense. The formalization and modular nature of the development process can lead to much more *maintainable* code. It is the same problem faced in making a transition from assembly language to a higher level language.

Is it worth the time and pain to learn an operating system? The answer depends on your situation. DCX and RTXtiny have very few commands to learn while USX, CMX, and Byte-BOS are much more complicated. In another sense though, there is less for you to do with the latter because the large variety of system calls can avoid the need for multiple calls in sequence.

# 3

# Build Your Own: Scheduler or Kernel

Up to now, the multitasking was all in general terms using commercial system calls. In this chapter, I want to take you inside several operating systems and show you how they work. First comes a *scheduler* and then two *kernels*—first a very small one and then a more complex one. None of them come close to the complexity of the larger commercial systems described in the last chapter, but they give you a taste for the techniques involved.

Some distinction is made between an operating system and a ***kernel*** or ***executive.*** The former, often *including* a kernel, is usually more elaborate, possibly including drivers for specific hardware such as disc drives and networking. The latter, when a distinction is made, refer to the most basic part of the system that tracks priorities and manages task switching. I will refer to the two operating systems that I write as *kernels* because they are primitive compared to what most people now expect of big computer operating systems.[1]

## MULTITASKING HARDWARE REQUIREMENTS

> **MULTITASKING:** *One* timer driving *one* interrupt is all you need for an efficient system

*All* members of the 8051 microcontroller family contain the two hardware features that are essential for real-time systems—timers and interrupts.[2] Be-

[1]Realize that it is only to avoid being confused with the "big boys" that I make the distinction. Actually, I suppose it sounds more powerful to say you are using a *kernel*, since *everyone* uses an *operating system*.

cause the details of these two hardware functions relate so directly to the systems that follow, we will review them here.

### Timer

A ***timer*** independently generates the regular intervals that allow things to happen on time independent of the software that might be running. With the 8051 family, a timer is actually an internal counter running off the microcontroller's clock. Since the counter keeps on counting independent of program flow, it can manage time delays while software is doing something else.

### Interrupt

Along with a timer, most real-time systems have at least one ***interrupt***. You may associate interrupts with signals from external hardware, but the most useful interrupt is one caused at regular intervals by the timer overflowing. Such an interrupt can break the processor out of any task so it can see if anything *else* needs attention. With such an interrupt, you do not have to do anything special in the individual tasks to be sure everyone gets a share of the processor's time.

Other interrupts or a task finishing its activity can also cause task switching, but this will be discussed later.

With multitasking, data processing and other computing can be done in background, lower priority tasks (or in the simplest form in the main program) while time-related things—refreshing a multiplexed display, scanning a keypad, watching for the arrival of a slow event—can be done in the interrupt routine.

> **SCHEDULERS ARE THE EASIEST:** For a small system, nothing beats a scheduler based on a real-time interrupt. You own all the code and have no special commands to learn

## SCHEDULER

By programming a timer and an interrupt found in any member of the 8051 family, you can generate a regular, timed interrupt, called a ***real-time clock***.

---

[2]Find details on setting them up in the Chapter 3 of *Volume I.*

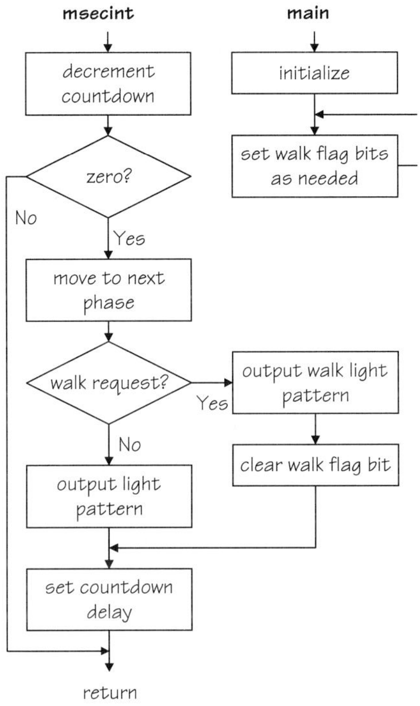


**Traffic Light Flowchart**

Using this simplest form of multitasking, you can make a ***scheduler*** such as the one revisited here from *Volume I*. It uses a real-time clock to run *tasks* at specific times on a *schedule*. The one shown here has no formal structure and there are no system calls. Everything repeats at specific numbers of system ticks. To make the structure more apparent, I could have the scheduled *tasks* be separate functions called from the interrupt routine, but, as I see it, the best way to grasp schedulers is just to study the examples.

## Example: Traffic Light Scheduler

In this first scheduler example (repeated from *Volume I*), the program runs a traffic light. This is a good job for a scheduler because the changes of the light should happen at specific times no matter what else the processor is doing.[3] A normal schedule consists of sending out the light patterns and

[3]For hardware, we can simulate the powerful, probably relay-driven lights with a few LEDs and the rugged pedestrian walk buttons with inexpensive pushbuttons. Later chapters (*Input* and *Output*) go into hardware details which are omitted here.

waiting the specified number of interrupts for each of the six basic parts of the light cycle. Having a real-time clock *ticking* at 1 kHz, the main program can ignore the lights. Pushing a walk request button can change the schedule, but all the state changes of the lights happen in the software that runs when the system clock ticks (real-time clock interrupts).

We can recognize the requests for a *walk* by polling in the main routine. You could have handled the *walk* requests with separate hardware interrupts, but this adds communication between the interrupt activities and the main (background) program. Set up a flag (*walk*) to share between the main program and the interrupt.[4] I arbitrarily choose to make the walk lights the most significant bit (msb) of each nibble of the pattern going to the LEDs.

**Traffic Light With Walk Buttons**

```
#include<reg51.h>
#define uchar unsigned char
#define uint unsigned int
#define LEDS P1
#define PBS P3
uchar walk=0;
struct {uint delay,uchar pattern}cycle[]=
                {{1000,0x12},{3000,0x41},
                {1000,0x21},{5000,0x14},
                {3000,0xc1},{5000,0x1c}};
                /*--R,-Y-;     -G-,--R;
                  -Y-,--R;     --R,-G-;
                  WG-,--R;     --R,WG-;*/
void initialize(void){
 TMOD=0x10;
 TH1=~(1000/265);
 TL1=-(1000%256);
 TR1=1;
 IE=0x11;
}

void msecint (void) interrupt 3 using 1{
 static uchar index;
 static uint countdown;
 uchar i;
 countdown=countdown-1;
```

[4]To be shared it will have to be global—declared outside the functions.

```
 if((countdown==0){ /*update at end of delay*/
  index=index+1;
  if(index==4) index=0;
5 switch(index){
6  case 1:if(walk&0x2>0){
7    walk=walk & 0xfd;
8   i=index+3;
    }
    break;
   case 3:if(walk&0x1>0){
9    walk=walk & 0xfe;
10   i=index+2;
    }
    break;
    default:i=index;
  }
  LEDS=cycle[i].pattern;
  countdown=cycle[i].delay;
  }
 }

 void main(void){
  initialize();
11 for(;;) walk=walk | PBS;
 }
```

## Nothing to Do

Let me repeat that the basic approach to programming for a scheduler (or *any* multitasking system) is to write so the program flow never waits.[12]

[5]Set index to next state.

[6]One of the features of C is the definition of *true* and *false*. *Any*thing other than zero is a value for *true*. Only zero is a value for *false*. In the line below, the *true* condition occurs when the bit is set high. It could have been written as *if (walk&0x02>0);* or *if ((walk&0x2)==2);* Note that the & grouping does not have to have the extra parenthesis because the > or == has higher precedence than the &. Be careful! If in doubt, always use extra parenthesis.

[7]Reset walk request flag.

[8]Use alternate light pattern for walk cycle.

[9]Reset walk request flag.

[10]Use alternate light pattern for walk cycle.

[11]Set bits for walk requests.

[12]For a scheduler that is specifically applicable to the interrupt function.

Rather it moves on through any task as quickly as possible and returns to the main program (or, in a kernel, encounters a system call). In short, you almost never write *while()* or *for(;;)* loops where the exit depends on an outside event. In this traffic light example, because *main* has nothing else to do and stops only briefly each millisecond for the interrupt, the program is actually polling the walk buttons at a fierce rate. It shows how, once you have taken the timing out of the hands of software delay loops, your programs will seldom tax the capability of the microcontroller.

## Communication by Shared Variables

The previous chapter introduced the concept of communication between tasks (semaphores and messages). In the traffic light example, there is communication of the most primitive sort—a ***shared variable***. The interrupt and the main program share two bits of information about the walk requests. In other situations, you might set a flag in one task to *request* processing and then clear it in another task to *acknowledge* that the processing is done. You can exchange not just binary information represented by flags, but also numbers, pointers to messages, or even messages themselves. This is where commercial operating systems shine, but they utilize more elaborate versions of the concepts presented here.[13]

## Example: Solenoid Cycler

Let's revisit the first solenoid cycler example of Chapter 2. It turns on and off two solenoids in a fixed-time pattern. Designate the solenoids as *S1* and *S2*. Here, I program a *fixed* cycle where, at the start they are both off. Two seconds later *S1* comes on. A tenth second later, *S2* comes on. *S1* stays on for 2.0 sec and *S2* stays on for 2.4 sec. For this example, the total cycle takes 4.5 sec.[14] Whenever the number of clock ticks matches the table value, the scheduler issues the next solenoid states. The match with the counter incremented in the real-time interrupt determines the intervals. The basic program flow is shown next.

---

[13]Illustrated in the next chapter.

[14]A dynamically assigned cycle would be more realistic, with user inputs for setup, but this shows the basic principles.

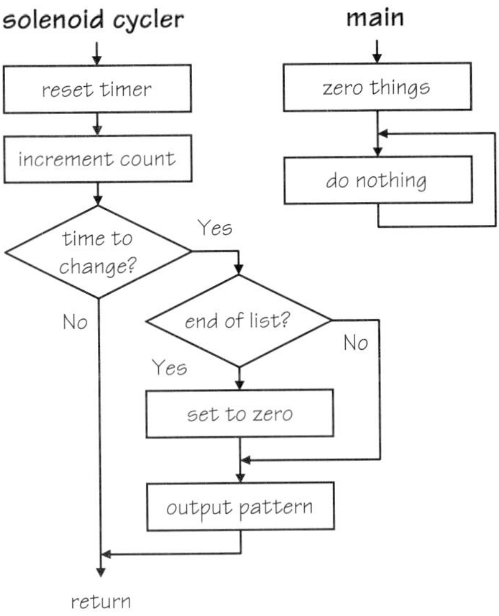


**Solenoid Cycler Flow Chart**

**Solenoid Cycler**

```
#include <reg51.h>
#define uchar unsigned char
#define uint unsigned int
uchar i;
uint nowtime;
[15]struct code {uint abstime;
uchar pattern;}next []= {{0,0x00},{200,0x01},
                {210,0x03},{400,0x02},{450,0xff}};
[16,17]void cycletimer (void) interrupt 1 using 1{
[18] TH0 = -8333/256;
  TL0 = -8333%256;
  ET0 = 1;
```

[15]This structure is a fixed-content schedule for the scheduler. In the next chapter, an expanded version is presented with a user-entered schedule. The first part of each structure element is the absolute time at which the corresponding solenoid states are sent out. Here only two solenoids are used, but the entire byte is available if desired.

[16]This interrupt is the key to the scheduler. Every 10 msec the interrupt causes this interrupt routine to run. The first step is to reset the timer and start it out again so the minimum of time is lost before the next time interval is started.

[17]Real-time interrupt routine happens every 10 msec

[18]10 msec with 10-MHz crystal

```
[19] nowtime++;
  if (nowtime==next[i].abstime){
[20]  if (next[i].pattern!=0xff)i=nowtime=0;
    P1=next[i++].pattern;
  }
}

 void main(void){
  nowtime=i=0;
[21] for(;;);
 }
```

## Example: Jam Detector

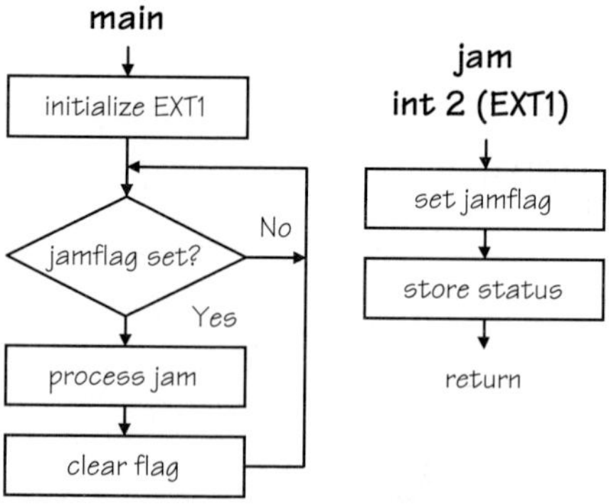


**Jam Detector—Flowchart**

The envelope system of Chapter 1 was shown using hardware interrupts in Chapter 2. This scheduler example only reports the particular sensor causing the interrupt (from subsequent polling) to another task (for further processing). The communication is by shared variables.

[19]This variable is the counter of 10-msec units into the cycle.

[20]The end of the cycle is indicated by a 0xff for the pattern.

[21]Endless do-nothing background task.

**Envelope detector**

```
#include <reg51.h>
#define uchar unsigned char
#define uint unsigned
uchar status;
bit jamflg;

void jam (void) interrupt 2 using 2 {
 jamflg=1;
 status=P1;
}

void main (void){
[22]IP=0x04;
[23]IE=0x86;
 for(;;);
  if (jamflg) {
   switch (status){
[24]    case 0: break;
    case 1: break;
    case 2: break;
    default:;
   }
   jamflg=0;
  }
 }
}
```

## Example: Pulse Generator

The final scheduler example, also repeated from the previous chapter, brings about a regular, variable pulse-width signal that you can use to drive a motor or as a "poor man's" D-A converter. Although you cannot change the values of *width* and *frequency* in *this* example, for a finished design you would probably change them from outside by some user-input task using shared variables.

---

[22]Set IE1 high priority.

[23]Enable IE1 and IT0.

[24]Handle jams here.

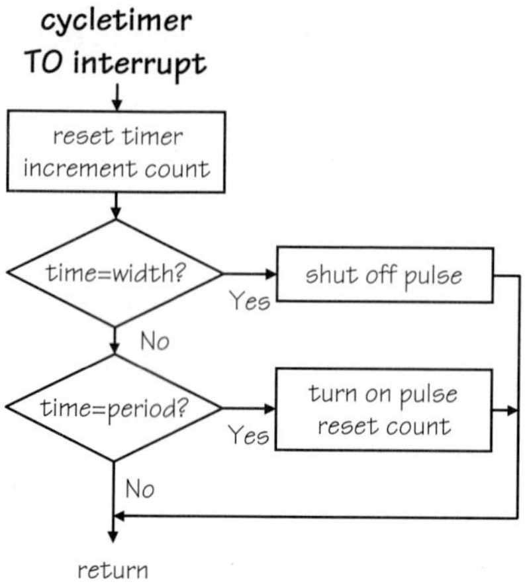


**Pulse Task Flowchart**

---

**Pulse Width**

```
      #include <reg51.h>
      #define uchar unsigned char
      #define uint unsigned int
      uchar i; uint nowtime;
25,26 uchar width=50;
      uchar freq=250;
27    void cycletimer(void) interrupt 1 using 1{
28     TH0 = -8333/256;
       TL0 = -8333%256;
       ET0 = 1;
29     if (++nowtime==width) P1=0;
       else if (nowtime==freq){
        nowtime=0;
```

[25]These variables are set to initial values so the program will do something on startup. In Chapter 11, a similar version of the pulse function is used with a varying input from a control function.

[26]20% duty cycle.

[27]Real-time interrupt routine happens every 10 msec.

[28]10 msec—10 MHz crystal.

[29]*Nowtime* is incremented before the test. If the time reaches the number in *width* the high part of the pulse is ended while if the time count reaches *freq*, a new pulse is started.

```
        P1=1;
    }
}

void main(void){
 forever{}
}
```

## Delayed Tick Due to Overload

If there is a lot to do during the interrupt (the system tick) the interrupt processing can go on until processing time exceeds 1 msec. At that point, there would be no time before the next interrupt. As configured up to now, any activity in the interrupt requiring longer than 1 msec will still run without interruption. This is because, in the 8051 family at least, all interrupts are automatically *disabled* by the hardware until the interrupt routine is exited (re-enabling of the interrupt is automatic at the return from the interrupt with a *reti* instruction). The hardware would delay the *next* interrupt: when the first, overly long interrupt finished, it would lead immediately into the start of the second interrupt. In other words, the system *delays* the second interrupt but allows it to start when the first one finishes. As long as the system never got *two* ticks behind, all the processing would happen and hopefully things would get back in step. If it got two ticks behind, the second timer overflow would set a flip-flop that it already set before and one tick would be lost.

## Catching All Ticks

If you *enabled* the interrupts right at the *start* of the interrupt, then the interrupt could interrupt itself! That way you could process successive ticks before the long processing of the first tick had finished. If you do this, however, be very careful. Remember that you may end up processing the variables in reverse time-order and, if you do not put them on the stack, they may be overwritten. Such an interrupt function must be ***reentrant*** so that the second invocation does not mess up something going on in the first invocation. At a minimum, to be reentrant, the function must have no static internal variables and should affect no global variables. Unless you specify the function as *reentrant* in the top line, to be more efficient and because of the limited stack space, some 8051 C compilers will still keep *automatic* variables in fixed locations.

## KERNEL AND OPERATING SYSTEM FOR MULTITASKING

All the RTOS of the last chapter are not magic! They all rely on a means to frequently check (even in the middle of a task's running) for *other* tasks that have become more urgent than the *current* activity. A clever programmer can do that *without* an operating system, but as programs grow larger and involve *several* programmers, this becomes very complicated to manage.

*Why* might you want to use a real-time kernel (or operating system)? The arguments were put forth in the last chapter, but let me repeat a few:

A formal system avoids some of the programming headaches that come with scratch-built embedded control applications. System calls formalize and standardize things.

Unlike other multitasking systems, *some* of the 8051 kernels are very small and control-oriented. They handle multiple real-time inputs and outputs quickly and efficiently. You do not give up as much as might be expected by using a kernel instead of developing similar functionality on your own.[30]

If your projects grow and change, you will not have to reconsider and scrap large pieces of code. With a kernel, the task interactions will not be buried in a subtle, clever (and obscure!) custom approach you have carefully developed from scratch. At the same time, your C programming should be fairly portable to some other higher-performance processor if necessary.[31]

The sections that follow develop two kernels somewhat like the commercial operating systems of the last chapter.[32]

---

[30]Not all the operating systems fit the 8051 well—some were originally for other processors that are more stack-oriented. If it is a C program, it easily ports over to the 8051, but that does not usually make an efficient system. There are a variety of approaches to context switching and keeping track of tasks. The size and speed requirements of your application are important in the choice.

[31]In the debate *between* operating systems, it will be pointed out that "pure" ANSI C should be easier to port to a different processor. You have to weigh speed and efficiency now against code portability in the future. There is no question that C is preferable to assembly if you anticipate any change.

[32]It would take a whole book to fully illustrate the practical uses of such systems—and I only introduce them. Bear in mind that I have only studied the user's manuals for the commercial RTOS (except DCX / BITBUS). I should also mention a book I recently read in more detail, *μC/OS the Real-Time Kernel* by Jean Labrosse (R & D publications, 1601 W 23rd St., Lawrence KS 66046 ISBN 0-13-031352-1). While it takes a different approach, a lot of the terms and concepts are discussed there and it gives you a code-level view of a much more elaborate operating system.

**BENEFITS OF FORMAL OPERATING SYSTEM:** While the code may not be more efficient, for large projects with multiple programmers an RTOS is more efficient overall

### When the Kernel Runs

Keep in mind that *all* kernels (operating systems) are really just additional program code, running every so often to keep track of tasks. *One* high-priority activity, usually not referred to as a task, *is* the operating system. This activity (the operating system) comes into play:

1. any time a running task issues a system call;
2. at each time-increment (system tick—usually an internal timer-caused interrupt); and
3. for most kernels, at any other interrupt.

At such times, the program flow returns to the operating system program. There is a sequence of activities that happens whenever a kernel takes control of the processor. The kernel code begins running (*wakes up*), recognizes what has caused its activation, and saves the return address pointer of the task that *was* running (found on the on-chip stack). The kernel *then* adjusts tables that keep track of task state, interval count, time-out count, etc. Having updated these tables, it *then* searches through various tables and queues to determine which task to next make the running task. Finally, if a different task is to run, it *replaces* the return address pointer in the microcomputer's stack (normally a dangerous practice!) and issues a *RETI* instruction.

From the perspective of any individual application task, a system call is simply a call to a function, which, after some period of time, finishes, at which time your task's program flow resumes. The fact that *other* tasks may have run in-between is invisible to your task's program flow.

### Choices for Context Switching

There has to be some way to move from running the code of one task to that of another. In a kernel, this is done with a set of instructions, called a ***context switch,*** executed every time a change of task occurs. This change can be because a task is ready to give up the processor and wait for a period of time or because something else more urgent needs to happen. The main benefit of multitasking is keeping the processor busy with other tasks.

When switching tasks it is important to know where the *RUNNING* task left off so flow can resume there when the wait is over. The program counter (*PC*) always holds the two-byte address of the *next* instruction to be executed. When a *CALL* is made or an interrupt is recognized, the 8051 hardware automatically pushes those two bytes onto the stack. When there is a return (*RET* or *RETI*), the top-of-stack values are automatically popped to the *PC*, and program flow resumes where it left off. To switch tasks, a kernel has only two choices—either swap the top-of-stack values or move the stack pointer *(SP)*. In either case, the next return brings the flow into the new task rather than the one that is now *ASLEEP*.

The kernel *could* pop the PC values off the top of the stack, save them in a data structure, and push the new task's previously stored PC values onto the stack. Unfortunately, in many cases the old task may have been using the stack, so the changing of *just* the PC on top can mess everything up. The better solution is to move the stack pointer and let each task keep its own stack out of reach of other tasks.

Sometimes a more urgent task may need the processor before the *RUNNING* task is ready to give it up. Rather than having to wait in a *READY* queue, a ***preemptive*** kernel (driven by interrupts) can override the running task and give control to the higher-priority one.[33] If the program flow *instantly* switched to the new task and the new task used the stack and the registers (*R0-R7*), it could destroy intermediate results that the first task was using.

As far as saving the registers, the original hardware developers of the 8051 included an *inherent* mechanism—the four register banks. Whenever an interrupt task runs, changing 1 or 2 bits in the *PSW* register shifts operation to a different set of 8 bytes (leaving the original 8 unchanged for when flow returns). This preserves the 8 bytes (*R0-R7*) in the one task and shifts to a different set of 8 bytes for *R0-R7* in the new task. At the start of the interrupt it may still be necessary to push the value of the *ACC, B, DPH*, and *DPL* registers onto the stack. Even then, a complete context switch on an 8051 can thus involve only four pushes and one bit-change instruction. That is fairly quick, which was the intent of the 8051 hardware designers from the start.

---

[33]If you avoid having preemption, you can still respond fairly quickly if you make sure that the tasks all run only a short time before checking if there is anything else waiting a turn. If you are familiar with Visual BASIC, *not* a preemptive system, the command *DoEvents* can be liberally inserted in any loops to make sure you don't lock out any other events.

The drawback that makes large RTOS designers hate the 8051 is the fact that only four register banks exist and they only hold 8 bytes each.[34] Most kernels must allow *more than* four priorities so they use only two register banks—one bank for the running task and one bank for the operating system. For a context switch, the kernel copies all the current registers' contents to somewhere else in memory. The contents for the next running task are brought back from wherever they were saved earlier. This allows for unlimited priority levels. You pay for the flexibility in the longer time it takes to copy the eight registers to memory.[35] While a linker would never overlay a *static* variables, if preemption is possible someone must ensure that neither variables nor the stacks overlap.[36] To avoid running out of on-chip stack space the kernel must either manage the on-chip stacks carefully or else copy each task's stack to off-chip RAM as part of the context switch.[37]

---

[34]The original 8051 designers intended essentially only three priorities. Interrupts could have two priorities to preempt the main program and each other. The higher priority masks off low-priority interrupts, so there would never be more than two levels of preemption. Such a scheme could never need more than three register banks for context switching, but it would never allow more than four levels of priority. A fourth bank came from a seldom-used third-priority scheme that is described in the 8051 user's manual. For our purposes it is sufficient to know that at most four different priority levels are inherent in hardware and only three tasks can in turn be preempted by other tasks If you will settle for a round-robin scheme, the hardware can still support multiple tasks of *equal* priority, because they will not preempt each other. The DCX operating system combined both preemption and equal-priority in a two-tiered approach.

[35]Almost certainly, these long-term storage areas would be in off-chip RAM since there is so little on-chip RAM (64 to 256 bytes) and the running tasks would need as much of it as possible. Remember that off-chip memory access requires both setting *DPTR* and using the *MOVX* command—a much slower process that a direct access to a register. Certainly a function repeatedly using a variable should keep it on-chip, while a store for context switching could afford the overhead or archiving off-chip.

[36]For all the available 8051 RTOS, as far as I know, tasks are always linked together into one module. You cannot take entirely separate, absolutely-located programs and pull them in one-by-one as needed in the way that PC systems load programs.

[37]A further word on other processors is in order. Most computer systems have one address space for code and variables. The space holds the program code, the variables, and the stack. Because C is a very stack-oriented language on most machines, most operating systems push the *entire* context on the stack. Then context switching merely involves swapping stack pointers to designate the particular stack areas maintained for the next task. The reason this doesn't work well with the 8051 family is the fact that the on-chip stack space is quite small—it competes for space with registers and on-chip variables in the 64 to 256-bytes of on-chip RAM. For the same reason, you should avoid ***reentrant*** functions—in both cases, the software tries to push everything. I seem to recall that some compilers solve this by automatically generating a ***synthetic stack*** in off-chip RAM. Instead of pushing and popping directly with 8051 instructions, a complex set of data pointer moves and *MOVXs* make a stack out of a sort of first-in-last-out buffer. I'm fairly sure that Keil's tools include this capability. Unfortunately, this is going to be much slower, so you have to decide whether it is worth it.

For big 8051-family RTOS where *many* priorities and *hundreds* of tasks are permissible, the kernel developers must *clean out* parts of on-chip memory between tasks. That way each task can have the full amount of on-chip resources—particularly stack. The specific details of the individual systems differ and you will not find the information readily provided. If the system uses *overlapping* on-chip stack, it *probably* copies the entire stack for the task to off-chip RAM. If it is going to overlay *other* on-chip variables, it also moves them to a safe place off-chip. In such cases, each task can operate as though virtually all the on-chip resources are at its disposal. The operating system may either blindly copy *everything* or else it cleverly saves only the parts being used by the present task (either by information derived from the compiler/linker files or else by specific inputs from the programmer).[38] Some of these more elaborate operating systems are ported over from other processors with only enough fix-up to run on the 8051—they were designed by *big-system* programmers with little appreciation for the limitations of the 8051 hardware. When you investigate commercial operating systems, be sure to ask if the system is available for other processors—sometimes a *bad* sign in terms of its fit with the (unusual) architecture of the 8051.

The drawback of some big RTOS is speed of context switching. It will take considerable time to move a couple hundred bytes to off-chip RAM. As a system developer, you have to choose whether you need *lots* of tasks and priorities with *lots* of variable space, or you need *lean and mean* operation having quick task switching. The answer can vary depending on the application and the size of the tasks.[39] The day might come with the 8051 family where large applications can fit, and still save costs over more advanced

---

[38]For such systems, you would not need to link together tasks if the on-chip space can effectively overlap, but the issue of conflict over off-chip RAM may be serious. For the 8051 family, that does not have relocatable final code, *that* space can *never* overlap. You should force locating of code and variables to different parts of off-chip RAM for different tasks if they individually locate and link them. I know of no linker that will locate multiple tasks non-overlapping in off-chip RAM and code space while overlaying *all* on-chip resources.

If you really want a huge multitasking system and insist on the 8051, how about a system where context switching merely entails *hardware bank-switching* both off-chip RAM and off-chip ROM? The operating system could reside in on-chip ROM and you could either move on-chip variables to the given task's RAM bank before a switch or write tasks that avoid all on-chip memory usage. About now you should be asking why you don't migrate to a much bigger processor with more suitable architecture!

[39]My own view is that it is inappropriate to use the 8051 for huge applications and that the practical embedded applications I envision can run nicely with only a few priorities, but that is admittedly a personal bias.

case, the task "wants" to run, while in the latter the task is waiting for an event. Once the event happens, the task gets in line in the *READY* queue. As the *RUNNING* task finishes (by waiting for an event or by deferring to the other ready tasks), the next task in the *READY* queue gets to run until it in turn chooses to wait for an event. Notice that there is no priority—the first task into the *READY* queue is the first task to run. If you want to have all tasks get a good chance to run, you must write either waits or defers into each of them so they do not *hog the processor.*

### *RTKS* System Calls

The system calls are the basic tools of an operating system. For *RTKS* there are system calls to wait for events as well as calls to set and check signals. They are described here:

| RTKS System Calls | |
|---|---|
| *WAITT (time)* | This call puts the task to sleep for the number of system ticks indicated. Assuming a 1 msec interrupt, the units are in milliseconds with the maximum delay being 255 (about ¼-second). At the end of the wait time the task goes on the ready queue to run when its turn comes. |
| *DEFER()* | This is a wait of zero time so the task goes to the end of the queue to allow ready tasks to run. This is the heart of cooperation. If there are no other tasks in the queue, the task continues running with no context switching. |
| *WAITS (timeout)* | This causes the task to sleep until its signal bit is set by another task. If that has already happened then the task goes on the end of the queue. If it has not yet happened then the task is flagged as waiting until another task sets the signal bit. |
| *SCHEK( )* | A task can check if its signal bit is set without giving up the control of the processor. A zero is returned if the bit is clear and a 1 if it is set. |
| *SIGNL (task)* | A task can set the signal bit for any task. The action does not stop the calling task's running. If the flag is already set no additional action takes place. |
| *INIT( )* | Sets up all the task storage locations and initializes for no active tasks. |
| *CREAT (task#, taskaddress)* | This is needed to put the program counter values into the tables. Each task goes into the ready queue in the order created. |
| *RTKSstart()* | This begins the interrupt and starts the first task running. |

These system calls are defined in a header file for C. By putting the name of the *.h* file in an *#include* line at the start of your task modules, you can make references to all the *RTKS* calls without having to prototype them yourself.

**RTKS.h**

```
#define uchar unsigned char
#define uint unsigned int
```

```
void WAITT(uchar msecs);
void WAITS(uchar msecs);
uchar SCHEK(uchar task);
void SIGNL(uchar task);
void CREAT(uchar task,uchar code *tskadd);
void DEFER(void);
void INIT(void);
void RTKSstart(void);
```

### Storage Tables for *RTKS*

Before getting down to serious coding, any software project should define the variables and their arrangements. Even this small kernel must keep track of a few things for each task. Keeping the number of things to store small will enable *RTKS* to cut space and increase speed. For *each task,* I define the storage structure to include (byte 1) a top-of-stack pointer—a byte since the *on-chip* stack address never exceeds 256 and (byte 2) a system tick counter.

In addition, I need (byte 3) a queue showing the numbers of the tasks waiting to run. The program counter (bytes 4 and 5) is actually kept on the separate stack of each task so that a task can resume running where it left off.[42] So operating system management needs 5 bytes for each task.

To include a simple signaling capability, I use 2 *bits* for each task—the signal and a bit to indicate if a task is already waiting for its signal. By limiting to 8 tasks, that takes up 2 additional *bytes.*

### Non-preemptive Context Switching

Before writing the *RTKS* context switcher, to determine if registers must be saved, I wrote a short test to see how the (Keil) C compiler treats registers. First, I did a math operation on a variable before passing it to a function. You can see that the compiler assumed the result of the math operation (in *R6, R7,* and stored in *x2*) was *still* in the registers at the next line.[43] In the second case, I inserted a dummy system call such as would be used to return to the operating system. The compiler then did *not* assume the register contents were unchanged—it did not expect the function to preserve or restore the registers.

[42]Two additional bytes of stack space would be needed for each nested call from within a task, but this is discussed more later.

[43]The math on the integer involved one of the integer (2-byte) math functions which returned the value in the *R6, R7* register pair. This is consistent with the parameter passing conventions of Keil C as discussed in *Volume I.*

**Test without Intervening Call**

```
x2=1456*x1;
 MOV R7,x1
 MOV R6,#00H
 MOV R4,#05H
 MOV R5,#0B0H
 LCALL ?C?IMUL
[44]MOV x2,R6
 MOV x2+01H,R7
bin_asciidec(x2,&x3,4);
[45]MOV R3,#00H
 MOV R2,#HIGH x3
 MOV R1,#LOW x3
 MOV ?_bin_asciidec?BYTE+05H,#04H
 LCALL _bin_asciidec
```

**Test with Intervening Call**

```
x2=1456*x1;
 MOV R7,x1
 MOV R6,#00H
 MOV R4,#05H
 MOV R5,#0B0H
 LCALL ?C?IMUL
[46]MOV x2,R6
 MOV x2+01H,R7
syscall();
 LCALL syscall
bin_asciidec(x2,&x3,4);
[47]MOV R7,x2+01H
 MOV R6,x2
 MOV R3,#00H
 MOV R2,#HIGH x3
 MOV R1,#LOW x3
 MOV ?_bin_asciidec?BYTE+05H,#04H
 LCALL _bin_asciidec
```

[44]The result of the multiplication is put in the absolute storage location for *x2*.

[45]The result of the previous operation is in *x2* but, since it is still also in *R6, R7,* the parameter passing preparations can omit the retrieval.

[46]Again, the values of *x2* are returned.

[47]The compiler reloads *x2* into *R6* and *R7* for passing to *bin_asciidec()*. Even though *syscall()* involves no parameters either way, the compiler makes no assumptions that the function preserved the registers.

So, as long as my kernel is *non-preemptive,* I can omit saving registers—the task can only get to the kernel with a call and the compiler will always have neatly restored any temporary register values before issuing the call. The *system tick* (timer interrupt event) can make an *ASLEEP* task *READY* by putting it at the end of the *READY* queue, but the flow still returns to the previously running task.

Aside from the preemption by the timer interrupt (which I carefully write to not disturb the *registerbank0* registers), no other task activity can take over the registers until the running task has intentionally relinquished control (which means it no longer depends on the registers' contents). In *RTKS,* I choose to forbid any other interrupts. If you intentionally unmask one of the external interrupts or the other timer, you are on your own—you should use *registerbank2* and manage priorities or you may lock the kernel's timer and defeat it entirely.

## Managing the Stacks

Since the test established that I do not have to preserve the registers for a non-preemptive system, the big context-switching issue for *RTKS* is the stack. The C compiler makes sure that the *PUSHes* at the *start* of a function are matched by the *POPs* at the *end.* There is no reason, however, why any function cannot call another function while the stack is deeper than at its start.[48] Since function calls are the only way a task returns control to the operating system, a second task, running in turn, might either add to or remove from the stack before control returns to the first task.

For *RTKS* I choose a *fixed stack area* for each task.[49] When the task is created, the storage structure includes a stack location and depth. When you configure the kernel, you have to anticipate how deep the stack might get for each task and request enough space. Usually, you would choose two plus two for each level of nested calls within the task. If you use floating-point variables or some of the more elaborate library functions, however, you could need much more stack—you would have to determine that empirically and probably shouldn't be using *RTKS* for this. A fixed stack for each task makes context switching fast and easy—I just set the stack pointer to the

---

[48]Incidentally, with this system you can call a non-reentrant function from any task because you will not enter the function again until you have left it to return to the main task code. You should never make any system calls from a shared function, however, because you then would be able to reenter it before the first invocation ended.

[49]That was the approach of DCX. It was fast but it put the burden on the user to set up the initial task descriptor (IDT) with the start addresses of the tasks, the stack sizes and several values related to the interrupts, priorities, and the location of the next task in the startup chain.

next task's last value and issue a return. There is no moving of stacks—each one stays put. While I prefer the elegance of the stack management described next that I chose for the development of *RTKB*, fixed stacks are much more efficient both in code size and time.

As you know from *Volume I,* the stack for the 8051 resides in on-chip RAM with between $32_{10}$[50] and $234_{10}$ bytes available for the stack if all four register banks are in use. If you have eight tasks on an 87C750 that leaves only 4 bytes for each stack *with no other variables.* To give you control, I write the kernel code in assembly with several global equates defining the size of the stack and the number of tasks.

It is usually not difficult to predict stack requirements in advance *except when library functions or reentrant functions are called.* The stack depth goes up by 2 bytes for each nested level of call—each call to another function from within a function. Math functions can be implicitly called—especially if you use floating-point variables. While you do not knowingly invoke a function, the compiler will bring in functions from the library that have to drive the stack quite deep during execution. To set fixed-stack depth in that case, you would have to run multiple simulations to be really sure what stack depth is needed then. Putting all that in rule form:

*Allow 2 bytes of stack for the task plus 2 bytes for each additional call to a function in the task code.* A simple task is possible with only 2 bytes of stack!

*Never use reentrant code or floating-point variables.* Almost all the libraries are reentrant and will use some more stack.

### *RTKS* Source Code

The code itself, for the bulk of the operating system was finally written in assembly. I struggled with doing it in C, but there were things the C compiler wouldn't allow (all those shortcuts that permit the non-structured disasters). Also, loving efficiency, there were things that a careful study showed could reduce the size of the code.[51]

---

[50]This number is for the 750/1 that has 64 bytes of RAM, The options are to either have fewer tasks or use fewer register banks. There is no room for sloppiness here since the call automatically pushes the 2-byte program counter value onto the stack.

[51]Actually, I wrote the stack move function intended for *RTKS* (and later moved to *RTKB*) in C and then stripped the assembly code listing down to just the mnemonics in an editor as the starting point for an assembly version. Most of the improvements involved overwriting registers when their use was finished to avoid some automatic variables that were being kept in directly-addressed on-chip RAM. I got the stack move code down to 98 bytes from 113. Is that worth it? It depends on the value of your time versus the importance of speed and size. I felt that for an operating system used over and over, the development time was justified.

The key features are that with assembly, I could avoid needing a stack—no calls *within* the operating system, and, aside from the permanent tables for the task data, I could use only *bank1* registers. By judiciously arranging the call entry points, almost all the jumps are short, saving a few more bytes of code.

When you look closely at the flowchart below, you see that almost all the system calls are entries to fairly simple code pieces. There are two categories of calls—ones that can lead to a change of tasks (*WAITT, WAITS, DEFER*) and those that set a signal or report back on a signal (*SIGNL, SCHEK*). Since this is a non-preemptive system, they can never cause a change of task. If, for example, you want to signal a task and then allow it to run if it is already waiting for the signal, the task has to first call *SIGNL* and then *DEFER*.

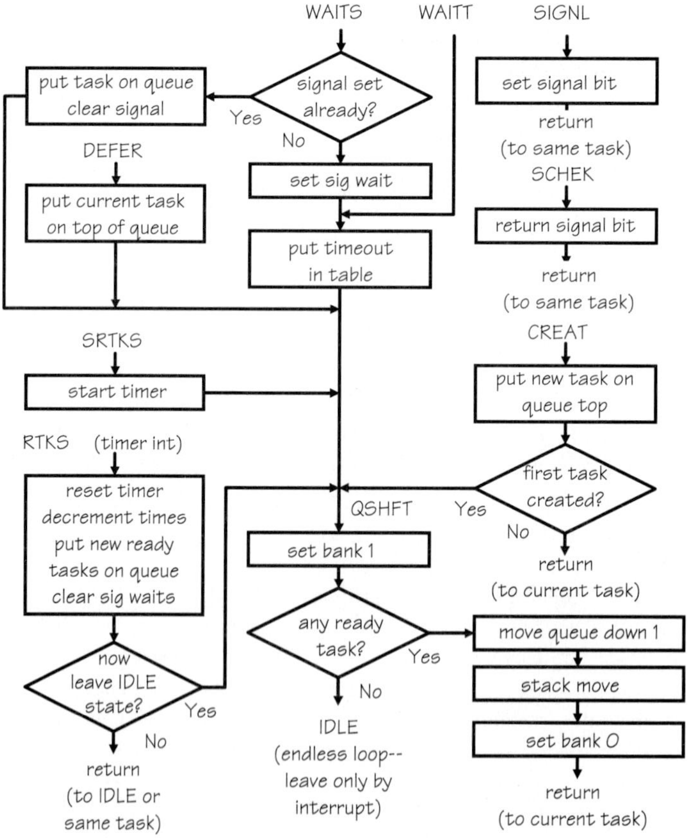


**Flowchart for RTKS**

I made some choices about system calls that could have just as easily been different. For example, if a task waits for a signal, but the signal bit has *already* been set, the task still goes to the end of the queue. If a task times out waiting for a signal or just waiting for time, it goes to the end of the queue.

One word of caution—if you put a task to sleep waiting forever for no event—*WAITT(0xff)*—there is no way to restart that task because the timer interrupt will never make it *READY* and no other task can change its delay value. It's a heck of a way to delete a task, but it works.

---

**RTKS**

```
PUBLIC RUNNING,READYQ,READYQTOP,SIGS,SIGW,TTS,SPTS
PUBLIC_WAITT,_WAITS,_SCHEK,_SIGNL,_CREAT,INIT
PUBLIC RTKSSTART,DEFER
THIGH EQU 4 ;SETTING FOR 1 MSEC
TLOW EQU 0
RAMTOP EQU 03FH ;SET FOR 87C750—64 BYTES
NOOFTSKS EQU 5 ;CAN BE UP TO 8 TASKS
STACKSIZE EQU 4 ;(SEE DISCUSSION) 2+2*#OF CALLS

MSKS SEGMENT CODE
RSEG MSKS
 MSK: DB 1,2,4,8,16,32,64,128;TO SELECT SIG BITS

TBLS SEGMENT DATA
RSEG TBLS
 READYQTOP:DS 1 ;ADDRESS OF LAST READY TASK
 RUNNING: DS 1 ;# OF CURRENT TASK
 READYQ: DS NOOFTSKS ;EMPTIES MARKED WITH NOOFTSKS
 SPTS: DS NOOFTSKS ;SP FOR EACH TASK
 TTS:DS NOOFTSKS ;FOR TASKS 0-7: REMAIN TIMEOUT TIME
                ;0 = READY, FF = NOT TIMED
 SIGS:DS 1 ;UP TO 8 SIG BITS—LSB TSK 0; MSB TSK 7
 SIGW:DS 1 ;TASKS WAITING FOR SIGNALS

ISEG AT RAMTOP-(NOOFTSKS * STACKSIZE)
 STKRESERVE: DS NOOFTSKS * STACKSIZE

CSEG AT 0BH ;INTERRUPT ADDRESS FOR TIMER 0
 LJMP TINT

RTOSTINY SEGMENT CODE
RSEG RTOSTINY
```

```
USING 1 ;ALL ABSOLUTE REFERENCES TO REGISTER BANK 1

_SCHEK:;SYS CALL ENTRY CHECK SIGNAL BIT FOR TASK
  ;RETURN 0 OR 1 IN R7 ;NO NEED FOR BANK SWITCH
  ;IMMEDIATE RETURN—NO CONTEXT SWITCH
  MOV DPTR,#MSK
  MOV A,RUNNING
  MOVC A, @A+DPTR ;FETCH SIG MASK BYTE
  ANL A,SIGS
  MOV R7,#0
  JNZ V1
  RET ;SIG NOT YET SET—R7=0
V1:INC R7 ;SIG SET—R7=1
  RET

_SIGNL:;SYS CALL ENTRY—SET SIGNAL BIT FOR TASK
  ;NO NEED FOR BANK SWITCH—NO CONTEXT SWITCH
  MOV DPTR,#MSK
  MOV A,R7 ;TASK# PASSED IN R7
  MOVC A,@A+DPTR ;FETCH SIG MASK BYTE
  MOV B,A ;SAVE MASK
  ANL A,SIGW
  JZ G1 ;TASK NOT ALREADY WAITING
  MOV A,B
  XRL SIGW,A ;CLEAR WAITING BIT
  MOV A,B
  XRL SIGS,A ;ENSURE CLEARED SIGNAL BIT
  INC READYQTOP
  MOV R0,READYQTOP
  MOV @R0,07 ;SET READY TASK ON READYQ
  RET ;DON'T GIVE UP RUNNING (MUST DEFER)
G1:MOV A,B
  ORL A,SIGS ;SET SIGNAL BIT
  RET

_WAITS:;SYSTEM CALL—WAIT TIMEOUT OR SIGNAL ARRIVAL
  ;TIMEOUT PASSED IN R7
  MOV DPTR,#MSK
  MOV A,RUNNING
  MOVC A, @A+DPTR ;FETCH SIG MASK BYTE
  MOV B,A ;TEMP SAVE MASK PATTERN
  ANL A,SIGS ;TEST FOR SIGNAL ALREADY THERE
  JZ U1 ;NO SIGNAL YET SO TASK MUST WAIT
  MOV A,B ;RELOAD MASK
```

```
    XRL SIGS,A ;CLEAR SIGNAL BIT
    INC READYQTOP
    MOV R0,READYQTOP
    MOV @R0,RUNNING ;TASK TO END OF Q (NO WAIT)
    SJMP QSHFT ;OFF TO START NEXT TASK IN Q
U1:MOV A,B ;RELOAD MASK
 ORL SIGW,A ;SET SIG WAITING BIT
_WAITT:;SYS CALL ENTRY POINT—WAIT FOR TIME
    ;_LEADS NAME FOR FUNC WITH REG-PASSED PARAMS
    ;TIME PASSED IN R7
    MOV A,#TTS
    ADD A,RUNNING
    MOV R0,A
    MOV @R0,07 ;BANK 0 R7—TIMEOUT PUT IN TABLE
    SJMP QSHFT ;TASK NOT ON Q—WAITING

RTKSSTART: MOV TH0,#THIGH
    MOV TL0,#TLOW
    SETB TR0 ;START TIMER
    MOV IE,#82H ;ENABLE TIMER0 INTERRUPT
    SJMP QSHFT

DEFER:;SYS CALL ENTRY
    INC READYQTOP
    ADD A,READYQTOP
    MOV @R0,RUNNING ;PUT RUNNING AT END OF READYQ

QSHFT:;SHIFT READYQ BY ONE
    ;SAVE PRESENT RUNNING TASK STACK PTR
    ;GET NEW RUNNING TASK FROM READYQ
    ;USE BANK 1—MAY HAVE ENTERED FROM INTERRUPT
    ORL PSW,#8 ;SET TO REGISTERBANK 1
    MOV R1,#READYQ
S1:MOV A,@R1 ;SHIFT Q DOWN 1
    DEC R1
    MOV @R1,A
    MOV A,R1
    INC R1
    INC R1
    CJNE A,READYQTOP,S1
    DEC READYQTOP ;THEY ALL MOVED DOWN BY 1
    DEC R1
    MOV @R1,#NOOFTSKS ;TOP SPOT NOW IDLE
RSCON:MOV A,#SPTS
```

```
    ADD A,RUNNING
    MOV R0,A
    MOV SP,@R0
    ANL PSW,#0E7H ;SET TO REGISTERBANK 0
    RETI ;MAY HAVE ENTERED FROM TIMER INTERRUPT
    ;OTHERWISE NO HARM ANYWAY

TINT:;INTERRUPT ENTRY ONLY FROM TIMER
    MOV TH0,#THIGH ;RESET TIMER
    MOV TL0,#TLOW
    SETB ET0
    SETB EA
    ORL PSW,#8 ;SET TO REGISTERBANK 1
    MOV R0,#TTS+NOOFTSKS ;R0 POINTER IN TIMEOUT TBL
    MOV R2,#NOOFTSKS ;DO ALL 8 TASKS
K2:MOV A,@R0 ;GET TIME FOR TASK
    CJNE A, #0FFH, K4
    SJMP K1 ;FF=TASK NOT TIMING
K4:DEC A
    MOV @R0,A ;SAVE DECRMTED TIME FOR WAITING TASK
    JNZ K1 ;TASK NOT TIMED OUT YET
    INC READYQTOP ;POINT TO TOP OF READY READYQ
    MOV R1,READYQTOP
    DEC R2
    MOV @R1,AR2 ;PUT TASK# IN READY READYQ
    INC R2
    MOV R6,#1 ;TELL IF ANY TASK COMES READY
    MOV DPTR,#MSK
    MOV A,RUNNING
    MOVC A, @A+DPTR ;FETCH SIG MASK BYTE
    XRL SIGS,A ;CLEAR SIGNAL BIT (IF SET)
K1:DEC R0 ;MOVE DOWN 1 IN TABLE
    DJNZ R2,K2 ;END OF 8 TASKS YET?
    MOV A,R6
    JNZ K3 ;NO TASK ADDED TO READYQ
    MOV A,RUNNING
    CJNE A,#0FFH,K3 ;NOT IN IDLE STATE
    LJMP QSHFT ;IDLE AND NEW TASK TIMED OUT
K3:ANL PSW,#0E7H ;SET TO BANK 0
  RETI

INIT:;SET UP VARIABLES
    MOV RUNNING, #NOOFTSKS ;IDLE TASK RUNNING
```

```
    MOV READYQTOP,#READYQ
    MOV R7,#NOOFTSKS
    MOV R0,#TTS
    MOV R1,#READYQ
  I1:MOV @R0,#0FFH ;NO TIMER ACTION
    MOV @R1,#NOOFTSKS ;IDLE TASK IN READYQ
    INC R0
    INC R1
    DJNZ R7,I1 ;SET UP ALL TSKS
    MOV READYQTOP,#(READYQ-1)
    MOV SIGS,#0 ;CLEAR ALL SIGNALS
    MOV SIGW,#0 ;NO SIGNAL WAITS PENDING
    MOV R0,#SPTS ;INITIALIZE ALL STACK POINTERS
    MOV R7,#NOOFTSKS
    MOV A, #STKRESERVE
  I2:MOV@R0,A
    ADD A,#STACKSIZE
    INC R0
    DJNZ R7,I2
    RET

  _CREAT:;SYS CALL ENTRY TO CREATE A TASK
    ;TASK # PASSED IN BANK0 R7
    ;TASK START ADDR PASSED IN BANK0 R4,R5
    ;TASK PUT AT END OF READYQ
    INC READYQTOP
    MOV R0,READYQTOP
    MOV @R0,07 ;BANK 0 R7—PUT NEW TASK END OF Q
    MOV A,#SPTS
    ADD A,R7
    MOV R0,A ;R0 POINTS TO TASKS SP
    MOV 01,@R0 ;FETCH TASK'S SP
    MOV @R1,04 ;PUT ADDR OF NEW TASK ON ITS STACK
    INC R1
    MOV @R1,05
    INC R1
    MOV @R0,01 ;SAVE TASK'S NEW STACK POINTER
    RET
  END
```

## How to Include *RTKS*

The operating system consists of a block of code to be assembled along *with* your tasks. The maximum number of tasks is pre-set at eight. An *include* line at the top of the file declares all the system call functions. An example is shown in the following section.

You must write each task as an endless-loop function with system calls to branch out within the loop. As you can see in the later examples, a *main()* function is needed to first create all the tasks and start the operating system.

## Size of *RTKS*

As you may guess, the goal for *RTKS* is to have a *very* small kernel that can run even on an 87C750 (having 64 bytes of RAM and 1K of code space all on-chip). The *.M51* file that follows (from the example on pg. 139) shows that the operating system requires $266_{10}$ ($10A_{16}$) bytes of code and, $14_{10}$ ($0D_{16}$) bytes for the tables with three tasks. Having 4 bytes of stack for each task also takes $12_{10}$ ($0C_{16}$) bytes of on-chip RAM for stack (the absolute module shown in the .m51 file).

Since all the tasks run in bank0 and the kernel runs in bank1, the space for banks 2 and 3 is occupied with variables. Even so, as you can see you have only $22_{10}$ ($09_{16} + 0D_{16} = 16_{16}$) bytes for variables on an 87C750, although the kernel leaves $758_{10}$ bytes left for task code.[52]

```
                 Test program allocations using RTKS
TYPE BASE       LENGTH RELOCATION    SEGMENT NAME

     * * * * * * * D A T A M E M O R Y * * * * * * *
REG    0000H    0008H ABSOLUTE       "REG BANK 0"
REG    0008H    0008H ABSOLUTE       "REG BANK 1"
DATA   0010H    000DH UNIT           RTKSTABLES
DATA   001DH    0009H UNIT           ?DT?REKSEX1
       0026H    000DH *** GAP ***
IDATA 0033H     000CH ABSOLUTE
IDATA 003FH     0001H UNIT           ?STACK
     * * * * * * * C O D E M E M O R Y * * * * * * *
CODE   0000H    0003H ABSOLUTE
0003H 0008H     *** GAP ***
```

[52]I finally determined that the software included with the book refuses to let the code start at $0000_{16}$ because it is a restriction of the free demo software—you have to buy the next step up!

```
CODE 000BH     0003H ABSOLUTE
     000EH     3FF2H *** GAP ***
CODE 4000H     0102H UNIT   RTKSCODE
CODE 4102H     008CH UNIT   ?C_C51STARTUP
CODE 418EH     002CH INBLOCK ?PR?CALCULATE?REKSEX1
CODE 41BAH     002AH INBLOCK ?PR?MAIN?REKSEX1
CODE 41E4H     0027H INBLOCK ?PR?TACHOMETER?REKSEX1
CODE 420BH     001CH INBLOCK     ?PR?PULSE?REKSEX1
CODE 4227H     0008H UNIT    RTKSBITMASKS
CODE 422FH     0005H UNIT    ?C_INITSEG
```

The very limited space for variables could be a serious handicap. If you switched to 2-byte stacks (no calls from within a task) an extra 6 bytes of space would be freed for variables. Going to the maximum permissible eight tasks requires $24 + 2 + 1 = 27_{10}$ bytes for the tables and at least $8 \times 2 = 16_{10}$ for the stack plus one for the starting stack or $44_{10}$ bytes total. With $16_{10}$ for the two banks, that leaves 4 bytes for variables! I do not think any significant tasks can be built around so few variables, so it really is not practical to have the full eight tasks on an 87C750. Even *RTKS* is not a kernel to use casually on such a small processor! On the other hand, on an 8051 (128 bytes of RAM) you could have eight tasks with stack for calls nested two-deep within the tasks and still have 52 bytes left for variables!

## RTKB—A BUILD YOUR OWN PRIORITY-BASED PREEMPTIVE KERNEL

What sort of features should go into a somewhat larger kernel? How much more elaborate (and therefore probably slower) should it be? What flexibility or additional capability is worth the sacrifice of space and speed? There are no sure answers to these questions, but entering into the development will help you to understand the tradeoffs.[53]

---

[53]You can, of course, proceed to revise my code and produce your own custom calls for other features. I certainly have no objections to your incorporating it in applications royalty free or and modifying it as long as you don't go into *business* selling a copy of my software as your own commercial kernel. If you do use it, I again remind you that I make no claims as to fitness or reliability of my code. It has not been thoroughly tested and is not warranted in any way. In fact, with *RTKB* I am sure there are bugs and have not taken time to really test it. ***Use it only as an educational example—not a "free" kernel.***

## Intervals

One of the first additional features for my second real-time kernel (I will call it *RTKB*) is repeating *intervals* for each task. Among the calls in *RTKS* was a wait for time and there were time-outs with some of the other calls.[54] I want to add a feature associated more with a scheduler—an event that can *regularly* wake up a task with no other intervention. You approximate the effect by immediately initiating a *wait for time* each time a task resumes, but an interval makes the time until the next interval event regular even when the task does not get to run immediately—the next interval begins as soon as the current interval is processed.

## Messages and Semaphores

Instead of binary semaphores, I choose to give *RTKB counting* semaphores and a *messaging system.* I expand the binary semaphore of the first kernel to counting semaphores. Each task has one semaphore. *Any* task can add or subtract units for a task's semaphore and a task can put itself *ASLEEP* until its semaphore reaches zero.

I add a configurable number of unattached single-message mailboxes. They can each hold either a pointer to a message or the number of a task waiting at that mailbox. The messages themselves must be configured by the tasks—size, order, identifying headers are all *their (your)* responsibility. The mailbox only holds a 2-byte pointer to the larger message in off-chip RAM. On request, the kernel reports back to a task that a message is in a given mailbox or that another task is already waiting there for a message to arrive. If a task was waiting at a mailbox before a message arrives, the task goes to *READY* when another task sends the message.[55]

## Dynamic Tasks

*RTKS* had a fixed number of tasks configured at the compile time. Here I make the maximum number of active tasks configurable. I want *RTKB* to allow a task to spawn other tasks and then delete them when their

---

[54]It is somewhat like Intel's DCX operating system that came with BITBUS, although I have never explored the assembly language innards of the that system. DCX is now in the public domain.

[55]One task can thus wake up another by sending it a message. If there is really no message to exchange, you may choose between use of a null message or a semaphore—there is a lot of freedom left to the people writing the tasks.

time is done. There can be inactive tasks, as well as dynamic creation and deletion of recognized tasks.[56]

Since each active task takes up RAM space for tables, by adding and deleting tasks I can keep the total table space smaller. For example, there could be forty possible tasks with no more than ten active at a time. Then the kernel need only search through lists of ten items and the storage space is less. As a task is needed, it can be created. When it is done, it can be deleted (not removing any code). Later it might be created again. For example, you might create X and Y stepper tasks each for one motion of a positioning table. When each task's motion is done, they could subtract a unit from a semaphore and delete themselves. You could then create new tasks to carry out some calculations.

The operating kernel assigns its own temporary *task number* to an active task and communication uses the assigned number. When a task wishes to talk to another task, it must already know its *function id* (reached by mutual agreement among the programmers). The task can then look up the current task number. The kernel might currently have assigned a 4 to the task with a function id of 27. The message then is sent to task 4. If by some chance 27 is not currently created, the lookup call would return a code indicating that no such task currently exists.

## Priority and Preemption

I am going to allow a large number of priorities (actually 127 since it is easier to use a byte and I utilize 1 bit for other things). Rather than trying to mix round-robin and priority scheduling (as in DCX), I will have only straight priority. Any time a task switch takes place, *RUNNING* status will go to the highest priority *READY* task regardless of which task has been *READY* longest. There is actually no queue—just a non-ordered list of all the *READY* tasks. I run the risk that a low-priority task in a heavily loaded system might never get to run, but that is a system design problem. Each task can have a priority from 1 to 127 with a lower number being higher priority. If two tasks have the same priority, which one will run first is not guaranteed. All that can be said is that the second *READY* one will run after the first has gone to sleep, but if the first one gets back on the queue before then, it may again go before the second one.

---

[56]This is not to say that the code for the tasks comes and goes—that is fixed—only the RAM is reallocated. This is unlike what you would have with a big operating system where programs are loaded from disk storage. I only have ROM code space in any small embedded system.

I am including preemption.[57] If an interrupt comes along, program flow goes to the interrupt routine immediately after the current instruction finishes. The latency is no more that a couple of microseconds worst case (the multiply instruction). That kind of preemption requires saving the context in more detail. The analysis for the *RTKS* showed that registers were safely tucked away before a task would issue any call. With a preempted task, you cannot assume things were wrapped up as cleanly. There can easily be important values left in *R0-R7, ACC, B,* or *PSW*. So to switch context in a way that a preempted task can eventually be safely restored, the kernel must either switch register banks or elsewhere save all the values for each task as it ends and restore them when it resumes. Intel's DCX developers chose to switch context using register banks, the inherent mechanism. That is fast, but it is limited to three priorities. I choose instead to follow the path of all the other developers and save the registers off to off-chip RAM, allowing many levels of priority.

One of the features I want is the ability to respond to other interrupts—specifically the two external ones. These are a thorn in the side of all the commercial operating systems because if you let the user write a nice long interrupt routine, they can delay the basic operating system timing. Also, if they change the masking, they can block it totally. Usually commercial systems come with big warning signs and forbid the interrupts handlers to make system calls since a system call could come in the middle of the operating system's running its own code and really mess things up. My solution, much like DCX, is to *allow* the external interrupts but make them events associated with one task that the *kernel* controls. You can wait for an external interrupt, but it is the kernel that takes the actual interrupt and processes it by its own rules and tables. It can bring about preemption in the same way that the system tick interrupt does.

## *RTKB* Task States

The state diagram here summarizes the task states for *RTKB*. Note that, unlike *RTKS*, the *READY* state is not a first-in-first-out queue. Instead,

---

[57] *RTKB* could have had priorities *without* preemption—then the kernel would just keep a FIFO *queue* of tasks but put a newly-ready higher-priority task at the head of the line. Even so, that task would run only when the running task *voluntarily* relinquished the processor. With preemption on the other hand, an urgent task can start up immediately, regardless of what task is running.

it is a pool of all the *READY* tasks and the next *RUNNING* task is always taken out based on the priority of all the *READY* tasks. In an overloaded system, the lower priority tasks could sit endlessly in the *READY* pool. Since there is no time precedence, it is pointless to have a defer instruction—as soon as a high-priority task deferred and went into the pool it would again become the *RUNNING* task. The only option is to have the task do a wait of some sort.

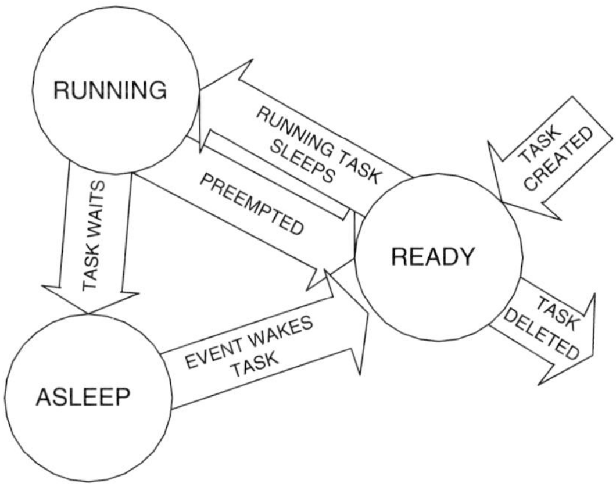


**RTKB Task States**

## *RTKB* System Calls

It gets to be a challenge to make all the calls five letters and still keep them logical. Some of the operating systems append a set of letters to identify the system calls, but I hate extra typing and it seems like a free advertisement for the company.

Notice that not all the calls can cause a change of running task. All the wait calls will try to make another task run unless the event for which the task waits has already come. Changing priority of a task can make it run if it is ready and gets raised to a higher priority than the running task. Subtracting semaphore units or sending a message can do it also if a higher-priority task is waiting on that event. Deleting a task will do it if the task deletes *itself.* Creating a task can do it if the new task is created a higher priority.

Other than that, it is always safe to check things, set an interval, or add units to a semaphore without risk of preemption. The other calls will not cause preemption as long as the tasks have a lower priority.

| System Call | Description |
|---|---|
| WAITT (time)* | sleep for a set time |
| WAITI ()* | sleep until interval event |
| WAITS (timeout)* | sleep until semaphore reaches 0 |
| WAITM (mbx#, timeout)* | sleep until message in mailbox |
| WAITX (int#, timeout)* | sleep until ext. hardware interrupt |
| INTVL (time) | set time interval for this task |
| ADDSU (task#, # of units) | add semaphore units of given task |
| SUBSU (task#, # of units)* | subtract semaphore units |
| SENDM (task#, pointer)* | send message to another task |
| uchar CHEKS () | return semaphore count |
| uchar CHEKM () | check for waiting message |
| SETPR (priority#, task#)* | change priority of a task |
| uchar FINDT (id#) | return task# of known id# |
| CREAT (task#, addr, priority)* | create a task |
| DELET (tsk#)* | delete a task |
| INITK () | set up tables for kernel |
| STRTK () | start operating system functioning |
| | *can cause different task to run |

**RTKB System Calls**

## Storage Arrays for *RTKB*

| | task 0 | task1 | ... | task n |
|---|---|---|---|---|
| status (msb 1 = preempted) | | | | |
| task's timeout count | | | | |
| | | | | |
| | | | | |
| task's interval count | | | | |
| semaphore (msb 1 = waiting) | | | | |
| priority (msb 1 = using interval) | | | | |
| task's stack pointer (SP) | | | | |

**Storage Structure for RTKB**

Before writing software you must decide what records to keep for each task. I allow the number of tasks to be configurable when you compile the kernel code. As you can see from the table of arrays, each task requires 8 bytes of (off-chip) storage for tables. In addition, each task needs 12 bytes for the registers (*R0-R7, ACC, B,* and *DPH/L*).[58]

***status*** holds 0x7f for any task that is asleep, 0 for the running task, and the priority number of any ready task. In this system, it is not a *queue*. In addition, the most significant bit (msb) is set to a 1 if the task was most recently preempted (indicating that restoring the task will have to include restoring the registers).

***timeout*** is counted down one for every system tick. If a task is not waiting for a timeout, the value is 0xffff—the counter will not be decremented. If the count reaches 0x0000, the task is marked *READY* by putting its priority number in the status array (*ASLEEP* is indicated by a 0x7f and *RUNNING* is indicated by a 0x00). Timeouts can be combined with other waiting—for example, a task could wait a little while for a message, but go on to something else if none is forthcoming. Since there is not a queue, a timeout of zero will have no effect, unlike *RTKS* where it can defer a running task to allow another to run if any are ready.

***interval*** is the value that can be loaded into the timeout counter each time it reaches zero. It provides a regular event for a task. Each time *timeout* reaches zero, if an interval is indicated by the priority msb, the timeout is automatically reloaded (otherwise it would be left at 0xffff. This can behave like a scheduler where the basic clock tick is counted to decide when to do various tasks.

***semaphore*** holds a counting semaphore function for each task. The msb indicates if the task is already waiting for the semaphore to reach 0. If it is, then the task is made *READY*; if it is not, then the semaphore is left at its new value. In the latter case, when a task issues a wait for a semaphore and finds it already at 0, it immediately becomes *READY*.

***priority*** holds the current priority of each task (1 to 127) with the lower numbers being higher in priority. The msb actually was appropriated to flag the interval function since 127 priorities is certainly enough for any practical application. Since the tasks do not queue up, the status table is scanned each time a context switch is planned and the lowest number one becomes the *RUNNING* task. Priority can be dynamically changed for a task. Two tasks can have the same priority, but then there would be an arbitrary favoring of one over the other; ignoring which one has been *READY* longer.

***stack*** holds the most-recent stack location of each task. There are other things involved with the stacks that are discussed later.

---

[58]I was initially going to make this an array of structures, but closer consideration of the way the code would have to work through the array picking out just one element of each structure suggested that having separate arrays is more efficient—just because C makes it easy to the programmer doesn't avoid the efficiency questions.

| | task 0 | task1 | ... | task n |
|---|---|---|---|---|
| R0 | | | | |
| R1 | | | | |
| R2 | | | | |
| R3 | | | | |
| R4 | | | | |
| R5 | | | | |
| R6 | | | | |
| R7 | | | | |
| ACC | | | | |
| B | | | | |
| DPTR | | | | |
| | | | | |

## Register Storage

***register storage*** is allocated in off-chip RAM for any active task that is preempted. As mentioned, 12 bytes are required if preemption happens, but no storage is needed if the task voluntarily relinquished the processor with a system call. The difference is indicated by the msb of *status*.

| | mbx status msb 1 = task waits | message pointer |
|---|---|---|
| mbx0 | | |
| mbx1 | | |
| ... | | |
| mbx n | | |

## Mailboxes

***mailboxes*** hold the addresses of the latest message that has been sent to it. To save space, they will be in off-chip RAM and the number of mailboxes is configured at compile time. This system will not have dynamic allocation of memory space for messages. Only 2 bytes will move into the mailbox—any long message remains in the space where it started and only the pointer is passed. In addition to the 2-byte message pointer, each mailbox has a third byte that identifies the task that is waiting for a message at that mailbox or the task that has left off the last message. If a task is already waiting for a message, another task cannot wait at the same mailbox. The sending and receiving tasks must agree on how to identify the messages and how to determine when they have been processed and can be reused. The real benefit of this message scheme is the ability for a task to sleep until the message arrives—no waking up periodically to check for the message.

## Preemptive Context Switching

As you can imagine, there is an almost continuous spectrum of possible operating systems. I chose to have *RTKB* add preemptive multitasking while still using the on-chip stack. By being preemptive, any task can be interrupted *anywhere* in its running. No guarantees can be made about the contents of registers, unlike the case with non-preemptive systems. Since the tasks will all be linked together, it isn't necessary to preserve the *entire* on-chip RAM space—you should set *NOOVERLAY* in the linker to disallow overlaying of any automatic variables you might accidentally have used in a task. The linker will then be sure no variables are overlaid on those of other tasks and the kernel doesn't have to worry about any variables being overwritten.

## Stack Options

***Fixed stack locations.*** As you know from *Volume I,* the normal stack for the 8051 resides in on-chip RAM with between 48[59] and 240 bytes available for the stack. If I want, say eight tasks that leaves between 6 and 30 bytes for each stack if no other variables are kept there. As with *RTKS*, I could just keep separate, non-moving stacks for each task. This makes context switching fast and easy—just swap out the stack pointer and return. There is no moving of stacks—each task's stack stays put.

***Off-chip stacks.*** A second option is to park each task's entire stack in *off-chip* RAM. This would make context switching slow because moving off-chip, with *MOVX, SWAP,* and *INC* commands, takes at least five instruction cycles for each byte, but it avoids the memory crunch—there can be up to 64 K bytes of off-chip RAM. I believe one commercial RTOS swaps the *entire* 128 or 256 bytes of on-chip RAM with an off-chip block for each context switch—very straightforward and not very code intense, but very slow.

Incidentally, I believe that the Keil C compiler supports a ***synthetic stack*** in off-chip RAM. It does not work for the calls, but some of the stack-intensive math and reentrant functions are written to work *directly* to off-chip space with separate pointers instead of using *PUSH* and *POP* instructions. If that is the case for the functions you use, you could still get by with a small "real" stack.

---

[59]This number assumes two register banks and no variables on an 87C750/1 (64 bytes of on-chip RAM). The options are to either have fewer tasks or use fewer register banks. There is no room for sloppiness here since the call automatically pushes the 2-byte program counter value onto the stack.

### Dynamically Shifted Stack

For RTKB, I am choosing a third option, as best I can tell adopted by Keil's RTX51. I allocate all the available on-chip RAM space for stack and *dynamically shift* the stacks to give the running task as much of the free space as possible. The diagram may help.

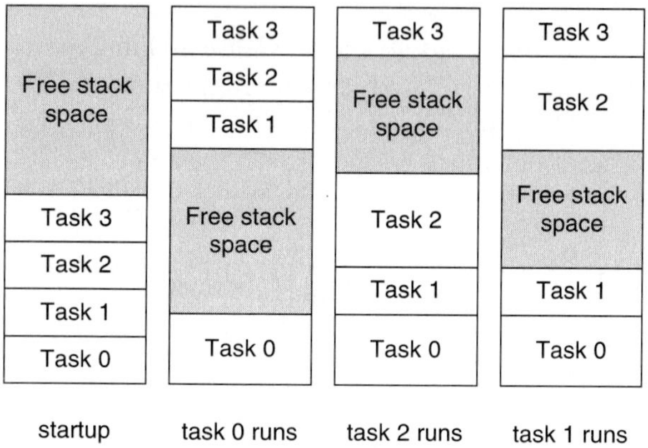


**Dynamically Shifting Stack**

Move the higher-in-number tasks' stacks to the *top* of the total space and put the lower-in-number tasks' stacks at the *bottom* of the space. The running task's stack is then at the top of the lower group and can expand up to the bottom of the upper stacks. Thus, during a task's running, *all* the available free stack space is at the disposal of that task. By eliminating unused space between *ASLEEP* tasks, I maximize the running task's potential stack space. This shuffling of stacks only involves moving one block of data up or down and adjusting the stack pointers. This makes the efficient use of the on-chip memory space to give the *RUNNING* task the most possible stack space.

Since stack usage by library functions is finished before a call, this does not tie up the space long term. A very real risk is when several tasks use floating-point math and can be preempted. Unless they use a synthetic stack, you could end up with both tasks needing, only temporarily, very deep stacks. In such a case the options would be:

1. Use 8051 family members with a full 256 bytes of on-chip RAM.
2. Check carefully that a synthetic stack is used.
3. Do floating-point math in only *one* specific task—do not have a situation where math could be underway in several tasks at the same time.

If you want *many* tasks, you will have to use the family members with 256 bytes of on-chip RAM—not the 8051—in order to have enough stack room. If you need more than ten or twelve tasks, you probably should go to a commercial operating system that dumps the stacks to off-chip RAM (and sacrifice context switching speed).

The function to move the stacks as described is shown next in C. (I switched it to assembly at one time and saved a little code, but it was very difficult to follow and, since the rest of the kernel is in C, I use this version.)

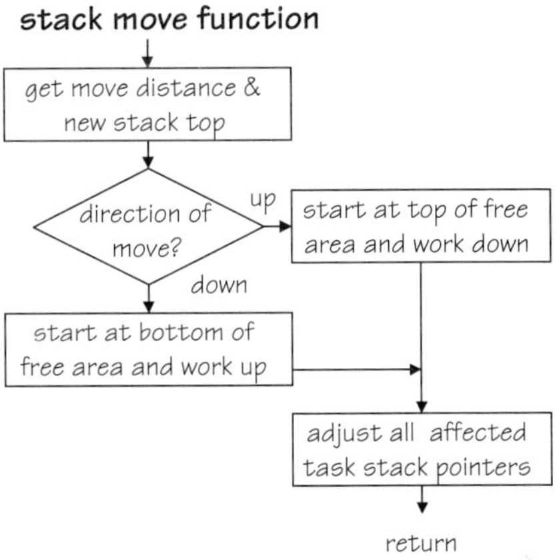


**Stack Move Flowchart**

**Stack Move Function**

```
#define uchar unsigned char
#define uint unsigned int
extern uchar idata *topfree;    /*address of top of
                middle free area*/
extern uchar idata *toptsks[]; /*address of top of
                each task's moved stack*/
extern uchar idata *tSP; /*saved task stack pointer*/
void stackmove (uchar oldtsk, uchar newtsk)using 1{
                /*pass number of each task*/
 uchar move; /*address shift of stack tops*/
 uchar idata *tfn; /*temp value*/
 uchar s; /*start task number of moved stacks*/
 uchar e; /*end task number of shifted stacks*/
 uchar idata *H;/*pointer to fetch actual stack to
                move*/
 uchar idata *L; /*pointer to store moved stack data*/
```

```
smv: toptsks[oldtsk]=tSP; /*save current task's stack
                pointer*/
 move=topfree-tSP;/*determine distance of move*/
 tfn=toptsks[newtsk];/*new top of free area*/
 L=tSP;
 H=topfree;
 if (newtsk>oldtsk){ /*moving stack(s)down*/
  while(H < tfn){L++; H++; *L=*H;}
  topfree=tfn; tSP=L;
  s=oldtsk+1; e=newtsk; move=-move;
 }
 else{ /*moving stack(s)up*/
  while(L > tfn) {*H=*L; L—; H—;}   /*move stack
                contents*/
  topfree=H; tSP=tfn;
  s=newtsk+1; e=oldtsk;
 }
 while(s<=e)toptsks[s++]+=move;/*update each task's
                stack pointer*/
}
```

In order to be sure everything was working, I tested the function with the following program on the simulator (DS51). I'm sure that, by now, you are not so naive as to believe what I show was what I started with and it worked the first time!

**Stack Move Test Program**

```
#define uchar unsigned char
#define uint unsigned int
sfr SP=0x81;
uchar idata *topfree;   /*address of top of middle
                free area*/
uchar idata *toptsks[5]; /*address of top of each
                task's moved stack*/
uchar tsp; /*saved SP for last-running task*/
void stackmove (uchar oldtsk, uchar newtsk); /*pass
                number of each task*/
void main(void)using 1{
 uchar idata *i;
 toptsks[0]=0x45;
 toptsks[1]=0x47;
 toptsks[2]=0x55;
 toptsks[3]=0x69;
```

```
  toptsks[4]=0x73;
  topfree=0x50;
  tsp=0x49;
  for(i=toptsks[0];i<=toptsks[4];i++)*i=i;/*fill stack
                        area*/
  SP=0x20;
  stackmove(1,3);
}
```

## *RTKB* Source Code

Before I could finish the source code, I had to investigate how the tasks could be created—I had to push the task address onto the right place in the stack even though, at the start, there is no call from the task. Like with *RTKS*, I had to run a sample piece of code to be sure of the way parameters are passed when they are pointers to code. It is shown next, with the C source in bold mixed in with the assembly. My conclusion is that the third parameter (the address) passes in *R2* and *R3*, to *CREAT*, but, being the only parameter, it passes to *settaskadd* in *R6,R7*. I had to be sure that a 2-byte pointer didn't somehow get treated like a generic one.

**Parameter Passing Test**

```
FUNCTION _settaskadd (BEGIN)
-- Variable 'tskadd' assigned to Register 'R6/R7' -
void settaskadd(uchar code *tskadd)using
  uchar code *x;
  x=tskadd;
   MOV      x,R6
   MOV      x+01H,R7
}
  RET
FUNCTION _settaskadd (END)

FUNCTION task (BEGIN)
void task(void){
  unsigned char x;
  x=5;
  MOV  x,#05H
}
RET
FUNCTION task (END)
```

```
FUNCTION _CREAT (BEGIN)
void CREAT(uchar task,uchar prity,uchar code
                *tskadd)using 1{
  PUSH     PSW
  MOV      PSW,#08H
-- Variable 'task' assigned to Register 'R4' --
  MOV      R4,AR7
-- Variable 'prity' assigned to Register 'R5' --
-- Variable 'tskadd' assigned to Register 'R6/R7' --
  MOV      R7,AR3
  MOV      R6,AR2
  priority[task]=prity;
  MOV      A,#priority
  ADD      A,R4
  MOV      R0,A
  MOV      @R0,AR5
  semaphore[task]=0;
  MOV      A,#semaphore
  ADD      A,R4
  MOV      R0,A
  CLR      A
  MOV      @R0,A
  settaskadd(tskadd);
  LCALL    _settaskadd
}
POP      PSW
RET
FUNCTION _CREAT (END)

FUNCTION main (BEGIN)
void main(void){
  CREAT (5,100,task);
  MOV      R7,#05H
  MOV      R5,#064H
  MOV      R2,#HIGH task
  MOV      R3,#LOW task
  LCALL    _CREAT
  for(;;);
  ?C0004:
   SJMP    ?C0004
}
  RET
FUNCTION main (END)
```

With the exception of a small amount of assembly to handle context switching and the stack, the entire kernel is written in the C language. I do not think C is as efficient as assembly could be if one had unlimited time to optimize, but having spent most of a week with *RTKS* in assembly, I am not sure it is worth it for *RTKB*. At the end of the C source I list the source for the assembly pieces as well.

**RTKS C Source**

```
#include <reg51.h>
#define uchar unsigned char
#define uint unsigned
#define nooftasks 3
#define noofmbx 3
extern xdata uchar CTXT[];
void banksave();
void bankrestore(uchar task);
void settaskadd(uchar code *tskadd);
uchar data running, noofactive; /*current number of
                active tasks*/
uchar waitex0;
uchar waitex1;
uchar xdata priority[nooftasks];
uint xdata time[nooftasks];
uint xdata interval[nooftasks];
uchar xdata ready[nooftasks];
uchar xdata semaphore[nooftasks];
uchar xdata catalog[nooftasks];
uchar xdata toptsks[nooftasks]; /*address of top of
                each task's  stack*/
uchar xdata mbxstat[noofmbx];
uchar xdata *msg[noofmbx];
uchar idata *topfree;  /*address of top of middle
                free area*/
uchar idata *tSP; /*saved task stack pointer*/

void stackmove (uchar oldtsk, uchar newtsk)using 1{
               [60]/*pass number of each task*/
 uchar move; /*address shift of stack tops*/
```

[60]This function shifts the stacks up or down to make the current active stack the top of the lower ones and all the higher stacks pushed up to the top of the available space. This maximizes the available stack space for the running task.

```
  uchar idata *tfn; /*temp value*/
  uchar s; /*start task number of moved stacks*/
  uchar e; /*end task number of shifted stacks*/
  uchar idata *H;/*pointer to fetch actual stack to
                move*/
  uchar idata *L; /*pointer to store moved stack data*/
  toptsks[oldtsk]=tSP; /*save current task's stack
                pointer*/
  move=topfree-tSP;/*determine distance of move*/
  tfn=toptsks[newtsk];/*new top of free area*/
  L=tSP;
  H=topfree;
  if (newtsk>oldtsk){ /*moving stack(s)down*/
   while(H < tfn){L++; H++; *L=*H;}
   topfree=tfn; tSP=L;
   s=oldtsk+1; e=newtsk; move=-move;
  }
  else{ /*moving stack(s)up*/
   while(L > tfn){*H=*L; L--; H--;} /*move stack
               contents*/
   topfree=H; tSP=tfn;
   s=newtsk+1; e=oldtsk;
  }
  while(s<=e)toptsks[s++]+=move;/*update each task's
                stack pointer*/
}

61/*return lowest priority# task in ready array*/
uchar highestpriority(void)using 1{
 uchar high=0xff; uchar task,i;
 for(i=0;i<=nooftasks;i++){
  if(priority[i]&0x7f<high){
   high=priority[i];
   task=i;
  }
 }
 return task;
}
62/*swap tasks voluntarily*/
```

[61]This function searches through the priority array to find the READY task with the lowest priority number. There is no queue as such.

[62]This function changes the stack and adjusts the READY “queue” to have a different task RUNNING.

```
void contextswitch(void)using 1{
 uchar temp;
  if((temp=highestpriority())!=running){
   if(ready[temp]&0x80){ /*was preempted*/
    ready[temp]&=0x7f; /*clear preempted*/
    bankrestore(temp);
  }
 ready[running]=0; /*old task gave up to wait*/
 stackmove(running,temp);
 running=temp;
 ready[running]=0; /*running task now not in array*/
 }
}

63/*swap tasks involuntarily*/
void preempt()using 1{
 uchar temp;
 if((temp=highestpriority())!=running){
  banksave();
  ready[running]=priority[running]|0x80;
  stackmove(running,temp);
  running=temp;
  ready[running]=0; /*running task not ready now*/
  if(ready[temp]&0x80){ /*was preempted*/
   ready[temp]&=0x7f; /*clear preempted*/
   bankrestore(temp);
  }
 }
}

64/*dec time, set ready any tasks that time out*/
void decrtime(void)using 1{
 uchar i;
 for(i=0;i<nooftasks;i++){
  if(time[i]!=0xff){ /*0xff indicates non-timing
                      task*/
   if(--time[i]==0){ /* time expired; set ready*/
```

[63]This function takes care of context switching where the task is not voluntarily relinquishing the processor. Such a switch requires the storing of the registers in off-chip RAM by the assembly language routine, *banksave*.

[64]This function is invoked every time there is a system timer interrupt (a tick). It adjusts the timeout time for any waiting tasks, marks READY any tasks that timeout, and restarts the count for any task that times out a repetitive interval.

```
      ready[i]=(ready[i]&0x80)|(priority[i]&0x7f);
      if(priority[i]&0x80){ /*running interval*/
       time[i]=interval[i]; /*reset time*/
      }
     }
    }
   }
  }

65/*picks up from assy interrupt routine
ff=timer tick, 0 external int 0; 1 ext 1*/
void FROM_INT(uchar event) using 1{
 if (event==0xff){
  TH0=4;
  TL0=0; /*choose for 1 mSec*/
  decrtime();
  preempt(); /*context switch if needed*/
 }
 else if (event==0){
  if(waitex0){
   if(ready[waitex1]&0x80) preempt();/*already
               waiting*/
   else waitex0|=0x80; /*set event occurred*/
  }
 }
 else if (event==1){
  if(waitex1){
   if(ready[waitex1]&0x80) preempt();/*already
               waiting*/
   else waitex1|=0x80; /*set event occurred*/
  }
 }
}

66/*wait for time
 always switch context
 return status:
 ff when time expires*/
```

[65]This function is reached from the assembly language interrupt handlers by a jump (strictly forbidden in C!). A passed number identifies which interrupt it was. As indicated, this kernel “owns” the external interrupts and handles them as events for specific tasks.

[66]This system call makes a task ASLEEP for a simple timeout.

```
uchar WAITT(uint timeout)using 1{
 time[running]=timeout;
 contextswitch();
 return 0xff;
}

67/*wait interval
 return status:
 01 if interval already arrived
 ff if interval comes later*/
uchar WAITI(void)using 1{
 if((priority[running]&0x80)==0){ /*no interval yet*/
  contextswitch();
  return 0xff;
 }
 else{
  priority[running]&=0x7f; /*clear intvl flagl*/
  return 0x01;
 }
}

68/*wait until semaphore reaches 0
 switch if not 0
 return status:
 00 if semaphore comes later
 01 if semaphore already zero
 ff if timeout*/
uchar WAITS(uint timeout)using 1{
 if(semaphore[running]&0x7f!=0){ /*wait for 0*/
  semaphore[running]|=0x80; /*set msb to 1=waiting*/
  time[running]=timeout; /*set timeout value*/
  contextswitch();
 }
 else return 0x01; /*if semaphore is already 0 then no
                        task change at all*/
}
```

[67]This system call either puts the task to sleep until the timeout (interval) or it puts the task in the *READY* array if the interval event has already occurred.

[68]The system call for signals makes a task *ASLEEP* until either the semaphore reaches zero or the timeout happens.

```
69/*wait for message or timeout
 switch if not 0
 return status:
 00 for delayed message
 01 or immediate message
 02 for another task already waiting
 ff for timeout*/
uchar xdata *WAITM (uchar mbx,uint timeout)using 1{
 if((mbxstat[mbx])&0x80){/*0x80=msg present*/
  mbxstat[mbx]=0; /*clear out message*/
  return 0x01; /*exit if message already there*/
 }
 else if(mbxstat[mbx]>0) return 0xff; /*other task
                 already waits*/
 else{ /*else wait for timeout or message to arrive*/
  mbxstat[mbx]=running; /*save to recall when msg*/
  time[running]=timeout; /*set timeout value*/
  contextswitch();
  return 0x00;
 }
}

70/*wait for external interrupt
 no task switch if already happened
 return status:
 00--waited and then returned due to ext int
 01--event already happened--no switch
 02--another task already waits
 ff--waited and returned from timeout*/
uchar WAITX(uchar intno,uint timeout)using 1{
 if(intno==0){
  if(waitex0&0x80){
   waitex0=0; /*already happened; clear event*/
   return 0x01; /*already happened*/
  }
  else{ /*prepare to wait*/
```

[69] The system call to wait for a message has the task *ASLEEP* "camped out" at a specific mailbox waiting for a message. The possibility of a timeout is there as well unless a 0xffff indicates to wait forever. A mailbox holds only one message and can have only one task waiting. In some ways, it is like an uncommitted semaphore.

[70] This system call makes a given task sensitive to an external interrupt. It avoids the need to write an interrupt handler and makes the event similar to the others, much like DCX did.

```
   if(waitex0&0x7f!=running)return 0xff; /*other task
                waits*/
   else{
    waitex0=running;
    time[running]=timeout;
    contextswitch();
    return 0x00; /*OK*/
   }
  }
 }
 else if(intno==1){
  if(waitex1&0x80){
   waitex1=0; /*already happened; clear event*/
   return 0x01; /*already happened*/
  }
  else{ /*prepare to wait*/
   if(waitex1&0x7f!=running)return 0xff; /*other task
                waits*/
   else{
    waitex1=running;
    time[running]=timeout;
    contextswitch();
    return 0; /*OK*/
   }
  }
 }
 else return 0xfe; /*bad intno value*/
}

71/*set interval value and repeat bit for given task*/
void INTVL(uchar taskid, uint intv)using 1{
 uchar task;
 task=FINDT[taskid];
 interval[task]=intv;
 priority[task]|=0x80; /*set interval bit*/
 time[task]=intv;
}

72/*add units to semaphore; never task switch*/
void ADDSU(uchar taskid, uchar units)using 1{
```

[71]This system call sets the repeat timeout that constitutes an interval.

[72]This system call adds units to a given task's semaphore. Note that this does not allow a task switch.

```
  uchar temp,temp2,task;
  task=FINDT[taskid];
  temp=semaphore[task]&0x80; /*msb=waiting*/
  if((temp2=semaphore[task]&0x7f+units)>0x7f)
   semaphore[task]=0x7f|temp; /*max at 0x7f*/
  else semaphore[task]=temp2|temp;
 }
```

[73]
```
/*subtract units to semaphore; freeze at 0 units; no
                 task switch*/
void SUBSU(uchar taskid, uchar units)using 1{
uchar temp,task;
 task=FINDT[taskid];
 temp=semaphore[task]&0x80; /*msb=waiting*/
 if(semaphore[task]&0x7f<units){ /*reach zero*/
  semaphore[task]=temp;    /*max at 0x7f*/
  if(temp){ /*task already waiting; set ready*/
   ready[task]=(ready[task]&0x80)|
               (priority[task]&0x7f);
  }
 }
 else semaphore[task]-=units; /*just decrement
                 semaphore*/
}
```

[74]
```
/*check semaphore; return # units; no task switch*/
uchar CHEKS(uchar taskid)using 1{
uchar task;
 task=FINDT[taskid];
 return semaphore[task]&0x7f; /*msb=waiting*/
}
```

[75]
```
/*return status of given mailbox; status:
                 00 mailbox free to wai
                 01 message waitin
                 02 task already waiting*/
uchar CHEKM(uchar mbx){
 if(mbxstat[mbx]&0x80)return 0x01; /*msg waiting*/
 else{
```

[73]This system call subtracts units from a task's semaphore.

[74]This system call returns the current number of units in the task's semaphore.

[75]This system call checks for the status at a mailbox

```
/*task to get revs/1/4 sec; send to calculate task*/
void tachometer(void) using 1 {
 static uint revn,revo;
 xdata uint msg;
 INTVL(250); /*count revs for 1/4 sec*/
 for(;;){
  revo=revn;
  WAITI();
  revn=((uint)TH1<<8)|TL1;
  msg=revn-revo;
  SENDM(2,&msg); /*report new reading*/
 }
}

/*task to do simple incr/decr of drive percent*/
void calculate(void) using 1{
 uchar rev;
 for(;;){
  rev=WAITM(2,0xff); /*sleep till new speed rdg*/
  if (rev < target){
   if (percent<100) ++percent;
  }
  else if (percent>0) –percent;
 }
}

void main (void){
 INIT();
 CREAT(0, pulse);
 CREAT(1, tachometer);
 CREAT(2, calculate);
 TMOD=0x11; /*T1 in endless 16-bit overflow mode*/
 TCON=0x54; /*T0,T1 running, IE1 edge triggered*/
 TH1=0;
 TL1=0; /*set initial count (unnecessary)*/
 percent=50; /*start at mid-range*/
 RTKBstart();
 for(;;);
}
```

## Keypad to LCD—*RTKS* Example

This example is broken into three tasks that show communication. I use the modular programming approach from *Volume I,* where all the details for one task are in one module, separate from each of the other tasks. I've also kept the DCX message size of 13 working bytes, deleting the rest of the address information that is part of a BITBUS message. The first task, *key* scans a 4 × 4 keypad and sends a message with the 16-bit key pattern to *keyinterp*. That task converts the bit pattern to ASCII chars in a message sent to *numdisplay*. That task displays the characters on the LCD.

Part of the *main* function (or another function, *initialize*) sets up the LCD display module and registers. The details of LCD setup are of no particular concern here—see Chapter 5, page 248 on Output for more details.

The *keyscan* task handles the reading of a user keypad. The keypad is a 4 × 4 matrix like the example from Chapter 3. The *keyscan* task takes care of the scanning at a 40-msec cycle time to eliminate bounce. When it is recognized that a button has changed state, the value is forwarded to the *keyinterp* task.

The *keyinterp* task allows a small amount of editing. The A key (accept) saves the numeric value in an array. The B key (back up) removes the most-recent digit entered. The F key (fail) erases the number without saving and clears the display.

The *numdisplay* task illustrates converting decimal to ASCII. Here you see that tasks can include several routines.

---

**Keypad to LCD Startup**

```
#include <reg51.h>
#include<RTKS.h>
void numdisplay(void);
void keyscan(void);
void keyinterp(void);

void main(void){
[96]INIT();
 CREAT(0,keyinterp);
 CREAT(1,keyscan);
 CREAT(2,numdisplay);
 TMOD=0x01;
 TH0=~(1000/265);
 TL0=-(1000%256);
```

[96]This is the expected startup procedure. The *INIT* sets up the tables, the *CREATs* put the tasks on record, and the *RTKS*start begins the actual operation.

```
 TR0=1;
 ET0=1;
 RTKSstart();
 for(;;);
}
```

---

**keyinterp control task**

```
/*keyinterp control task*/
#include <reg51.h>
#include <RTKS.h>
#define forever 0xff
extern uint keypattern; /*raw code form keyscan*/
extern uint binarynum; /*accumulated by numdisplay*/
uchar array[10];
uchar key; /*shared with other tasks*/

uchar keycodeconvert(void){
 static uint oldpattern;
 uint newpushes;
 uint i;
 newpushes = (keypattern^ oldpattern) & oldpattern;
  /*keypattern have 0 for pushed key;
  this gives a 1 for each newly pushed key position*/
 for (i=0;i<=15;i++){
  if(newpushes & 1){
   return i;   /*quit at first new*/
  }
 }
}

void keyinterp(void){
 static uchar j;
 j=9;
 for(;;){
  WAITS(forever);
  key=keycodeconvert();
  if(key<10) SIGNL(2); /*add numeric to display*/
  else if (key==10){ /*A-save and ready for next*/
   array[(j=++j%10)]=(binarynum);
   key=0xff;
   SIGNL(2); /*clear display*/
  }
  else if (key==11){ /*B--back up one digit*/
   key=0x0f;
   SIGNL(2);
```

```
  }
  else if (key==15){ /*F--erase-no save*/
   key=0xff;
   SIGNL(2); /*clear display*/
  }
 }
}
```

---

**numdisplay task**

```
/*numdisplay task*/
#include <reg51.h>
#include <RTKS.h>
#define forever 0xff
extern uchar key; /*data from keyinterp*/
static uchar temp[5]; /*accumulating characters*/
static uchar tempi;
uint binarynum; /*binary of accumulating number*/

void wcmd(uchar cmd){ /*sends cmd to LCD display*/
 P1&=0xdf;      /*set cm/da to*/
 P3=~(cmd); /*put on bus*/
 P1|=0x10;      /*strobe in data*/
 P1&=0xef;
 WAITT(1);
}

void wdat(uchar dat){ /*sends dat to LCD display*/
 P1|=0x20; /*set cm/da to da*/
 P3=~(dat); /*put cmd on bus*/
 P1|=0x10; /*strobe in data*/
 P1&=0xef;
 WAITT(1);
}

void dspwrite(void)using 1{/*converts, sends to LCD*/
 uchar j;
 for(j=0;j<=tempi;j++){
  wdat(temp[j]+'0');
  binarynum=10*binarynum+temp[j];
 }
}

void dspinit(void)using 1{
 WAITT(14);
 wcmd(0x38); /*function set*/
```

```
 WAITT(4);
 wcmd(0x38); /*function set*/
 wcmd(0x38); /*function set*/
 wcmd(0x38); /*function set*/
 wcmd(0x08);/*dsp off*/
 wcmd(0x01);/*dsp clear*/
 WAITT(1);
 wcmd(0x06);/*display stat,cursor ->*/
 wcmd(0x02); /*cursor home*/
 WAITT(1);
 wcmd(0x0e);/*display on,cursor off*/
}

void numdisplay(void)using 1{ /*display task*/
dspinit();
 tempi=0;
 for(;;){
  WAITS(forever);
  if (key==0xff){ /*erase*/
   wcmd(0x01); /*clear display*/
   WAITT(1);
   binarynum=0;
   for (tempi=0;tempi<5;tempi++) temp[tempi]=0;
   tempi=0;
  }
  else if (key==0x0f){ /*back up by one*/
   temp[--tempi]=0;
   dspwrite();
  }
  else { /*add digit to number*/
   temp[++tempi]=key;
   dspwrite();
  }
 }
}
```

---

**keyscan task**

```
/*keyscan task*/
/*task to scan the 4x4 keyboard every 40 msec it sends
                a message holding the scan pattern
                whenever there is a change of any
                key.*/
#include <reg51.h>
#include <RTKS.h>
uint keypattern;
```

```
void keyscan(void){ /*keyscan task*/
 uchar code scanp[]={0x1,0x2,0x4,0x8};
 uchar i;
 uint newsc;
 static uint oldsc=0;
 for(;;){
  newsc=0;
  for(i=0;i<=3;i++){ /*4 scans*/
   WAITT(10); /*10 msec -> full scan every 40 ms*/
   P1=P1 & 0xf0 | scanp[i];
   newsc=newsc<<4 | (P3 & 0xf);
  }
  if(newsc != oldsc){
   keypattern=newsc; /*shared variable*/
   SIGNL(1); /*changed*/
  }
  oldsc=newsc;
 }
}
```

## Keypad to LCD—*RTKB* Example

Finally I present an example using *RTKB* to do the same keypad to LCD operation.

**Keypad to LCD—Main (Startup)**

```
#include <reg51.h>
#include <RTKB.h>
void numdisplay(void);
void keyscan(void);
void keyinterp(void);

void main(void){
 INITK();
 CREAT(0,100,keyinterp);
 CREAT(1,20,keyscan);
 CREAT(2,150,numdisplay);
 TMOD=0x01;
 TH0=~(1000/265);
 TL0=-(1000%256);
 TR0=1;
 ET0=1;
 STRTK();
```

```
 for(;;);
}
```

---

**keyinterp control task**

```
#include <reg51.h>
#include <RTKB.h>
#define forever 0xffff
#define mbx0 0
extern uint keypattern; /*raw code form keyscan*/
extern uint binarynum; /*accumulated by numdisplay*/
uchar array[10];
static uchar key; /*shared with other tasks*/

uchar keycodeconvert(void){
 static uint oldpattern;
 uint newpushes;
 uint i;
 newpushes = (keypattern^ oldpattern) & oldpattern;
  /*keypattern have 0 for pushed key;
  this gives a 1 for each newly pushed key position*/
 for (i=0;i<=15;i++){
  if(newpushes & 1){
   return i;  /*quit at first new*/
  }
 }
}

void keyinterp (void){
 static uchar j;
 j=9;
 for(;;){
  WAITS(forever);
  key=keycodeconvert();
  if(key<10) SENDM(mbx0,&key); /*add numeric to
                  display*/
  else if (key==10){ /*A-save and ready for next*/
   array[(j=++j%10)]=(binarynum);
   key=0xff; /*clear display*/
   SENDM(mbx0,&key);
  }
  else if (key==11){ /*B--back up one digit*/
   key=0x0f;
   SENDM(mbx0,&key);
  }
```

```
  else if (key==15){ /*F--erase-no save*/
   key=0xff; /*clear display*/
   SENDM(mbx0,&key);
  }
 }
}
```

---

**number display task**

```
#include <reg51.h>
#include <RTKB.h>
#define forever 0xffff
static uchar *key; /*data from keyinterp*/
static uchar temp[5]; /*accumulating characters*/
static uchar tempi;
uint binarynum; /*binary of accumulating number*/

void wcmd(uchar cmd){ /*sends cmd to LCD display*/
 P1&=0xdf;        /*set cm/da to*/
 P3=~(cmd);   /*put on bus*/
 P1|=0x10;        /*strobe in data*/
 P1&=0xef;
 WAITT(1);
}

void wdat(uchar dat){ /*sends dat to LCD display*/
 P1|=0x20; /*set cm/da to da*/
 P3=~(dat); /*put cmd on bus*/
 P1|=0x10; /*strobe in data*/
 P1&=0xef;
 WAITT(1);
}

void dspwrite(void){/*converts, sends to LCD*/
 uchar j;
 for(j=0;j<=tempi;j++){
  wdat(temp[j]+'0');
  binarynum=10*binarynum+temp[j];
 }
}

void dspinit(void){
 WAITT(14);
 wcmd(0x38); /*function set*/
 WAITT(4);
```

```
 wcmd(0x38); /*function set*/
 wcmd(0x38); /*function set*/
 wcmd(0x38); /*function set*/
 wcmd(0x08);/*dsp off*/
 wcmd(0x01);/*dsp clear*/
 WAITT(1);
 wcmd(0x06);/*display stat,cursor ->*/
 wcmd(0x02); /*cursor home*/
 WAITT(1);
 wcmd(0x0e);/*display on,cursor off*/
}

void numdisplay(void){ /*display task*/
 dspinit();
 tempi=0;
 for(;;){
  *key=WAITM(0,forever);/*use mailbox 0*/
  if (key==0xff){ /*erase*/
   wcmd(0x01); /*clear display*/
   WAITT(1);
   binarynum=0;
   for (tempi=0;tempi<5;tempi++) temp[tempi]=0;
   tempi=0;
  }
  else if (key==0x0f){ /*back up by one*/
   temp[--tempi]=0;
   dspwrite();
  }
  else { /*add digit to number*/
   temp[++tempi]=key;
   dspwrite();
  }
 }
}
```

---

**keyscan task**

```
/*task to scan the 4x4 keyboard every 40 msec
 it sends a message holding the scan pattern whenever
 there is a change of any key.*/
#include <reg51.h>
#include <RTKB.h>
uint keypattern, /*passsed to keyinterp by shared
                 variable*/
```

```
void keyscan(void)using 1{ /*keyscan task*/
 uchar code scanp[]={0x1,0x2,0x4,0x8};
 uchar i;
 uint newsc;
 static uint oldsc=0;
 ADDSU(1,1); /*set semapnore so interp waits*/
 INTVL(10);
 for(;;){
  newsc=0;
  for(i=0;i<=3;i++){ /*4 scans*/
   WAITI(); /*10 msec -> full scan every 40 ms*/
   P1=P1 & 0xf0 | scanp[i];
   newsc=newsc<<4 | (P3 & 0xf);
  }
  if(newsc != oldsc){ /*changed*/
   keypattern=newsc; /*shared variable*/
   SUBSU(interptask,2); /*semaphore stops at 0
              units*/
   ADDSU(interptask,1);
  }
  oldsc=newsc;
 }
}
```

# 4

# Inputs

## OVERVIEW

This chapter has examples of both circuitry and software involved in bringing information *in* to a microcontroller from things such as switches and keypads.[1] The next chapter goes into ways to get *out* from a microcontroller to external devices such as LEDs, stepper motors, relays, and DC motors. Many interfacing problems can be easily avoided with an understanding of the principles that are discussed here.

I emphasize designs that use simple devices such as transistors. While you may prefer to use more highly integrated devices, designs for high-volume products should be concerned with *all* aspects of efficiency, including circuit board real estate, the variety of parts to stock,[2] availability of parts from multiple sources, along with the usual concerns about reliability and user-friendliness. Real estate and availability may fight each other—the most compact solution may be a brand-new IC barely on the market yet and useful only in one place in the product. This new IC may also carry a premium price at the start and, of course, you run the risk of it going out of production before your product's end life. On the other hand, cost includes the labor to insert one part versus a handful of discrete or small-scale parts. The suggestions that follow can *also* lead to good choices for prototype projects—especially if a hard-to-get part fails just before a deadline.

---

[1]For lack of a better home, the last part of this section discusses power supplies—whether this is input to a micro is debatable. It is certainly not an input signal, but it is surely necessary somewhere.

[2]The variety of parts can directly affect purchasing volumes and resulting economies. If your design uses a 7400, a 7402, a 7404, and a 7432, it might be much better to use four or five 7400s instead. Of course, for that mix, the use of a small PLD for the glue logic might be even better.

## DIGITAL INPUTS

This section discusses inputs that are digital—either "high" or "low." These inputs can be standard TTL levels, discussed first, or significantly different levels, discussed under "level shifting." I include keypads and keyboards, since they usually are either on or off for any one key.

### TTL Compatible

To understand the problem of going into TTL logic, you need to visualize the currents and voltages involved. Since logic "1" is a voltage above 2.4 V and a logic "0" is a voltage below about 0.9 V, any circuit that can maintain those voltages at an input can work. But while a logic "1" involves virtually no current flow, a logic "0" (depending on the specific TTL family) requires about 1.6 mA of sinking current (*from* the TTL input to ground).

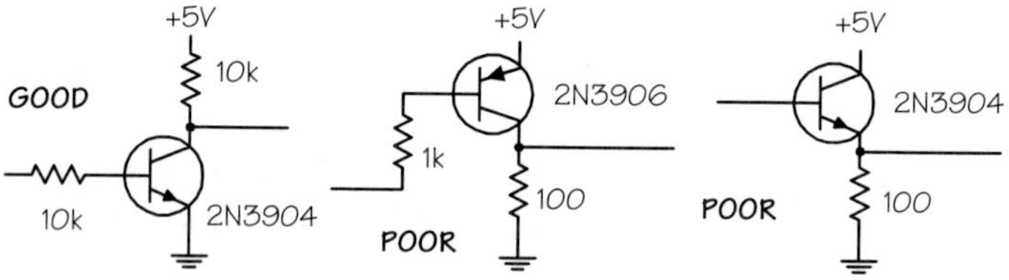


**Interfaces to TTL**

The circuits here show ways to convert external digital inputs into TTL-compatible outputs.[3] With a limiting resistor ahead of the base the inputs are not even limited to 0 to 5 V—as long as the base current is not exceeded, the transistor will serve as a safe buffer.

> **TTL INPUTS:** Driving any input should always favor current sinking—pulling input to 0.

A comparison of the three circuits will show that it is far more efficient to drive a TTL gate from an NPN device. When the device is driving a "0," it sinks the requisite current, but when it is driving a "1" almost no current is

---

[3]If you can convert an input to TTL-compatible form you can drive most embedded microcontroller ports

used (wasted).[4] The PNP device or the emitter-follower, on the contrary, require that the output resistor be small enough to sink the 1.6 mA without rising above 0.9 V (or more current for multiple inputs) and yet be driven above 2.4 V for a logic "1." The current required in the latter case is quite large and totally unnecessary since virtually no current flows *into* the TTL input.

### Microcontroller Inputs

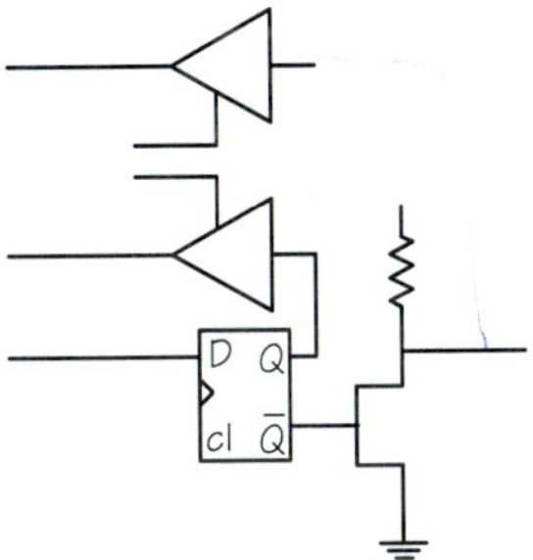


**Basic Internal Port Pin**

Internal ports of the 8051 are unusual in that they are only *partly* bi-directional (the term is *quasi*-bi-directional). You can see the hardware in the schematic for one pin here.[5] You can drive *out* any time but, for *input*, the output circuitry must be off. Otherwise you will be reading your own output latch rather than the signal coming in from the outside.[6] The key software aspect is the requirement that ***any port you are using as input must have a 1 written out to it***.[7] There are other functions for some of the port

---

[4]The 10-k resistor will pass ½-mA. It could be even larger in value but the speed would start to slow because of stray capacitance. The value is not critical and can depend on whether power is at a premium as with battery-powered equipment. Incidentally, you can often distinguish electronic designers by this resistor choice—there are the 10 k (conservative) engineers and the 1 k (bold or wasteful) engineers.

[5]Technically the schematic shows a P1 pin. The pins on P3 can have alternate output and input functions and P0 and P1 can function as address or address/data lines when off-chip memory access is called for by a *MOVX* instruction. The additional feature of reading the input off the latch or off the input was added for the *read/modify/write* commands.

[6]When two logic outputs are tied together, if there is a "winner" at all, it is the one pulling low to zero.

[7]On reset the latch is set high, so if you never write out, you do not need to worry. If you write out a 0, when you read in the port you will see the 0 being sent out rather than the condition of the external device.

pins not shown here. Suffice it to say, if you are using a pin for other functions you should not use it as a port. There is no special disabling or setting of port direction.[8] Also, P0 when used as an I/O port (as opposed to the data bus for off-chip memory expansion) does not have a pull-up resistor—the external device cannot count on an open floating high.

The numbers for 8051 inputs are: a 0 must be below 0.9 V (0.2 $V_{cc}$-0.1) while sinking 50 μA.[9] And a 1 must come in above 3.5 V (0.7 Vcc)

### Single Contacts

**Switches.** The circuit below is good way to bring switch inputs into a TTL input. The pull-up resistor supplies the very small current needed for a logic 1 input, and the short to ground by the switch supplies a logic zero very close to 0 V and with all the current sinking capability required. Thus the circuit wastes very little current in either state—only what is needed by the TTL. The one drawback, of course, is that the closure of the switch gives logic 0 and people like to think of a switch closure (or button push) as giving logic 1. Hopefully the software portion of this book has convinced you of the ease of complementing the input if you must have positive-true logic.

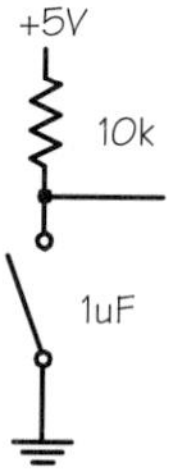


**Switch Interface**

The very worst arrangement, tried by legions of students over the years, is to tie a switch to 5 V and feed the other side into the input. The logic 1 when the switch closes is undeniably solid, but when the switch is

---

[8]Input operations read either the latch or the port pin, depending on the instruction. The instructions that use the latch when a port is designated are the read-modify-write instructions: *ANL, ORL, XRL, JBC, CPL, INC, DEC, DJNZ,* and three of the bit instructions. They should never have an *input* port as a destination for the result anyway. The distinction protects from reading a false state of an output port due to pin loading.

[9]This is a far cry from the 1.6 mA of standard TTL so the fan out *to* other microcontroller inputs could be very great. You could go to at least thirty other micros, steady state, but there is a much larger transition current (750 μA) sourced as the inputs cross the threshold. It is difficult to imagine such an application anyway—you might go out to many devices, but seldom would they be other microcontrollers without some sort of buffer or line driver.

open the input floats to a noise-sensitive high rather than a low—from high to floating high! Again, you *could* add a pull-*down* resistor as in the POOR circuits on page 152, but that has the same inefficiency.

**TIE SWITCHES TO GROUND:** It is always best going into TTL to run the switch to ground with a pull-up resistor

The arrangements here all use single-pole switches with a single-throw. You can use a double-throw switch with one contact tied to 5 V and the other tied to ground, but you still have the possibility of the contact floating during the time it moves between posts (break-before-make). Besides, double-throw switches are more expensive and not available in low-cost pushbuttons.

**Debouncing**. Except for very special devices such as mercury-wetted switches, there is a momentary period when the inertia of the switch contacts causes them to bounce apart several times before the motion stops. This bounce can be too fast for a human to notice (a few msec), but much to slow to be missed by electronic counting circuitry—flip-flops and count chains. Note that a switch only bounces when it *closes*—when it opens, as soon as the surfaces part, they never touch again.

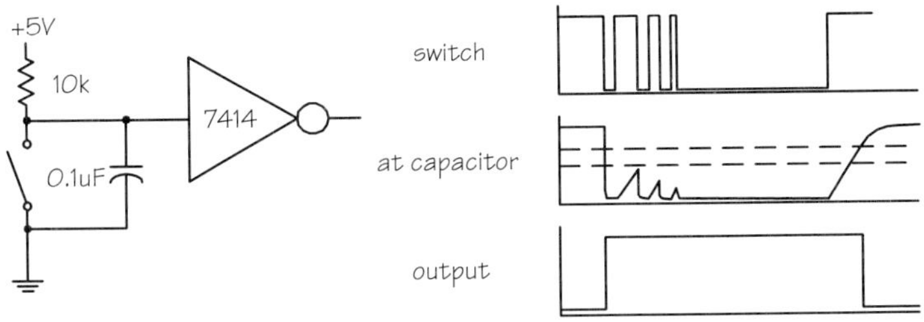


**Debouncing Circuit**

The one place for debounced switches is where counting is involved—especially directly into a counting chip. The circuit above will debounce any ordinary switch. Be sure to have the switch or button tie to *ground* rather than to +5 V because it allows a larger resistor and saves power. The closure pulls the capacitor quickly to zero where it hovers because the RC rise is too slow to get above the Schmidt trigger threshold before the contact has bounced up and returned to make contact. The capacitor is there to keep the input from jumping high when the contact bounces open. The Schmidt trigger is there to avoid extra edges at the end due to noise on the input as the capacitor slowly passes through the threshold—the hysteresis in the 7414 keeps small variations from putting out more than one negative going edge. The bounces on small switches end within 10 msec, so be sure to choose a capacitor big enough to keep the bounce from rising above the *upper* threshold of the 7414.

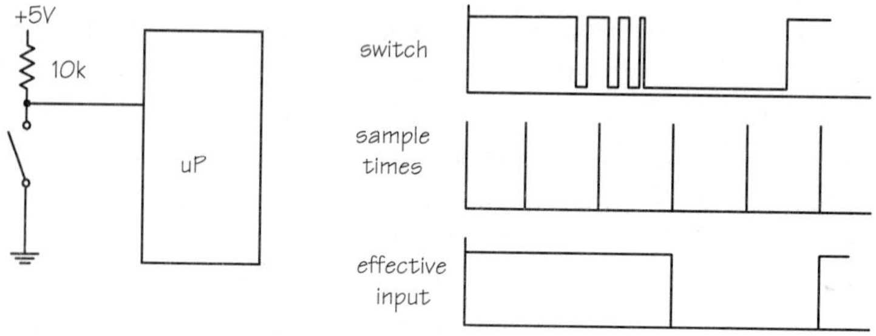


**Debouncing by Sampling**

Particularly when tying switches (and keypads) to microcontrollers, it is usually inefficient to use debounce circuitry. In software, it is quite easy to scan a switch or button once every 40 to 50 msec and miss any bounce entirely. The diagram above shows this. In this case, if the scan happens to fall in the middle of a switch's high-to-low transition it may catch the high bounce and decide the state hasn't yet changed, or it may catch the low part of the bounce and decide there has been a change. In the former case, the change won't be recognized until the next scan 40 msec later (which will cause no problem to a human user). The basic principle: *Small switches don't bounce more than 20 msec, and humans do not get upset by delays of even several hundred msec.*

> **AVOID DEBOUNCING:** Unless you are counting directly in hardware, it is always more efficient to debounce by the inherent timing of the scheduler or operating system.

**Buttons.** Inputs from buttons are much the same as inputs from switches. Give careful attention in your design to which is most "user friendly" and appropriate to the application.

## Level Shifting

Sometimes you have a digital input coming from a different voltage level. You can always convert between levels if speed is not an issue. There are ICs specifically for conversion between logic families—particularly to and from TTL, ECL, and CMOS logic families. A newer level shift problem is to transfer from 3.3 V logic and 5 V logic.[10]

Non-standard level shifting can require more thinking. If you are converting an incoming logic 1 at, say 24 V, you could use a simple voltage divider to get the 5 V range you need as shown here. Remember that TTL inputs require a significant current sink for a logic 0—to get the impedance low enough you might have to use very low resistance values in your divider, wasting what may be significant power. If the resistor to ground is to sink 1.6 mA from the TTL input that follows while not rising above 0.8 V,[11] it cannot be larger than 500 ohms. If the output to the TTL is not to rise above 5 V when 24 V is input, given the 500 Ω lower resistor, the ratio must be at least 19/5 or 1.9 k Ω. That means the circuit sinks 10 mA from the 24-V input signal and about ¼-watt is dissipated. The design values might be 1.8 k @ ½-watt and 470 @ ⅛-watt.[12]

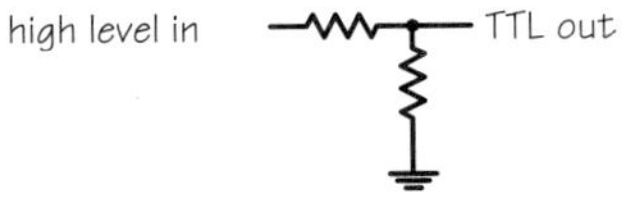


**Basic High-level to TTL Interface**

The better option is to add a transistor to reduce the current in the divider as shown here where a single NPN transistor allows very low-power divider while giving adequate TTL drive current. Where power is in short supply or where waste heat is to be avoided, an extra transistor might be a bargain! Remember that you have added an inversion to the logic.

---

[10] I believe that 3.3 to 5 V transitions are nearly automatic since a logic 1 for 5 V TTL is anything above 2.4 V which is easily be supplied by 3.3 V "TTL." A quick check of the low threshold for the 3.3 V logic is in order, but I suspect that the value is close to the 1 V of the normal TTL. Probably the only concern is to be sure that you don't overload a 3.3 V output so its logic high doesn't go high enough—it may be necessary to have a loading of only one device to be safe.

[11] *One* standard TTL input—LS inputs need only 0.4 mA at .8 V and multiple inputs multiply the current requirement.

[12] For carbon composition resistors, conservative design calls for no more that half the power dissipation of the resistor's rating.

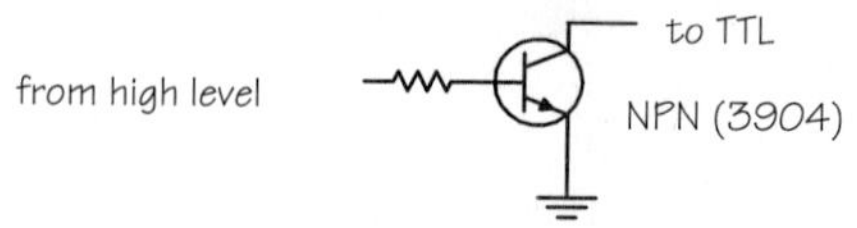


**Better High-level to TTL Interface**

Specifically, suppose you are getting a 24-V DC signal. The most efficient way is with the NPN transistor shown above. In computing the resistor, I will assume that the circuit is only driving one gate and must sink no more than 2 mA. With 2 mA of collector current, I assume the β (current gain in the transistor) would be at least 100, so I only need 20 μA of base current and a base-emitter voltage of about 0.7 V. Just for safety, to be sure the transistor stays in saturation, I'll plan on 40 μA through the resistor. The resistor would then be 23.3 V/40 μA = 582 k Ω. Now *that* is saving power. I would probably be pick the next smaller value in the 20 percent tolerance steps—470 kΩ — and an ⅛-watt size, but the power dissipation would only be 1.2 mW—a lot less than the ¼-watt in the first case.

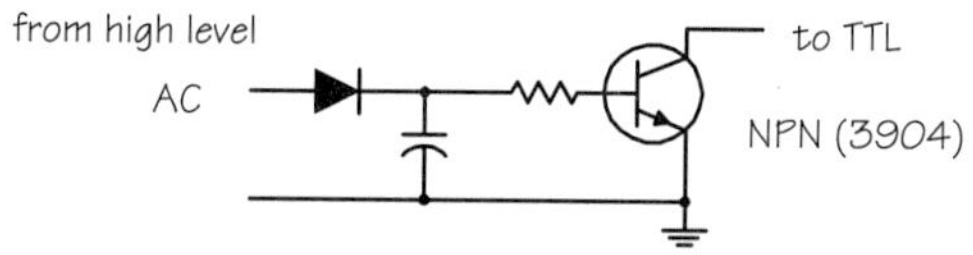


**AC to TTL Interface**

Finally take a case where you are getting in either a 110 V or 24 V AC signal as a "true" input. Perhaps you are interfacing a micro to a system where AC activates everything such as common relay logic. You need to be careful about the commons so you don't fry circuitry, but otherwise it is fairly easy.[13] Remember that the AC is running at 50 or 60 Hz, meaning that the signal peaks positive every 16 to 20 msec. All you need to do is rectify and filter the input to get a DC level. Then, if the voltage is too high, run a divider. It may be a good place to add a transistor as shown above if the amount of current needed would involve too much power.

Specifically assume the circuit above is to give a logic 0 when 110 V 60 Hz is present at the inputs. I'll assume the same 40 uA needed into the base of the transistor. Since the diode delivers charge to the capacitor only during the peak of the positive cycle every 17 msec, there must be enough capacitance to drive the transistor for an entire cycle. Since the 110 V is rms,

---

[13]You may want to look at the opto-isolated section for more protection ideas.

the peak is about 1.7 × 110 = 187 V. There isn't any hysteresis in the circuit so it could give some trouble to the following TTL if it is used for counting—a Schmidt trigger would be required or a comparator with positive feedback. Assuming that is not a problem, I would suggest that the capacitor voltage be kept to at least half at the end of a cycle so the time to sense a loss of signal not be too long while the continued presence of the signal not be succeptible to dropout. I'll say that the RC value should give a time constant of 17 msec. If 40 μA is to flow when the voltage is half (94 V) then R = 94 V/40 μA = 2.3 MΩ. Then C = 1/RT = 1 / (2.3 MΩ × 17 msec) = 25 μF. That's awfully big for a 200-V capacitor, but it would hold the signal long enough to remain for the entire cycle.

My love of efficiency, given a microcontroller on the other end, would call for *eliminating* the capacitor and going right into the transistor. Then the signal would be present only for part of the positive cycle, but a clever programmer would already have a real-time interrupt (perhaps every msec) that could control scanning of this AC input and define a signal present if it is found at least once in the 17 msec.

## Numeric Keypads

Keypads are a common input device for microcontroller systems—perhaps too common. While they allow easy numeric entry, they are often used when much simpler device would suffice or even be preferable. For more on this look ahead to *Knobs, Dials, and Sliders* under *Analog Inputs*. Also, in *Algorithms* (Chapter 6) look at the ways you can combine four buttons and a display to set parameters using two buttons as up/down adjusters.

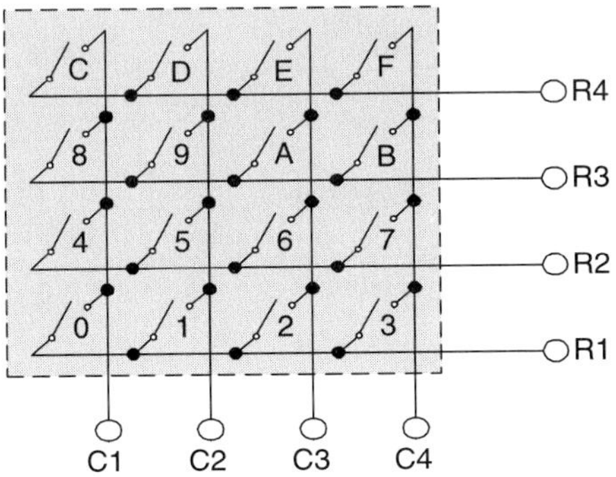


**Matrix Keypad Internal Circuitry**

The most common sixteen-key keypads have eight I/O lines. Pressing a key shorts two of the lines (a row and a column) together. Such ***matrix***

***keypads*** are very common for user input to an embedded controller because they are inexpensive and fairly easy to use.[14] This section focuses on interfacing a sixteen-key (4 × 4 matrix) keypad to a micro, first ***with a keypad decoder chip***, and then by ***connecting the keypad directly*** to an I/O port on the micro.

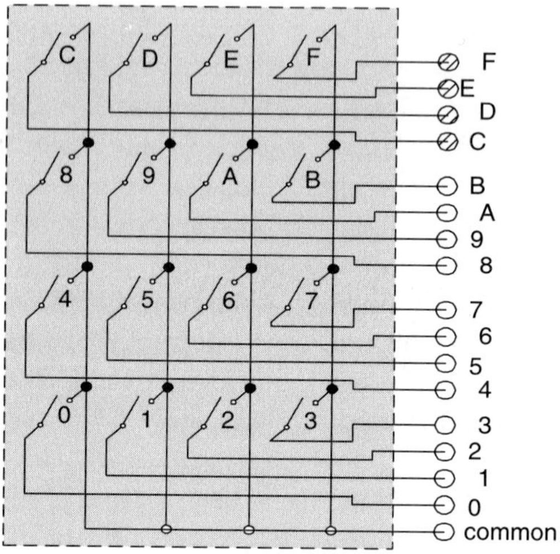


**Non-matrix Keypad Internal Circuitry**

An alternate type of keypad, not discussed further, brings out *all* the keys and a common line. With embedded controllers, this is poor because it takes up so many I/O pins, but it can be appropriate where the keys have direct actions on hardware rather than acting through a micro.

Another feature of some more expensive keypads is the addition of a diode with each key, called "diode isolation," discussed at the end of the section.

What is the most efficient *hardware* depends on how busy the processor is and whether saving three port bits is worth adding an additional chip and two capacitors.[15] The most efficient *software* relies on either a real-time

[14]The other very common keypad has ten or twelve keys and seven lines although some of the telephone keypads have a different arrangement to activate the Dual Tone, Multi-Frequency (DTMF) (Touch-tone®) encoder circuitry.

[15]I fear that many students, at least, use the encoder because they think it makes the software easier. As you can see, the scanning is extremely simple if you have a real-time interrupt already (as in Chapter 3).

interrupt—to scan directly or to poll an encoder chip—or an interrupt *from* an encoder chip when there is a new key. It is inefficient to tie up the processor waiting endlessly for a key to be pushed. At the fastest, input from a keypad can't come more often than every 100 msec, and, as discussed earlier, a scan rate of 40 msec eliminates any double counting caused by key bounce. Is there an extra external interrupt input available? Even if there is, that approach mandates an encoder chip—more hardware. If a real-time interrupt is already involved, scanning the keypad involves negligible software overhead.

The first example that follows uses an encoder chip ***without interrupts***. A call to the keypad routine halts program flow until a key is pushed because the routine can't leave the polling loop until there is data available.

In the second encoder chip version, the data available line is still ***polled*** to see if a new key has been pushed, but if no keys are pushed, the routine returns *immediately* with a "no key" code (FF). It is up to the main program to decide what to do if there is no valid data. Possibly the polling routine would be used by a scheduler. The main program would then be independent of the polling and would be notified of new characters by a shared flag or by observing that buffer pointers indicate that there are new characters.

In the third encoder chip example, the data available line is tied to a ***hardware interrupt*** line so that when (and only when) a key is pressed, the program flow at once stops to process the new key.

In the direct-scanned example that comes later, the keypad is tied directly to a port and a ***periodic, interrupt-driven scan*** (somewhat like the second encoder version) recognizes changes that happen at the keypad and converts them to coded values.

## Example: Encoder-Chip to Micro Port—No Interrupt

From a software standpoint, the first approach is the simplest. A 74C922 encoder chip takes care of the keypad scanning and encoding. The scan rate is set by the value of an added capacitor. Whenever a new key push is recognized the *Data AVailable (DAV* or *DA*) line goes high. The code for the particular key ($0000_2$ through $1111_2$) is then available on the other four output lines until a new key is pushed. Based on the configuration of the chip, the DAV signal either remains until the key is released, or times out after a delay determined by the other capacitor.

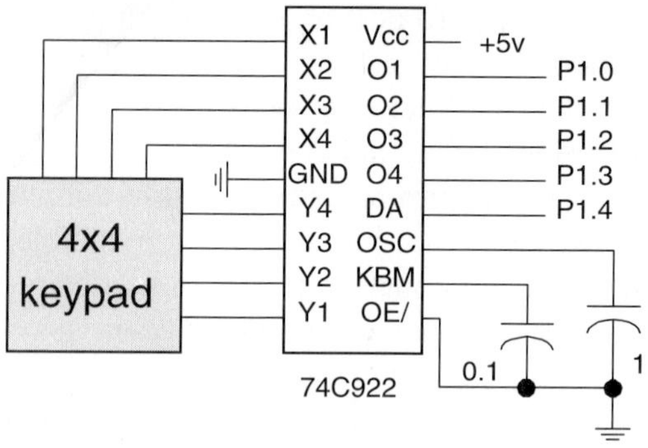


**Encoded Keypad Schematic**

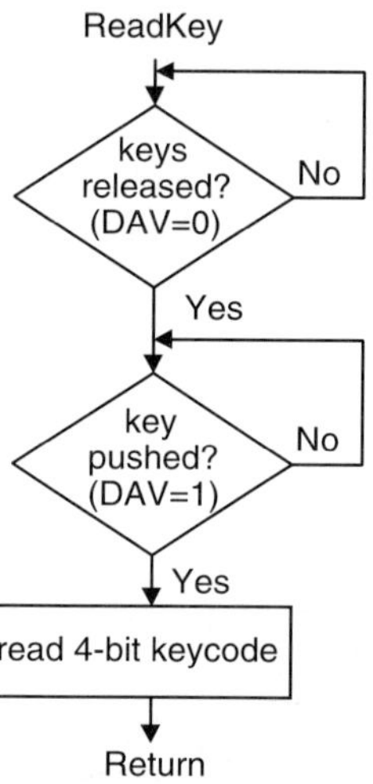


**Encoded Input Flowchart—Simplest Software**

The example assumes you are interfacing to the 74C922 using an on-chip port (*P1*) without any interrupts. The *readkey()* function waits until a key is pushed, so this routine is only useful if there are no other processor activities to take place until the key is pushed (or if other activities are interrupt-driven).

**KEY INPUT—ENCODED INPUT**

```
KEYM SEGMENT CODE
RSEG KEYM
 READKEY:
  K1:JB P1.4,K1 ;WAIT FOR LOW
  K2:JNB P1.4,K2 ;WAIT FOR HIGH
    MOV A,P1 ;GET 4-BIT DATA
```

```
  ANL A,#0FH ;RETURNED IN REG A
  RET
END
```

**Key Input—Encoded Input**

```
#include <reg51.h>
#define uchar unsigned char
uchar code ASCIIbtn()={"0123456789ABCDEF"};

/*prototype for undefined display function*/
void display(uchar ascii);

uchar readkey(void){
 while(P1.4); /*must go low first*/
 while((P1.4)==0); /*high for new push*/
 return P1 & 0x0f;
}

void main(void){
 for(;;){
  display(ASCIIbtn(readkey());
 }
}
```

**DON'T WAIT FOR INPUT:** It is much better, if there are other things to do, to fall through the key fetch routine if there is nothing new. This is consistent with multitasking and fuller use of the microcontroller.

### Encoder Chip with Added Port and Interrupt

A second circuit using the 74C922 chip uses the interrupt of the controller. When the *DAV (DA)* signal goes high it now clocks a flip-flop that triggers an interrupt. The microprocessor can then *read* the keypad decoder chip (via the 74LS244 tri-state buffer that serves as an input port) to determine which key is pressed. *Writing* to this same address will clear the flip-flop to allow the next interrupt. The example shows a memory-mapped input port chip at address F600 (decoded elsewhere from the address bus).[16]

[16]This design was adopted from the boards sold by *Iota*, but Gary Muhonen has since sold the business and trained in psychology. Who says you can't change career directions?!

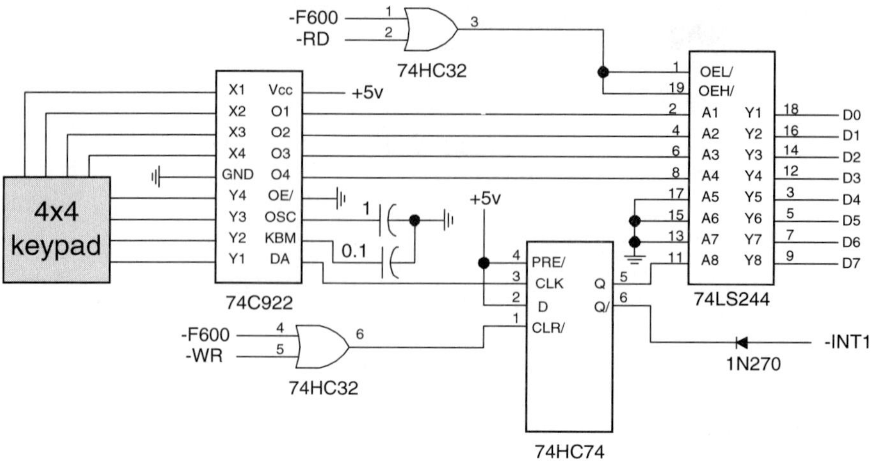


**Keypad Interface Using 74C922 with Interrupt**

This circuit is designed to bring multiple interrupts into a single external interrupt pin. The low-going signals pull the *INT* line down through the diode. To be safe, a low-voltage-drop (germanium) diode should be used so the low is below the maximum voltage for a logic 0. If the interrupt might come from one of several sources, after the interrupt signal arrives your interrupt routine will have to check around (*poll*) to find out which device caused the interrupt. As long as your software doesn't have to read each keypress code more than once, the circuit could be simplified to have the *read* of the address both bring in the data *and* clear the interrupt. By bringing the ORed *RD/* and *F600/* signal to the *CLR/* of the 7474, the high-going pulse would both enable the tri-state buffer for the read and clear the latch on the falling edge just after the read is finished.

If you have no more than two devices using external interrupts, you can dispense with the latch and diode and have them go directly to the *EX0* and *EX1* pins. You will still have to invert the signals, however, because the *DAV* line goes *high* while the *INT* signal must go *low*. You should set the interrupt mode to be *edge* triggered because the *DAV* signal won't go away until the button is released. If you have an on-chip port available (or already have added ones with an 8255), you can eliminate the additional read-only port (made here with the 74LS244 tri-state buffer) and the OR gates.

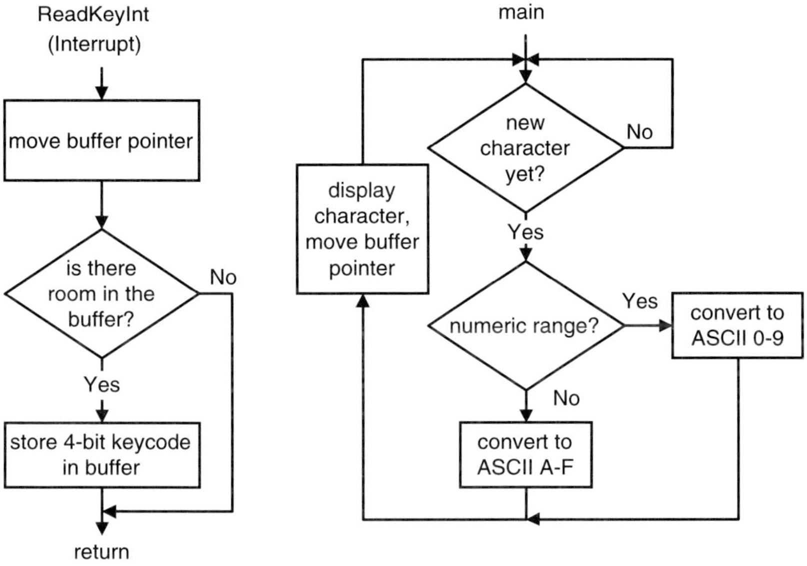


**Encoded Scan with Interrupt**

The C software of the encoded/interrupt example that follows assumes you are using the hardware shown in the previous schematic to latch the data and to manage an interrupt from the *DAV* line of the 74C922. It is a little more complicated because it buffers multiple keystrokes. When a keypad interrupt arrives, the interrupt routine captures the decoder output and saves it in a buffer, *key()*. The two counters, *kbin* and *kbout,* keep track of where the most recent character has been put and where the oldest unprocessed character resides. Notice the use of the *mod (%)* operator to fold the end of the array back to the start. Once some keystrokes have been processed, the "beginning" and the "end" of the array can be anywhere in the 20 bytes. If the two pointers are equal, then the buffer is empty and there are no characters to process. If inserting another character would make the counters equal, then the buffer is full. In such a case, it was decided to not accept the newest character. Alternately you could move both pointers and thus discard the oldest character. Other approaches could get into dynamic memory allocation, but that is beyond the scope of this book.

**Encoded Scan with Interrupt**

```
#include <reg51.h>
#include <absacc.h>
#define keydecoder XBYTE(0xf600)
```

```
#define uchar unsigned char
uchar code ASCIIbtn()={"0123456789ABCDEF"};

uchar key(20); /*a place to keep the readings*/
uchar kbin=0
uchar kbout=0;
void display(uchar x); /*fn prototype*/

void keyfn (void) interrupt 2 using 1{
 /*captures successive key entries and stores in an
               array*/
 uchar i=kbin;
 if((i=++i%20)!=kbout){ /*there is room in the
               buffer*/
  kbin=i;
  key(kbin)=keydecode|0x0f; /*comes in lower 4
               bits*/
 }
 keydecoder=0; /*clear the latched interrupt*/
}

void main(void){
 uchar key;
 for(;;){ /*endless loop*/
  if(kbin!=kbout){ /*process new key*/
   display(ASCIIbtn(key(kbout)));
   kbout=++kbout%20;
  }
 }
}
```

**TO BUFFER OR NOT TO BUFFER?** The question of using an external keypad interface vs. using the micro itself has several aspects:

1) Scanning with a *separate* chip requires extra hardware for *each* unit built and increases the cost and complexity of each unit built.

2) But avoiding the extra chip requires a bit more code space (software) and keeps the processor busier.

Just be sure you don't use an encoder chip out of a fear of software, since software represents a one-time development cost and processors are rarely too busy to scan the keypad.

## Direct Keypad Interface—Computer Scanned

If you have ports free and want a bare minimum of hardware, you can scan the keypad yourself in software and eliminate the 74C922 chip. This approach requires the entire port—4 bits in and 4 bits out—and involves more software, but it can save hardware complexity and cost. The approach is to drive one row at a time low and look to see if any of the column lines have gone low due to pushed keys connecting the low row to their particular column.

Pull-up resistors are advisable on the column line inputs if the port doesn't reliably read in floating lines high. With TTL, an open is *usually* read as a high, but this is not recommended practice because of the chance of picking up noise on what is essentially a high-impedance "antenna."

By spreading the scan over 40 msec, the problem of key bounce is eliminated since in that time the bounce for any reasonable keypad will have died out. The scan that comes in the middle of a bounce will either read the key closed (in which case the next scan will still see it as still closed) or else it will catch the open of a key bounce (in which case it will detect the closed key the next time when it has settled). A delay of 40 msec will be undetectable by a human.

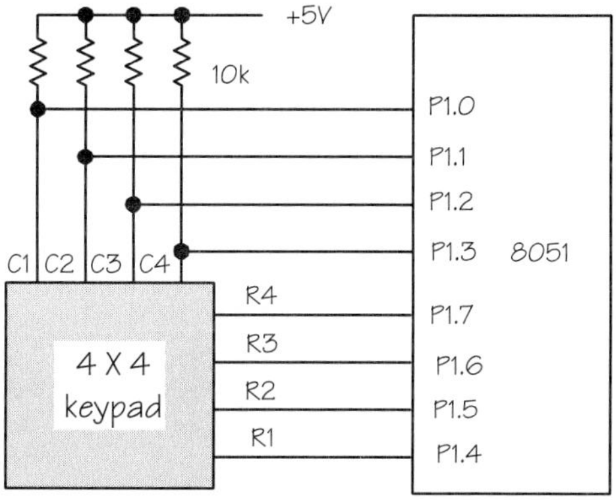


**Direct Scanned Keypad Circuit**

The C software shown on page 169 is more elaborate than before with the addition of buffering and a real-time interrupt. The timer provides a regular (every 10 msec) interrupt to cause the program flow to jump out and scan the keys. Every fourth time (40 msec) the *keyscan()* function decodes the accumulated scan pattern into 16-bit patterns of newly pushed and released keys. The software also includes buffering of all the newly pushed keys in their ASCII coded form. The function only changes the global variables when there has been keypad activity, and in such cases it sets the *keyschanged* flag.

The scan algorithm outputs 1's to all the rows except the one being scanned and takes in the four signals from the columns. For example, one-fourth of the scan would involve sending out 1101B (row 2) and reading back the column nibble. If, say, keys “5” and “7” were being pushed, then the row input would be 0101B. The pushed keys would pass the low from the micro on row 2 to the second and fourth column lines. By sending a succession of 0’s to the rows and storing the column readings, you can determine which keys are being pushed at any given moment. No software delay need be added between issuing the row scan and reading back the columns since the microcomputer cannot execute successive instructions quickly enough to encounter any settling problems. Two 4-bit readings are combined into a byte or the entire scan is combined into an integer with each of the 16 bits corresponding to the state of one of the sixteen keys.

Where the rest of the program needs to know only of any *changes* to the inputs—new buttons pushed or old buttons released—logical operators allow changes to be sorted out easily. By “exclusive ORing the most recent scan with the previous value, any change in keys can be recognized. The next table may help to visualize how the logic works here. The logical operations easily and efficiently mark the changed keys.

| | |
|---|---|
| `oldscan (previous: 0->pushed)` | 11010010 |
| `newscan (most recent: 0->pushed)` | 10110011 |
| `chgscan (oldscan ^ newscan: 1->changed)` | 01100001 |
| `newpush(chgscan & oldscan: 1->pushed)` | 01000000 |
| `newrel(chgscan & newscan: 1->released)` | 00100001 |

**Logic to Sort Key Changes**

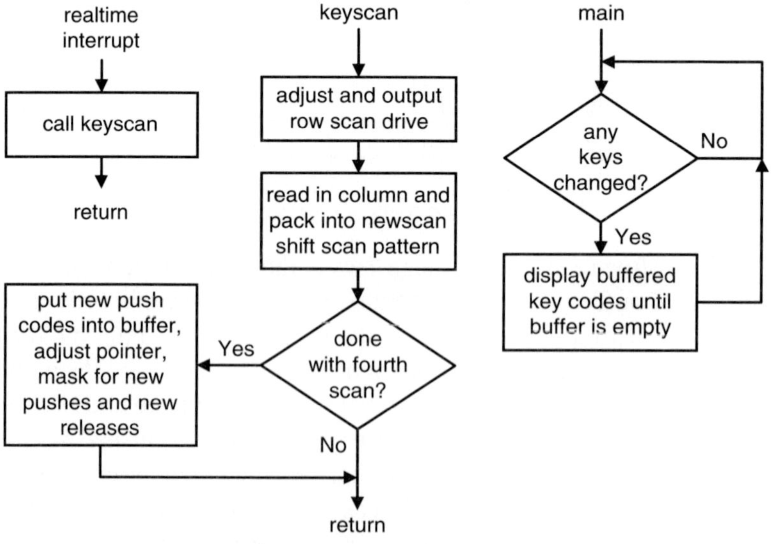


**Direct Scan—Flowchart**

**Direct Scan**

```
#include <reg51.h>
#define uchar unsigned char
#define keypad P1
/*#define keypad XBYTE(0xf600)*/
#define tenmseccnt -10000
uchar pattern=0x10;
uint newpush=0;
uint newrelease=0;
sbit keyschanged=0;
uchar newkeys[10];
uchar newkeysin=0;
uchar newkeysout=0;
uchar code ASCIIbtn[]={"0123456789ABCDEF"};

void display(uchar ascii); /*prototype for
              undefined display function*/

void realtimeint(void) interrupt 2 using 1{ /*every
              10 mSec*/
 TH1=tenmseccnt>>8;
 TL1=tenmseccnt&0x00ff;
 keyscan();
}

void keyscan(void) using 1{
 static uint oldscan,newscan;
 uint i,chgscan; uchar j;
 P1=pattern | 0x0f; /*drive out one row scan*/
 newscan=(newscan<<4)|(P1&0x0f);/*4 bits from
              columns*/
 if((pattern<<=1)==0){ /*end of 4-scan cycle*/
  pattern=0x10;
  if(newscan != oldscan){ /*some key changed*/
   keyschanged=1;
   chgscan=oldscan^newscan;
   newpush=chgscan&newscan;
   newrelease=chgscan&oldscan;
       /*process into buffer*/
   if (newpush)for(i=1,j=0;i>0;i<<1,j++){
    if(newpush|i)newkeys[newkeysin=++newkeysin%10]
            = ASCIIbtn(j);
   }
```

```
    oldscan=newscan;
   }
  }
}

void main(void){
 for(;;){
  while(keyschanged){
   display(newkeys[newkeysout]);
   if((newkeysout=++newkeysout%10)==newkeysin)
             keyschanged=0;
  }
 }
}
```

**Direct Scan**

```
PERMANENTDATA SEGMENT DATA
RSEG PERMANENTDATA
  OLDSCAN: DS 2

KEYM SEGMENT CODE
RSEG KEYM
KEYSCAN: CLR A
     MOV R6,A       ;NEWSCAN IN R6,R7
     MOV R7,A
     MOV R2,#10H    ;START COLUMN 1
KS1: MOV A,R2
     MOV P1,A      ;DRIVE COLUMN
     CALL ROT4      ;READY THE RESULT
     MOVX A,P1       ;GET RESULT
     ANL A,#0FH
     ORL A,R7
     MOV R7,A
     MOV A,R2
     RL A           ;MOVE TO NEXT COLUMN
     MOV R2,A
     CJNE R2,#10H,KS1 ;DONE SCANNING?
     MOV R2,#0      ;USE AS ='S COUNTER
     MOV A,OLDSCAN
     CJNE A,6,KS2   ;DANGER-ASSUMES BANK 0
     INC R2         ;ONE 'NO CHANGE'
KS2: MOV OLDSCAN,R6
     MOV A,OLDSCAN+1
```

```
        CJNE A,7,KS3   ;ASSUMES BANK 0
        INC R2         ;ONE 'NO CHANGE'
   KS3: MOV OLDSCAN,R7
        CJNE R2,#2,KS4 ;BOTH BYTES UNCHANGED?
        CLR A
        MOV R7,A
        MOV R6,A ;RETURN 0
   KS4: RET

   ROT4:MOV R3,#4     ;ROTATE R6,R7 LEFT BY 4
    RO1:CLR C
        MOV A,R6
        RLC A
        MOV R6,A
        MOV A,R7
        RLC A
        MOV R7,A
        JNC RO2
        INC R6        ;CARRY AROUND
    RO2:DJNZ R3,RO1 ;4-BIT ROTATE
        RET
    END
```

## Multiple Key Presses—Diode Isolation

Usually current can flow in either direction across the contact made by a closed switch of a key. As you can see from the diagram on the next page, this can cause trouble if the wrong combination of three or more keys is pushed at once. When R1 is being scanned, for example, if keys "3," "6," and "7" are being pushed at the same time, the row sinking drive will show up as a low on *two* columns, implying to the column-reading circuitry that key "2" is being pushed as well.

One solution is called "diode isolation" where diodes are added in series with the keys as shown below. If you follow the circuit you will see that the current flow is blocked by the diode at key "7." Thus the R1 scan will not cause a low on the C3 column. The impact of sixteen diodes on keypad assembly cost, however, is too high for many applications—particularly for membrane-type pads that are made up of layers of material glued together—no place to insert diodes! Frequently the system designer assumes that pushing more than two keys at once will have to be the user's problem!

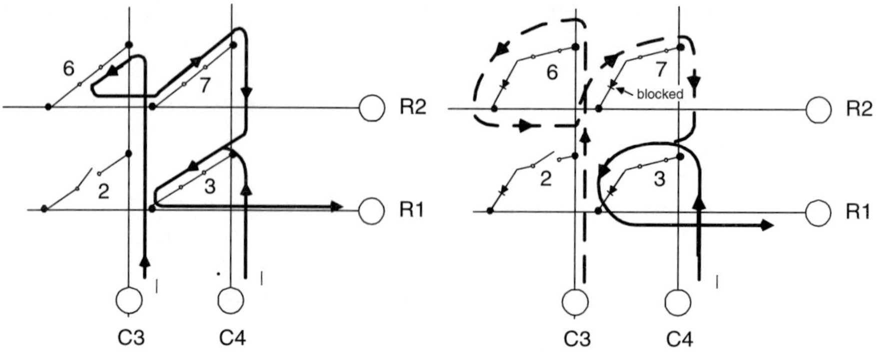


**The Need for Isolation**

If you should use a diode-isolated keypad, be careful of the voltage drop across the diode. Ordinary silicon diodes have about a 0.7 V forward drop. Since the threshold for a logic "0" into a TTL device is about 1 V, if the low driver for the scan doesn't go low enough, the input might be in the forbidden voltage region, or at least susceptible to noise. One option is to use germanium diodes which have about a 0.3 V drop. Failing that, it might be necessary to use comparators or a high-going scan drive. In any case, be careful and don't assume a diode-isolated system will work into TTL without special attention. The author had an unfortunate experience using a music keyboard that way in assuming that the key closures were direct shorts when they were actually made with conductive rubber that dropped to no lower than 100Ω and wouldn't go directly into the micro or TTL inputs.

The scan sequence of the 74C922 has not been checked, but for the direct interface example, row drive was chosen because users might be more likely to simultaneously push two side-by-side keys than two keys in a column. Since the scanned row is driven *low*, shorting two drive rows together by pushing two column keys will still allow the scan to go low because TTL-compatible drive has more power on a low than a high—the low "wins." This is true for "TTL-compatible" CMOS ports on a micro, but take care on the special ports with higher drive capabilities.

### Alphanumeric Keyboards

These are generally the same as the matrix keypads only with more keys. Unless I lost count, I am typing at one with 105 separate keys. That would require an 11 × 10 or 8 × 14 matrix. Since interfacing is usually done with 8-bit micros, the latter makes good sense even though it is one extra port pin. The process is no different than what we've just done with the keypad—just more codes.[17]

## Music Keyboards

Electronic music keyboards are another instance of digital inputs. For some student lab experiments I obtained replacement keyboards for electronic keyboard synthesizers.[18] The systems are *only* diode-isolated matrixes of switches. The basic approach is to read the *B* lines of the matrix while driving high all the *N* lines but one. The *B* lines tied to the given *N* line by the key closure will be read in as a zero while the other lines will be high.

A feature of the more elaborate keyboards is the presence of *two* closures for each key—one closure is made near the top and one near the bottom of the key travel. For basic use, you could ignore one of the closures, but for more elaborate effects you can easily measure the time *between* closures to determine a volume level. Much like pushing a piano key harder (and faster) produces a louder note, you can electronically determine how "hard" the keys are being pushed. The range in times between "loud" and "soft" will have to be determined empirically. You can produce all sorts of loudness and duration effects in software. The amount of sustain can also be based on when they key is released. If the resulting sound is generated independently by a special chip (see the next chapter on the programmable sound generator chip), it should be quite feasible for an 8051 to scan the keys often enough while doing other things.

---

[17]PC keyboards almost all have a built-in encoder (used to be a ROMed 8048 micro—the predecessor of the 8051). The processor on the motherboard communicates using a simple serial protocol so the PC is not actively involved in watching for keystrokes. Oddly enough, it may well be more efficient and less expensive to design in a commercial PC keyboard since there are so many on the market and they have become an inexpensive commodity. Commercial keyboards can easily be replaced at some future date if there is a failure. A quick scan of *PC Shopper* magazine shows a 104-key key*board* with electronics for $14! Many key*pads* cost about as much with no electronics.

[18]Specifically we used keyboards for the Yamaha PSS-460 and PSR-6300. These keyboard assemblies have none of the synthesizer electronics, but we have attached a 26-pin ribbon cable for easy interfacing to parallel ports. All scanning must be done by student programs. Unfortunately the contacts of the keyboards are not low enough resistance to supply a TTL low, so comparators had to be added to supply a TTL return from the scanning of the keyboard.

## POSITION INPUTS

This section comes before the analog inputs section because some of the position inputs are binary—*far* or *close* or have already been converted to serial data before they reach the microcontroller. . . For the others you will need to read the section on analog inputs that follows.

### Optical Devices

**Opto-interrupting.** Opto-interrupters are usually single packages holding an optical emitter (light-emitting diode—LED) lined up with a photo-detector with a slot in-between so an opaque object can block the passing of the light. The light is usually infrared (IR) that is not quite visible to humans.[19] Most of the interruptor packages have a very small slot for the object that blocks the light so that not very much ambient light can get in. Some detectors have comparator circuitry built in so they provide a direct logic signal. Others leave you to add the resistors as shown here. Review the digital inputs section (page 152) in designing the detector stage. If the transfer efficiency is not enough you will have to add a transistor to increase the gain, as shown to the right of the figure below.

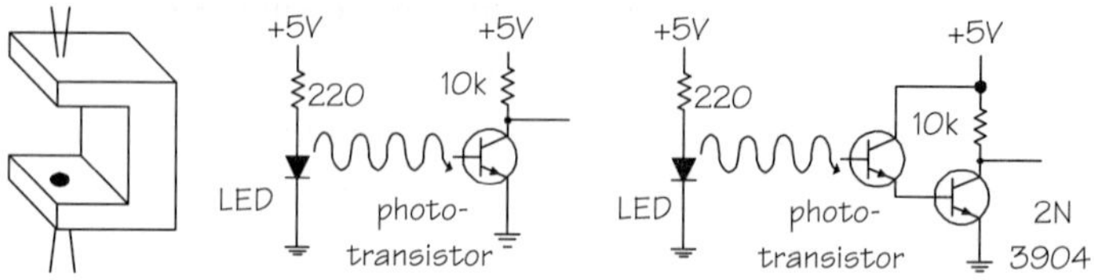


**Optical Interruptor**

If you need to detect over a larger distance some of the chopping techniques described later might be helpful. Remember that at long distances, ambient light easily becomes a problem and you must electronically or mechanically protect from its effects.

Opto-interruptors are quite useful where the thing being detected is not heavy[20] or where high reliability and long life are important. They can, for

[19]It happens that IR emitters are slightly easier to make and have a higher efficiency. I suppose the detectors can distinguish IR from visible light somewhat, but most silicon detectors are sensitive well into the visible red range.

[20]Obviously heavy objects can be detected with mechanical switches, but things like screws and nuts on a conveyor belt would not have enough mass for most switches.

example, be used in a coin-detecting mechanism with possibly longer life than mechanical switches.

**Optical Isolators.** I include these here for lack of a better place. They are not technically input devices, but rather opto-interruptors without the slot. In a *totally opaque* 6- or 8-pin, DIP is an LED and a phototransistor or photo-Darlington lined up with each other to allow an optical signal to be transmitted between them. Both sides could be at ordinary potentials but powered separately and the device would stop the spread of catastrophic voltages (from lightning or inadvertent connection of low-level inputs to power voltages). Alternately, the input LED could ride up and down sensing a signal on a 110 V 60 Hz-power line, while the detector is safely connected into a ground-referenced logic circuit. The two sides can safely be hundreds or even thousands of volts apart. The signal is basically digital—when the light is turned on, the detector saturates, pulling the output to ground or to the power supply. The frequency is limited by the speed of the detector to perhaps 100 kHz, and it is generally necessary to drive 10 to 20 mA into the LED to get a good output.

You can work the devices in the linear region if you can compensate for drift and lack of repeatability. In particular, I have data on a device by Texas Instruments (TIL300) that isolates analog signals by combining two photodiode detectors and a single photodiode emitter in one package. One detector is used to provide feedback on the LED brightness, while the other diode works on the isolated output side. Variations due to temperature or time can be compensated for and the detectors, being on a common substrate, experience the same drifts. TI claims gain transfer stability of 0.05%/°C. There are also some quite sophisticated analog/digital devices that convert an incoming analog signal into a digital pulse train through the opto-isolator and then put it back to analog, as well as systems that provide optically isolated feedback to stabilize the system. Analog designers must lie awake nights struggling with these issues!

### Proximity/distance

Knowing how close something is can be important in many applications. In one of my project classes, each group had to build a robot that would navigate a maze (much like IEEE's mouse contest). The biggest challenge was to determine how close to the walls the robot was. Three basic approaches were used—mechanical switches, optical reflective sensors, and strain-gauge sensors (normally used in the fingers of computer-input gloves).

**Optical Reflecting.** For proximity sensing, if you combine an optical emitter having a narrow emission angle with a detector having a narrow detection angle, you can angle the two devices so the spot of light is in the field of view only when the reflecting surface is at a specific distance.

As you can see from the drawing below, the typical optical proximity detector gives an analog level based on distance that peaks at a certain point and drops off when the surface is either closer or further away. Perhaps the most deceptive thing about the picture is the scale—the common proximity detector peaks at around 0.1 in. This is very close—don't expect to detect a surface inches or feet away with the standard optical pairs.

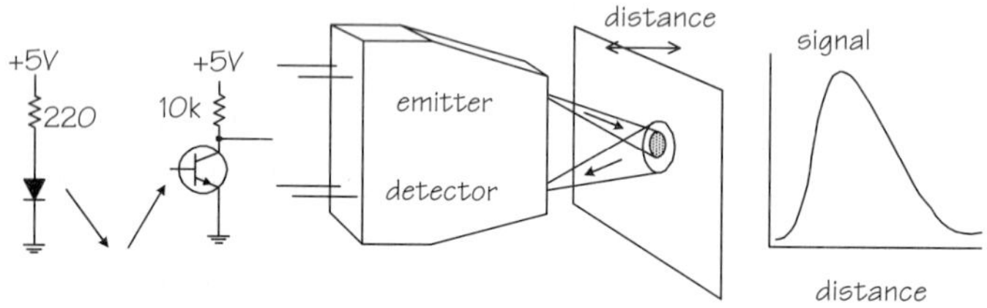


**Optical Proximity Sensing**

In addition, there is some constraint on the nature of the surface. A purely reflective one could fail to reflect at the right angle to get the light to the detector and a totally absorptive surface (flat black) could have the same problem. The best choice, if you have one, is a flat, diffusive surface. Or, for shaft encoding, the surface can be dull except for a polished area serving as a mark. The key is to have a change in reflected light to detect.

> **OPTICAL REFLECTOR MODULES:** Standard emitter/detector pairs are configured to work only for *very short* distances. Longer distances require more expensive systems or more complicated design.

**Separating Optical Signals from Noise.** You can extend the distance if you use separate emitters and detectors. If so, you have to start distinguishing between the sensing light and any ambient. Even though many of the systems use IR which you can't see, the detector is reasonably sensitive into the visible spectrum and sunlight or incandescent light can particularly cause problems. If you want to detect over a long distance, you can start by trying a long tube or lens to narrow the field of view. But if the ambient is an integral part of the surface illumination, you have to go to sorting out the signal from the noise. Trying to optically reject the visible is pointless because sunlight has lots of IR.

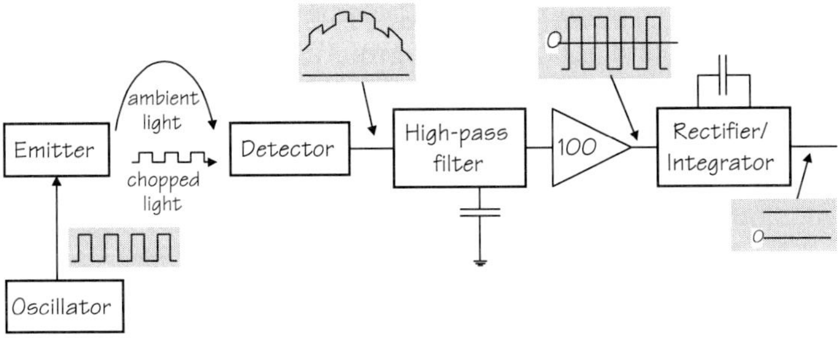


**Chopped Light—Filtered Detector**

The simplest approach, used on most IR remotes, is to ***chop the emitted light*** at around 40 kHz as shown above. Keep the first-stage gain after the detector low so ambient light doesn't swamp (saturate) the stage, and then *high-pass* filter before encountering further gain stages. That way all the noise (light from the sun or low frequency fluorescent light—120 or 100 Hz) is cut off after the detector. *Then* the actual signal can be amplified as much as needed. Most detectors specifically for IR remotes include amplification and filtering right in the module (giving a binary logic output), but you can do quite well with a simple R-C filter. Since the highest frequency to eliminate is the 120 or 100 Hz from fluorescent lights, you have 2½ decades for your filter rolloff.

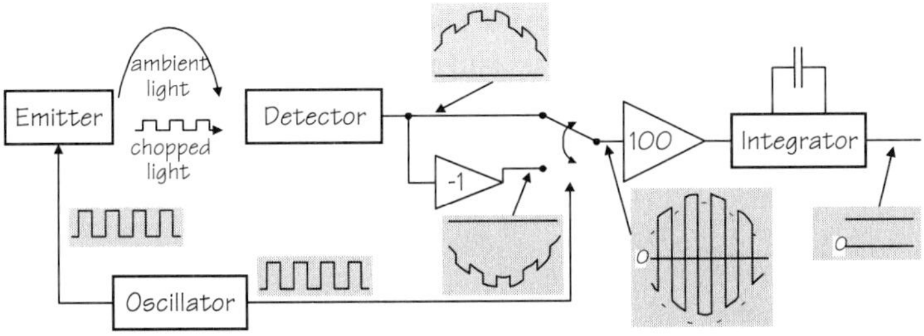


**Synchronous Detector—Functional Diagram**

If you *really* want to separate signal from noise, you can use a ***synchronous detector*** as shown above.[21] Rather than just switching the light on and off at a high frequency and filtering, it controls an analog inverter at the detector in synchronism with the chopping of the source. Since, unlike a remote control, you have ownership of *both* the emitter and the detector, you can design

[21]I think this is related to what is called a boxcar integrator.

a circuit where the signal when the light is on is added and the signal when the light is off is subtracted.[22] Feeding the result into an integrator, anything that is not in sync with the signal from the emitter will add and subtract more or less equally and integrate out. Only a signal in sync will add, and then only if it is the same frequency. Any signal, *even at nearly the same frequency*, will tend to be added about as often as it is subtracted. A 20-kHz signal would add for one cycle and subtract for the next. A synchronized 80-kHz signal would both add and subtract for one cycle. The toughest would be a 39 kHz that would come in and out of adding 1,000 times a second. If you filter the result with a low-pass filter cutting off at, say, 100 Hz, you can easily block out such a signal. Depending on how close a frequency you want to reject, you can drop the filter corner and separate your signal from 39.9 kHz or 39.99 kHz.

Synchronous detecting allows a signal to be orders of magnitude smaller than the ambient light to be detected and even measured for amplitude. The price you pay is response time—if you have a 1-Hz corner on your filter, you have to wait a few seconds to be sure the reading has settled. You can't turn the 40 kHz on and off rapidly to send information (as is usually done with an IR remote). In tracking the approach of a distant object, say from 20 feet away, you can probably tolerate a 100-Hz time constant since a significant reading ten times a second is probably reasonable for approaching physical objects (unless it is a speeding bullet).

*If* you can get enough signal, an efficient design would probably avoid synchronous detection and capitalize on the availability of IR detector modules—pre-filtered detectors are readily available for a very low price since you ride on the high production volume of such applications. Many remotes work over a range of 30 feet at all sorts of angles—for a proximity or range detector your job is to narrow the angle and derive an *analog* signal relating to distance.

**Capacitive Sensing.** For much more information in this field, I suggest Chapter 6 of *Capacitive Sensors.*[23] This technique can detect both conducting and non-conducting objects. One commercial sensor consists of

[22]There are two devices to use for this. The less expensive one is to use a conventional op-amp configured for unity gain inverting followed by an analog switch. For fewer parts, you could use an analog multiplier chip with the chopping signal fed in AC-coupled so it multiplies by +2.5 and then by –2.5. It becomes very important that the logic amplitude is controlled and for either approach the chopped signal must truly square—equal durations—best done by feeding the multivibrator/oscillator into a flip-flop to divide by 2.

[23]Larry K. Baxter, *Capacitive Sensors* (IEEE Press, 1997 Piscataway, NJ, ISBN 0-7803-1130-2).

concentric plates where the presence of a higher-than-air dielectric material increases the coupling between the plates and causes a Colpitts oscillator to break into oscillation. A detection circuit then gives a logic level output when the oscillation is going on.

On the other hand, if you want to get a range of distances, you could have a continuously running oscillator and measure the frequency.

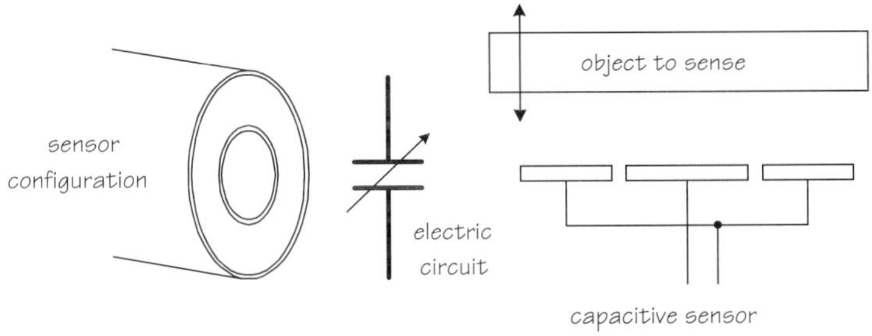


**Capacitive Proximity Sensor**

**Inductive Sensing.** In the same way that an object can change capacitance of a sensor, so a ferrous one can change the inductance of a magnetic circuit. With the frequency of an oscillator dependent on the inductance (an L-C oscillator), the arrival of ferrous material in the flux path changes the inductance and thus the frequency. A simple example of this was used in the flow-sense circuit of the last chapter of *Volume I* where a trapped ball bearing was allowed to move in and out of the center of a wire loop. The circuit here gave an oscillation that changed frequency by about 10 percent. The same could be used for proximity.

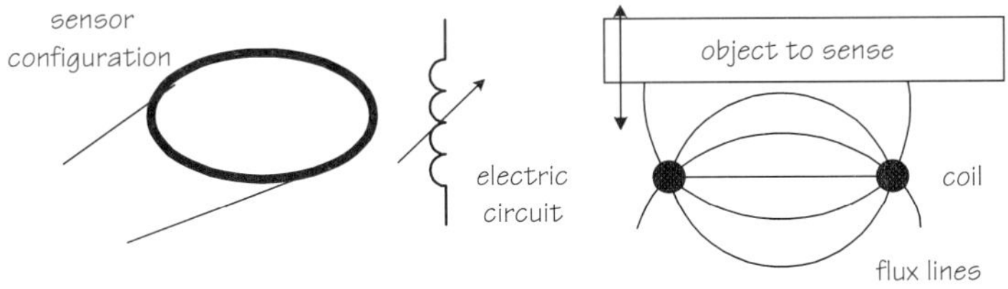


**Inductive Proximity Sensing**

**Inductive Proximity Sensor.** The following circuit is a simple oscillator where frequency accuracy is not important. It is efficient where you have multiple circuits because a 74LS14 has six gates in a very inexpensive package.

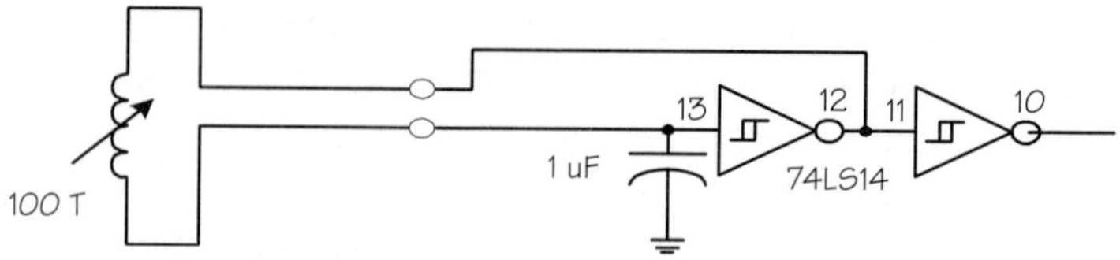


**Inductance to Frequency Circuit**

**Ultrasonic Ranging.** Ultrasonic distance sensing is somewhat like radar. Conventional radar relies on timing the return (the reflection) of an RF pulse. The longer the return echo, the further away the object. For air traffic radar systems the pulse is sent out in a highly directional beam that is rotated—hence the display that shows the sweeping beam and leaves the bright dots where airplanes are located. The return time is plotted as distance from the center.[24] Ultrasonic ranging is more like the air traffic system—the time it takes for the echo of an acoustic wave front (usually 40 kHz) to arrive back at the source corresponds to the distance to the object. Often the same transducer is used as both the loudspeaker and the microphone. Commercial ultrasonic ranging units (shown diagrammatically below) include the electronics and TTL interface to either give a pulse lasting the duration from the emission to the detected echo or else a separate pulse when the echo is detected. Polaroid Corporation has sold such a unit for many years as an offshoot of the focussing system for their cameras. Their modules are, of course, interested in the *first* reflection from the closest object.

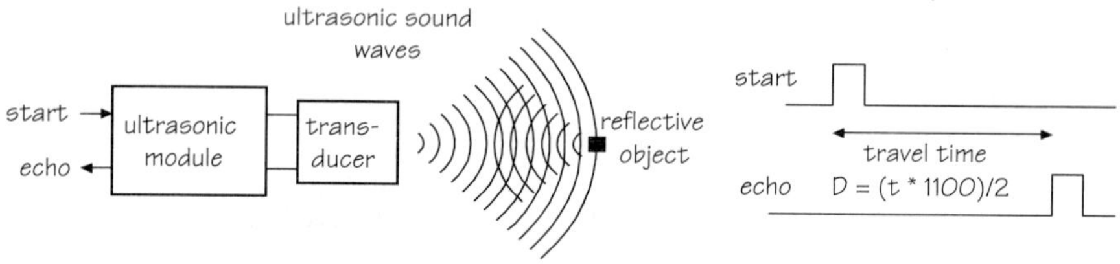


**Ultrasonic Ranging**

A variation on that principle is used underwater for fishing and navigation. By directing the ultrasonic pulse *down*, the echoes can be plotted to show both the reflection off the bottom and smaller reflections off fish.

[24]The radar associated with automobile *speed* (not position) measurement is more complex and involves the *shift* in reflected frequency (the Doppler effect, usually explained by the analogy to the shift in perceived sound pitch as one is approaching or leaving the sound source.). The police officer is not interested in how far away your car is—just how fast it is travelling!

Since the intensity falls off with the square of the distance, the better interface modules include an increasing gain function with time to offset the effects of dispersion—the longer it takes for the echo to arrive, the further away the object is and the weaker will be the magnitude.

**Image Processing / Focussing.** The problem of distance to an object has challenged all the auto-focus camera manufacturers. Those who have not relied on ultrasonic ranging have relied on the fact that they already have an image system and use sensors to look for maximum sharpness of the image. As the lens is rotated, the point at which the dark-to-light light difference reaches a maximum for the area of interest is the best focus (out of focus tends toward a uniform gray). Knowing the mechanics of the lens, the focal distance is determined.

Humans determine distance by parallax, since the creator graciously provided two eyes to view from different angles.[25] The optical sensing of an earlier approach more closely resembles this. Other focussing systems rely on the convergence of split image parts as in manual-focus camera viewfinders.[26]

Both of these techniques might be adapted for a custom system, but they are inherently much more complex than the earlier approaches. In very unusual applications, you might use one of the CCD camera modules combined with image processing, but such signals are generally too fast to process directly—some sort of video buffer would be required and the processing load would rule out continuous real-time processing. Efficiency might use such an approach to distance measurement as a by-product of an application that already was involved with image processing.

### X-Y Position

In addition to the techniques discussed next, it is possible to determine position with imaging systems—linear or array cameras, for example. The whole area of image processing has become very complex with a heavy theoretical base, but there are still very specific, constrained industrial applica-

[25]There are several other ways as well. If the object is known, its relative size or orientation is used. That is how the view of trick rooms can give a false perspective. Also, the fact that things become blurred and lighter due to atmospheric effects can suggest distance—the distant hills are progressively more light and blue. Finally, the focus of an individual eye can give distance information, but parallax is the key as our one-eyed cat could attest.

[26]Both these methods assume there is a discernable contrast—consider how even a human has a difficult time determining distance where everything is a uniform color or where there is a repetitive pattern so parallax can lock on the wrong image alignment.

tions where the desired information can be easily extracted from a video stream at low cost.[27]

**Joystick.** Some joysticks are actually just four switches—definitely a digital input. The digital joysticks must of necessity control a direction of change. If your game is shooting down enemy space ships or asteroids, you move from side to side by pushing the joystick one way or the other. In the middle, you stay in the current position. If you wish to be a little more sophisticated,[28] you can make the rate of position change dependent on the time you keep holding in a direction—the *longer* you hold to the right, the *faster* you move over. You will have to develop an empirical feel for this.

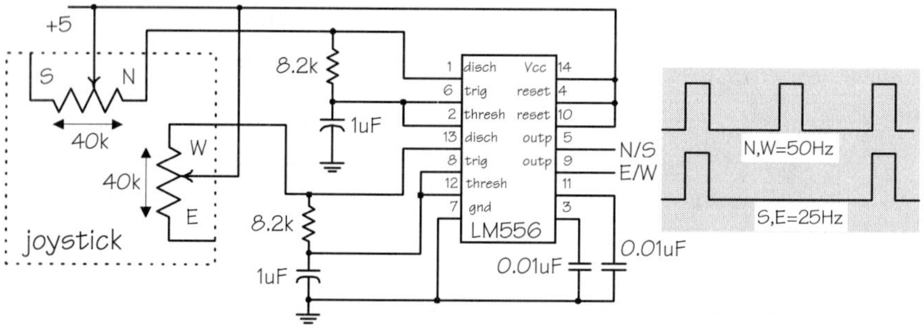


**Analog Joystick Interface**

Other joysticks are an arrangement of two potentiometers, which can give an analog voltage related to position. The variable resistances of the X and Y position pots can either be connected to A-D converters or to some sort of V-F converters. For our student lab hardware I used the common, very inexpensive 555-timer/pulse-generator chip (actually an LM 556—dual 555s) to provide variable-frequency 5-V pulses which interface directly to a

---

[27]For example, if the object being imaged were bright white against a black background, a simple pair of gated one-shots could produce pulses whose width related to the horizontal and vertical scan time to the object. A latched counter could capture the horizontal distance and the microcontroller's internal timer could resolve the vertical distance. The key is to "stack the deck" in terms of the image so the elaborate, general-purpose image-processing algorithms are unnecessary.

[28]You can do the same with pushbuttons, by the way, to have an up/down button pair increase the value by 3 per second for the first interval, 30, the second interval, 300 then next, and so on. When the value is close to the desired one, by releasing and then pushing the key again, the slow adjustment resumes Here is an excellent place to develop an empirical shift of rates.

microcomputer port or to a timer chip.[29] To simplify the software interface for the students, the pulse rate is made quite slow—approximately ranging from 25 to 50 Hz. The duty cycle of the resulting waveform (about 15 to 30%) is still large enough to easily not miss a pulse in software loops.[30] It would be possible use analog voltages to feed an A-D, but pulse-timing methods are generally much less expensive to implement and are closer to what is done for computer and video game inputs.

A fundamental choice you have in programming with analog joysticks is whether you control position or rate-of-change of position. Games modeling aircraft control the *rate* in the same way that the real thing responds to pilot controls. If you are modeling an input device such as a mouse or a touch-pad, you determine *position*. Some of the older laptops had a pointing device that controlled change of position (rate). Personal preference enters in, but it is inherently possible to control position to a higher resolution with a rate input.

Assume you have an analog joystick with a frequency output—how do you interface such a signal? First, with a system having a real-time clock you can measure the time *between* pulses by polling at the system tick as long as the pulses are always wide enough to last at least one tick. Here is an *algorithm*.[31]

**Timing Algorithm**

```
loop while joystick input = 1
counter = 0
while joystick input = 0
  wait 1 mSec
  increment counter
while joystick input = 1
  wait 1 mSec
  increment counter
now use counter as the delay
```

[29]The 556 timer does not give a linear relation between resistance and frequency so the midpoint may not be halfway between the extremes. A calibration lookup table may be in order. Also, different units may not be identical in frequency output. Probably the analog input should not be taken as a high-resolution input. If the joystick is spring loaded there may need to be a zero-calibration step. If there is no spring return, but it provides a rate signal, then a dead-band around the center is necessary.

[30]For more on measuring fast and slow pulses, see pages 298–302 of *Volume I.*

[31]In a software loop, be careful to allow a "double fall through" so the same "1" is not seen as both the start and the end of the loop. Or, if using the hardware with an interrupt, be sure the interrupt is edge triggered so the extended duration of the "1" does not cause multiple interrupts.

Second, you can measure the time between pulses with hardware. Given an interrupt and a timer device, the pulse from the joystick can cause an interrupt. An interrupt routine could capture the value of a (continuously running) counter. At the next interrupt, the difference between the new count and the old count will be the time between pulses. That serves as a period/frequency indication that can be related to joystick position.

Third, you can have hardware count the number of pulses in a fixed period of time. Connect the pulses from the joystick to counters (now you need two) and read the counters at regular times. The higher the count is, the higher the voltage. You could then speed up the pulse rate to get a reasonable count over a short time. Always you have system tradeoffs. *How often do you need a new reading? Is it important to have reading at a constant rate? What resolution is necessary?*[32]

**Mouse.** Usually there are two optically-encoded shaft encoders producing pulses as the mouse ball presses against the shafts and rotates them. (Optical-shaft encoders are discussed in the next section on page 186.) Unlike more elaborate shaft encoders, the absolute position of the mouse is immaterial—it is only the direction and increments of position that matter. It seems likely that an A-B encoding scheme is used based on two sensors spaced ¼-cycle apart. See the section on shaft encoding for more detail.

Like the PC keyboard, the mouse has become very standardized. And like the keyboard, there is some electronics somewhere to process the signal. My current mouse has a single chip (a Sysgration SYS2005) that takes the optical interrupted inputs and converts them to a signal into a serial port. The four-wire interface includes the power and ground lines to power the photo emitters and detectors as well as the chip. A large capacitor suggests to me that the chip may include its own switching regulator to get the necessary voltage from the 12 V that is supplied on the serial port. I suspect the communication is bi-directional because PCs can detect if the mouse is present.

If you are going to directly interface a mouse, at this time I suggest that you interface over the serial port using the commercial mice already available.[33] It is interesting to consider how much of Windows' activity is involved in interfacing with the muse. It is a very high priority part of the kernel to recognize the motion and update the display to show cursor posi-

[32] This basic problem is addressed in Chapter 6 (page 340).

[33] A quick scan of *Computer Shopper* shows mice for as little as \$8 U.S. It makes little sense to try to do your own thing unless there are very special requirements.

tion.[34] If you intend to use a mouse, you will probably want to set up an interrupt-driven serial interface.

**Digitizer**. A digitizer is a device that reports the position of a physical pointer or cursor. A user can trace the outline of an object on a paper drawing or follow a route on a paper highway map. It could be part of a system that manually measures the area of objects. A digitizer could be the input device for a computer drawing system (much closer to conventional drawing and painting than, say, a mouse or joystick).

There are systems that issue a sharp noise and time the travel of the sound wave front to long microphones on the X and Y axes (and Z if you wish). There are systems that induce magnetic fields in a back plane grid that are picked up by a sense coil at the desired coordinate.[35] There are CRT screen digitizers using photo-detectors that pick up the scanned beam timing relative to the retrace of the video. There are touch screens and touch pads that work on capacitive principles.[36]

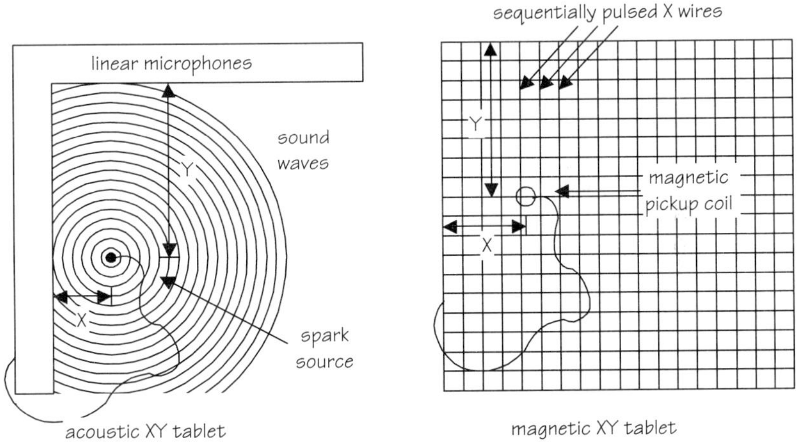


**Two XY Digitizer Techniques**

[34]Witness when the applications have crashed—often the cursor continues to move as the mouse is moved.

[35]The distance is resolved more finely than the coarse wire-grid spacing by interpolating based on the relative amplitude of the pickup from wires one-over on either side of the sensor.

[36]For a very practical coverage of capacitive sensors of all sorts, I recommend *Capacitive Sensors* by Larry K. Baxter (IEEE Press, 1997 Piscataway, NJ, ISBN 0-7803-1130-2). I may reflect some of his ideas in other input sections. The position input for laptop computers has been a problem and lately the capacitive touch-pad has provided another option for user input. The principle is an X-Y grid where pulses on one X wire after another are coupled capacitively by the user's finger to the Y grids. In addition to finding the X-Y combination with the largest coupling, it is possible to interpolate to a higher resolution by considering the coupling to adjacent crossings.

Most of the *complete* systems put out a string of numeric ASCII characters over an RS-232 serial port, so interfacing should be the same as any normal serial input.

## Angular Position

**Shaft Encoder.** A common use of opto-interrupters is to put between the LED and detector some disc with marks attached to a motor shaft. As the shaft turns, the light is blocked or allowed to pass, giving a signal related to the shaft turning. A single mark gives one pulse per revolution, or multiple marks can indicate the shaft position.

Optical shaft encoder assemblies usually use transmission through a glass disc to get very fine resolution easily up to 1/400 of a turn. There are two kinds of shaft encoders—absolute and incremental. The former use a series of detectors lined up behind different tracks on the disk so that there is a unique code for each angular position. The innermost track might be blocked for half the circumference while the next track might be blocked for ¼ of the circumference in two places. Actually the pattern is often what is called a "Gray code" so the alignment of the detectors is not so critical and so only one channel makes a transition at any one place. The point to remember is the fact that there is a unique and immediate readout of position.

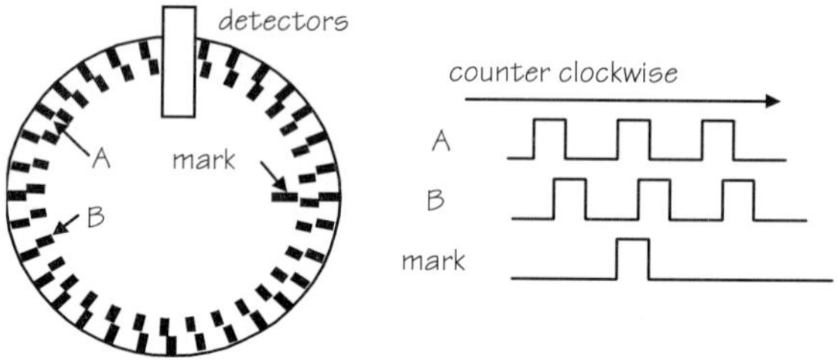


**A-B-mark Shaft Encoder**

An incremental encoder (sometimes called an A-B encoder) has only two or three channels, so it costs less to manufacture. The third channel, called the mark channel, gives a pulse only at one place in the revolution. But the A and B channels each have many equally-spaced marks surrounding the perimeter of the disc (these are usually printed photographically to get the precision and resolution needed). These marks and clear spaces are of about equal width so the detector signals are square waves. The two sets of marks are rotated relative to each other by about half a mark's width so the two square waves are shifted by a quarter cycle. If you consider the diagram you can see that the A channel signal goes high *before* B if it turns counterclockwise, but goes high *after* B if it turns clockwise. With a little bit of digital logic, it is possible to get signals for an up/down counter so the count goes up when the shaft turns one way, and goes down when it turns the other way. The mark pulse can reset the counter to give a known starting count. As you can imagine, an incremental encoder is no good when first turned on—it has to pass the mark once before it knows where it is.

There are some shaft encoders that are priced as low as about $20, but the high-resolution ones are precision instruments that can cost hundreds of dollars. An interesting new input device is a shaft encoder packaged like a potentiometer. It can turn continuously while putting out the A-B signal that can be used in tuning a microcontroller-controlled radio or anywhere else where the direct digital inputs are a savings. While more expensive than typical resistive potentiometers, it costs under $10 U.S.

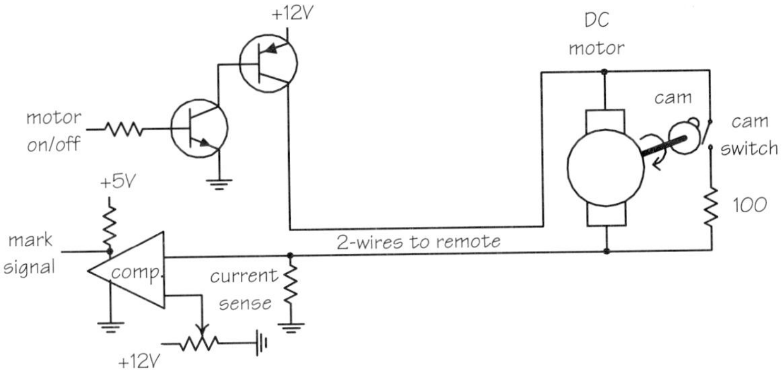


**Remote Angle Sensing System**

**Remote Mark Sensing.** A low-cost way to combine remote angular position sensing with the drive to a DC motor is to add a mechanical switch and cam at the motor. The cam switches in a parallel resistor at one point in the motor's rotation–a *mark.* Include a current-sense resistor in the drive electronics and the mark can be detected by the sudden increase in drive current—the current going into the resistor.[37] Incidentally, the same current sense can be used to tell if the motor is stalling or disconnected.

## Navigation

Orientation and position are important pieces of information for mobile systems. At Purdue, there have been student projects to produce a robot (the IEEE mouse maze) and ones to fly a model helicopter, unaided, on a mission. In both cases, serious navigation problems resulted.

---

[37]The comparator shown here is discussed in the next section (page 191).

**Compass.** Electronic compasses are discussed in the later section on Hall-effect magnetic sensors. Such systems rely on the earth's magnetic field. If high precision is necessary, corrections need to be made for north not being exactly the direction the field is oriented, depending on the location on the earth.[38]

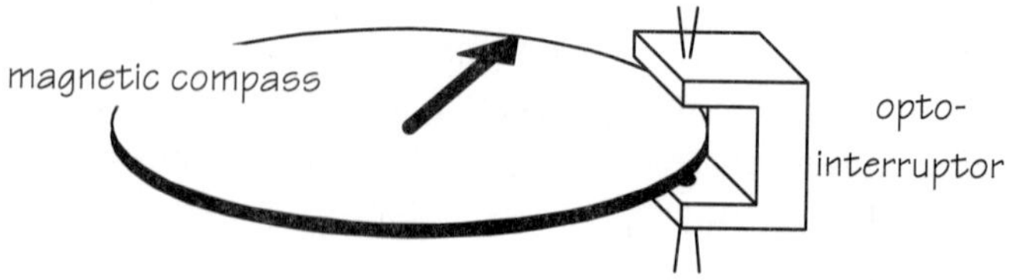


**Low-cost Single-Position Electronic Compass**

Thinking of efficiency, a cheap-and-dirty approach to an extremely low-cost electronic compass would be to take a conventional magnetic compass and add an opto-interruptor or proximity sensor. By judicious painting of the needle or disc, a black mark could only come into view at one specific orientation. It would not give a continuous readout of position, but a vehicle could rotate until it found the mark and then return to the desired orientation. Instead of the $30 to $50 that the electronic compass might cost, the total outlay would be no more than a few dollars.

Or, if your budget allowed, a shaft encoder with optical marks on a disc could be added and an A-B-mark encoder could give high resolution. How much rotational inertial is appropriate is another question. Clearly the mechanical systems suffer from vibration, oscillation and lack of damping while the electronic system are probably rugged enough and quick enough for use in missiles and the like.

> **USE ONLY THE MINIMUM:** Efficient design carefully assesses the information needed and avoids more precise sensors or calculations.

**Triangulation.** If you can find the relative ***direction*** to *three* known objects, even not knowing how far away they are, you can know where you are in three dimensions (for two dimensions only two objects suffice as long

[38] I understand that the orientation changes slowly with time over centuries or millennia, but with all the furor over the year 2000 deficiency (no one in 1960 worried about forty years away in the days when a memory was a scarce resource), it should merit consideration.

as you distinguish left and right). With RF you need highly directional antennas, but with light you can use a lensed detector or some sort of camera. The directional detectors would either have to be rotated (scanned) or be aimed at a rotating reflector that would direct the signal to the detector.[39]

Conversely, if you know your ***distance*** to three known objects you have a position fix. You might have three objects each emit an ultrasonic pulse and pick them up at the moving object. Given either a distinctive signal from each or a widely separated but precisely timed relation between the firings, you can determine the distance much like the ultrasonic distance measuring system. If you intend to use either of these approaches you will have to dust off your trigonometry!

**Gyroscope.** Gyros originally built on the principle that a spinning object resists a change of orientation because of rotational inertia. Perhaps you have played with one of the toy gyros that had a massive wheel in a globe-shaped cage that were spun by pulling on a wrapped string. As you moved the pivot point resting on your finger, the gyro resisted any change in orientation. The principle was, if something can stay in a fixed orientation, the device undergoing change in orientation can refer back to the reference to determine its new orientation. With a gimbaled support, the output could be determined from shaft encoders servos.[40]

One of my very advanced students worked with electronic rate gyros. The principles have escaped me, but some clever folks came up with a way to avoid all the large mechanics and still get the signal. I have a vague recollection that something very tiny spins inside (with magnetic support?), but it is quite expensive and accurate. If you enjoy such clever uses of silicon ICs, you should investigate solid state pressure transducers (page 217).

**GPS.** Global positioning satellite (GPS) systems are tied in with earth satellites and complex, government-run electronic communication systems to provide longitude and latitude information. I believe it is determined by individual modules communicating both ways with the satellites and then by triangulation calculations in the host system that knows the positions of at least three separate satellites above the earth. GPS is, to my knowledge,

---

[39]If you used an array camera with a 180° fisheye lens, you could process the video to determine where the bright spot is on the perimeter of the image. Similarly, you might generate some optical system that would project onto a linear CCD.

[40]I am relatively ignorant in the avionics field. I hope you know all this already or find someone else for more information—remember you bought this book because of the *microcontroller* emphasis!

*not* done by your module processing signals passively sent out by the satellites.[41] For purposes of this book, *magic happens* and out comes coordinate information from the several $100 (or more) devices you bought. The lowest cost devices provide only an LCD display, but I expect there are ones that provide the information over a serial output much like a digitizer.

## ANALOG INPUTS

Binary (digital) signals are characterized as either *1* or *0*, represented by either a high voltage or a low voltage (above 2.4 V or under 1 V for TTL). A signal should never occupy this in-between region for long. The binary nature allows for correction if the signal starts to drift—a 2.7-V input can be "corrected" back to 5 V with no loss of information. An *analog* voltage that should be 6.2 V, if it shifts to 6.3 V, can never be "fixed" down the line. There is no way to correct for small drift, because every value represents a valid reading. A long analog signal chain accumulates errors or noise, while a cascade of digital processing can be kept in the right range at each successive stage. That is why digital electronics has become so pervasive.[42] Needless to say, a major focus of analog circuit design is in minimizing inaccuracies. Likewise, the most efficient approach for most microcontroller systems is to convert analog voltages to digital codes as early as possible to reduce the chance of drift or cumulative errors. Nowadays digital approaches are generally less expensive than precision, low-drift analog approaches, particularly if the application has to have a microcontroller anyway. In the algorithms chapter, I discuss several analog-processing options, but in this section I focus on the ways to get the analog voltage into a digital value in a microcontroller.

### Knobs, Dials, and Sliders

While the input from these devices will have to become digital codes before they can be used in your software, in many cases they are much more

---

[41]It is my understanding that this involves all sorts of national security issues. I believe any transponder has to be registered and that the *highly* accurate information is not generally provided. Not just anyone can have the coordinate information—I suppose because an enemy invader should not attack courtesy of the host country's resources.

[42]*Digital* is not the same as *perfect*—it is possible to have signals get so far off or have a noise spike come up at the wrong time so the next stage comes up with the wrong value. Both in memory systems and in communication there are ways to add extra bits of information to a binary value or message (such as *parity* or more elaborate *error-correcting codes*) to first recognize an error and second, restore the right value. This rapidly gets beyond the scope of this book.

user-friendly than buttons, keypads, and switches. Take an example from a student project I recall where the user had to enter the time of day to set when the lights came on in a terrarium. The student's first approach was to have a keypad and LCD so the owner could key in the *exact* times—9:57 A.M. or whatever. But then you must allow for the possibility of pushing a wrong key and having to re-set or reenter values. Most likely, the owner doesn't care about the times that precisely anyway. Instead, the eventual solution was to have a potentiometer (pot) calibrated in hours over the entire 24-hour day—marked off in dark and light halves. An 8-bit A-D converter reads the voltage from the wiper of the pot. If the time is set to the nearest 15 minutes, you only need to resolve ninety-six different positions—more than enough for this application. If there are several multiplexed A-D channels (as in the 552, in this case) separate pots (or slide pots) could be used for the on time and off time. Since the current time must be set as well, the decision was to have a single pot and three buttons labeled "Time now," "On time," and "Off time." The user sets the pot to the desired value and pushes the button. If an LCD is available, the current reading off the pot can be shown to help in setting the desired value.

Other forms of user input can rely on a spring-loaded center-return pot (as on some joysticks). A digitally displayed value can be set very precisely by having the rate of change controlled by the pot. At the center, nothing changes, but, as the pot is moved to the left, say, the value goes down at a rate proportional to the rotation of the pot away from center (or the arm of the joystick slid to the left).

This whole process can be used for setting desired temperatures, levels, alarm limits, etc. Users often do not want to be bothered with fine details and many find digital number-setting very frustrating. Unfortunately, many of the designers who configure these systems (and happen to love pushing buttons and keying in numbers) judge all users by themselves. Efficiency relates to ease of use as well as the amount of hardware and software required. In this case, though, a pot is dramatically less expensive than a keypad if the A-D conversion can be managed.

**CONSIDER YOUR USER:** Many technically-oriented designers overlook the bent of the users in choosing precise digital inputs instead of analog pots for input.

### Comparators

Comparators are the most flexible way to interface *from* analog levels to binary (on/off) levels.

Comparators such as the LM 311 shown here can be powered by the ±12 V or ±15 V typical of many analog circuits while the output transistor can use the digital ground for the emitter and add, say, a 10 kΩ or 1 kΩ collector resistor to +5 V for TTL compatibility.[43] The circuit here shifts from an analog level to TTL logic levels. Basically there is an NPN transistor from the right-most pins. In the output configuration shown, the emitter is grounded and the collector is tied to +5 V through the resistor—the suggested TTL interface of page 152. As configured here, the analog input can span a 24-V range, but the output transition from a high to a low occurs as the input crosses the threshold (here as high as 3.3 V or as low as 1.7 V depending on the setting of the potentiometer). If you wanted a very wide threshold range, use only the potentiometer tied between +12 V and –12 V. Note that reversing the + and – connections will cause the output to go *high* as the input drops below the threshold.

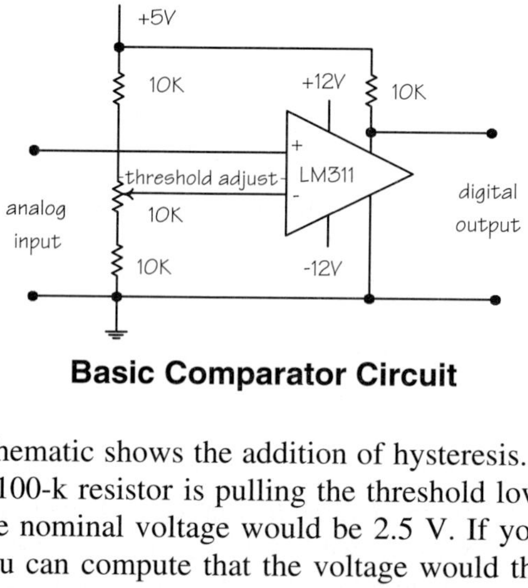


**Basic Comparator Circuit**

The next schematic shows the addition of hysteresis. When the output is low (0 V) the 100-k resistor is pulling the threshold low. Assuming it is set mid-range, the nominal voltage would be 2.5 V. If you have had basic circuit theory, you can compute that the voltage would then become 2.33. When the output is high, the threshold voltage rises to 2.66 V. What that means is that on the way down, the input has to go below 2.33 V to have the output go high, but on the way up it has to pass 2.66 V for the output to go back low. That says that there is a 0.33 V hysteresis. The benefit of this is

---

[43]The LM 311 can work from +5 V as long as the inputs stay between the supply voltages plus a margin. There have been significant changes recently with products for cellular phones. The problem for small prototype projects is that many of these new devices come in tiny surface-mount packages. If your budget allows, you can either develop printed boards or you can purchase adapter headers that convert the devices to DIP headers that can be inserted in conventional sockets.

that as a slowly changing input nears the threshold, small voltage fluctuations (noise) will not cause the output to shift rapidly back and forth.

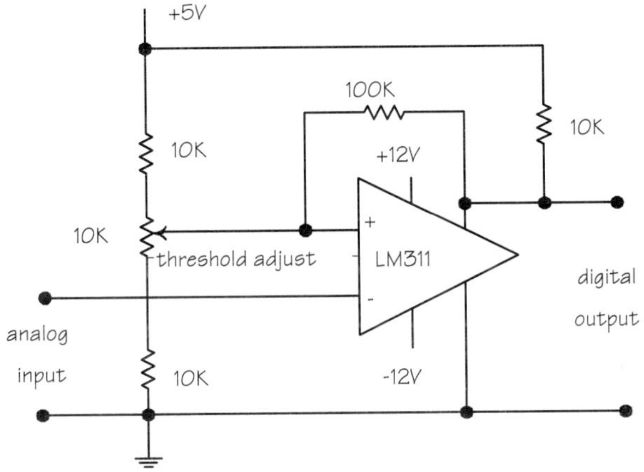


**Comparator with Hysteresis**

*Please* do not use an op-amp as a comparator to go to TTL. Unless the op-amp runs single-sided on 5 V, you should limit both the positive swing of the output and any travel below ground to avoid damaging TTL devices. Either you must clamp the output (which puts the op-amp in a current-limit mode, slows it down, and may exceed device limits) or you have to detach the TTL with diode isolation. Some op-amps *can* run on only +5 V, but usually these misguided interface efforts are due rather to using what's in your parts kit. Even a 5-V-only op-amp may not sink the current to go to TTL.

> **OP-AMPS MAKE POOR COMPARATORS:** They require extra voltages and do not interface well to digital logic.

## A-D Converter

One of the most common electronic tasks is to get information in from the outside world to the microcontroller. Most physical parameters such as pressure, temperature, humidity, light intensity, or even flow rate or water level are analog.[44] Aside from some of the more integrated sensors (see sec-

[44] If you want to quibble about quantum mechanics and the discrete nature of very small things, you should be theoretical physicists rather than an engineer. The real world we work in is close enough to analog as to not matter.

tion on Light), the information from a sensor ends up being an analog electrical voltage. That is where an analog to digital (A-D) converter comes into play.

**D-A with Comparator.** The most primitive way to get a digital equivalent of an unknown voltage is to digitally make one that just toggles a comparator. A simplified basic circuit is shown below, although the D-A converter to make the analog voltage is not discussed until the next chapter. For practical purposes, the A-D 558 puts out 0 to 5 V—5 V when the digital input is $11111111_2$, 2.5 V for $10000000_2$, and 0 V for $00000000_2$. If you connect one 8-bit output port to the "digital voltage code" inputs, you can have the software control the output voltage fed to one side of the comparator. Then, depending on whether the voltage you produce is higher or lower than the unknown, you will get a digital 1 or 0 out of the "digital feedback" pin, which you would wire to a single bit on an input port.

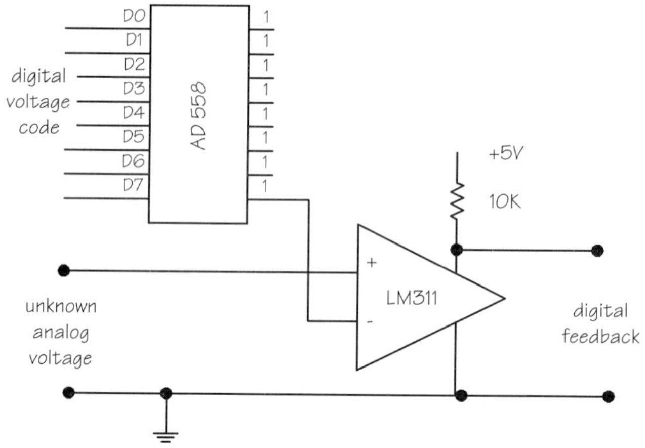


**A-D Using D-A and Comparator**

Several software algorithms are possible to determine the voltage from the photodetector.

1. Start with zero's into the D-A converter and count up (***ramping***). The count which causes the comparator output to go low is the code for the unknown voltage. For an 8-bit resolution, this takes 255 steps, worst case, to find the unknown voltage.

2. Starting at halfway (most significant bit high) and successively include or exclude that bit and then, one-by-one, the less significant bits based on the comparator output (i.e., include the current bit if the comparator is high). This routine can narrow in on the unknown voltage in eight tries (for an 8-bit converter). This approach is called ***successive approximation***.
3. Assume the unknown voltage has not changed much since last time you measured it. Start with the answer last time. If the comparator is putting out 0, count down (if it is putting out 1 count up) until the comparator reverses. Depending on how often you are taking readings and how rapidly the unknown is likely to change, this algorithm may get a new value in less than eight steps. This is a ***tracking converter***.

The approaches here can be carried out by electronic circuitry separate from a microcontroller. In addition, there are several quite different all-electronic methods discussed in the following sections.

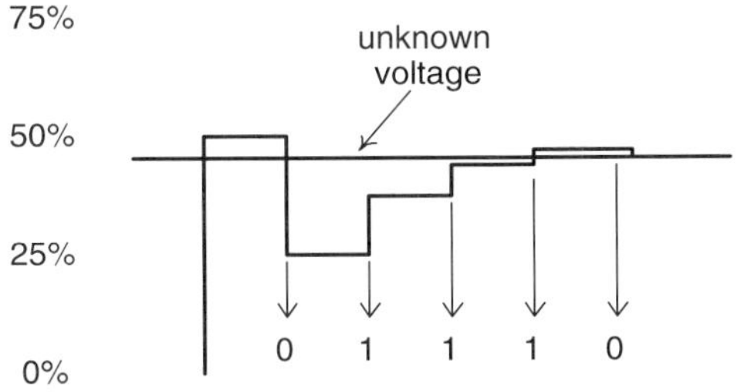


**Narrowing in by Successive Approximation**

**Successive Approximation.** *Relatively fast conversion time. Moderate cost. Noise problems.* This most common conversion approach is the type built into some of the 8051 family microcontrollers. The basic approach was introduced in the last section. If the converter can produce a digital code into a D-A converter to produce an analog voltage matching the unknown, that is the digital value of the unknown. Successive approximation is done much like you would play the game "I'm thinking of a number between 0 and 100—try to guess it." You would start by guessing 50 and then, if the answer is that "too low," you would guess 75 and so on. The converter does the same. For an 8-bit conversion ($00000000_2$ to $11111111_2$ possible), it would start by trying $10000000_2$ and seeing if the comparator says "too low." If so, it includes the first 1 and then tries $11000000_2$. If not, it then tries $01000000_2$. The divide by two is extremely simple in binary math! It successively, bit-by-bit narrows in its approximation until it runs out of bits to try hence the name "successive approximation."

*Several points to notice:* It takes eight approximations to get an 8-bit value. Ten or twelve would take correspondingly more cycles. The conversions require enough time for the D-A voltage to settle each time, so A-D conversion this way is inherently eight to ten times slower than a corresponding D-A conversion. Typical time for 8-bit conversions for typical, low-cost converters, (including those built into microcontrollers) is between 20 and 50 μsec.

Notice that the input voltage ought to hold still during the conversion time. If it changes during conversion, the answer may represent the voltage at the end or it may be hung up if the threshold of an earlier bit is crossed during the process.[45]

**The Analog Devices.** AD 573 is one common A-D converter. It is a monolithic 10-bit converter which puts out a code of $0000000000_2$ for 0 V and a code of $1111111111_2$ for 10 V, so each bit change represents a voltage change of about 10 mV. The internal logic (not shown) loads the successive bits one by one into a (serial-in, parallel-out) shift register to build up the answer. The shifting and comparison is gated by the clock beginning when a latch is set to start the process. When the 10 (or 8) shifts and comparisons are done, an output bit (end of conversion or EOC) goes high. The digital output remains there until the next conversion starts.[46]

---

[45]Successive approximation is not very noise immune during the measurement period because a glitch can cause an erroneous comparison and produce errors all through the rest of the conversion. If a significant change is likely during the conversion, either due to noise or rapid change in signal, a ***sample-and-hold*** circuit should be put ahead of the input. This is an analog switch (field-effect transistor-FET) that is normally closed to keep a capacitor charged to the current input voltage and then opens during the conversion so the signal from the outside can't change the capacitor voltage. The capacitor must be large enough that its charge can't leak off enough for the voltage to change by half of the voltage of the least significant bit (lsb) during the conversion time. At 20 to 50 μsec, conversion times that isn't difficult.

[46]Some A-D converters are buffered so the answer is latched to remain at the output even when another conversion has begun. That way you never catch a partially completed answer.

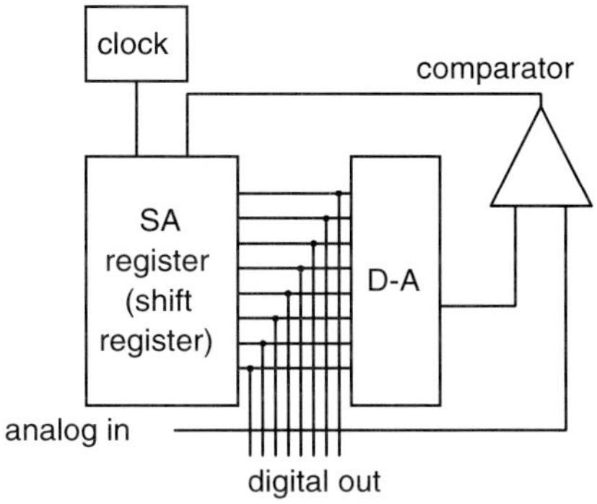


**Successive Approximation Converter: Functional Diagram**

The start-convert command can be sent from an I/O port using two instructions to drive the line first high and then immediately back low (producing with any 8051 member at least a 500-nsec pulse—more than long enough for any A-D chip). Following the start of conversion, you can read correct new data only after the EOC line goes high (up to 30 μsec after the start of conversion). The simplest way to do this in software is to toggle the start-convert line and then immediately enter a polling loop looking for a high on EOC. This is the method shown in the next program examples.

---

**A-D Interface**

```
/*assumes d2-d9 into P1; d0-d1 into lsb of P3; strobe
                from msb of P3; datardy/ to bit3 of
                P3*/
uint readad (void){
  P3=0x80;
  P3=0; /*issue start convert*/
  while((P3&8)==0);/*wait for EOC=1*/
  return((uint)P1<<2|(P3&3));
}
```

---

**A-D INTERFACE**

```
;ASSUMES D2-D9 TO P1
;D0,D1 TO P3.0,P3.1
;STROBE TO P3.7
;EOC TO P3.3
;RETURN INTEGER IN R6,R7
```

```
47READAD:SETB P3.7
   CLR P3.7 ;ISSUE START-CONVERT
 AX: JNB P3.3,AX ;WAIT FOR EOC=1
   MOV A,P1 ;PUT 2 BYTES TOGETHER
   RL A
   RL A
   MOV R7,A
   ANL A,#03H
   MOV R6,A
   ANL AR7,#0FCH
   MOV A,P3
   ANL A,#03H
   ORL A,AR7
   MOV R7,A
   RET
```

It would be more efficient for the software to go off and do something else and return sometime later to check EOC. You could use an external interrupt driven off the edge of the EOC line to avoid polling. The data remains on the A-D output lines until the next conversion is started, so there may be cases where you might not collect the data for some time.

If you wired the A-D directly to the bus you could issue a *start-convert* by writing an address decoder output to that pin. Then starting conversion is a simple matter of writing (anything) to the address. You could connect EOC to an interrupt or else read back and mask to check that signal. Many of the "microprocessor-compatible" A-Ds include the necessary buffering and decoding to require no extra chips.

Some even more "compatible" devices include a buffer and can continuously convert and upgrade the buffer value. Then any time a reading is desired, the microcontroller can immediately pick up the most recently completed one. The one I'm most familiar with includes eight multiplexed inputs and can go around one after the other in sequence and storing the most recent result. Essentially, then, the microcontroller just reads data from an 8-byte RAM that just happens to be the latest reading for that input.

---

[47]Note how easy the "bit twiddling" is in assembly—clearly what the language was optimized for. On the other hand, see how messy it is to configure the 2 bytes into a right-justified integer. A good hardware design would be changed to bring the bottom 8 bits into P1 and the top bits into the bottom of P3 so no rotating would be needed. You might be able to beat my assembly program with some rotates. The hardware configuration here is so you could choose to ignore the bottom 2 bits and just work with the top 8 if you didn't need the resolution.

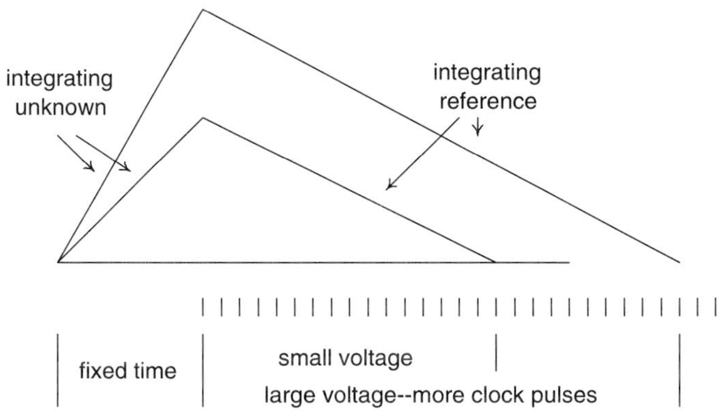


**Dual-Slope Integrating A-D Conversion**

**Integrating Converters.** *Slow but noise immune. Only reference accuracy matters.* A quite different type of A-D converter relies on comparing the incoming signal with a reference signal. Integrating converters measure the incoming voltage over a long time in a way that averages out noise. For reasons more of interest to analog designers, the technique cancels out the effect of component variations and temperature drift.[48] It is particularly useful for slow readings of steady signals in lots of noise, and can give high precision for relatively low costs. Two (patented) variations of the basic integrating conversion method are called dual slope and charge balancing.

The integrating takes time. This is the technique commonly used on digital meters where faster conversion is pointless because the human user couldn't even begin to recognize readings if they changed faster than a few times a second. Typical integration times for the unknown are chosen to be 16 or 20 msec—about 1,000 times slower then successive approximation. Signal fluctuations or noise during the integration time are averaged out. Picking the right fixed time also cancels out any 60 (or 50) Hz—by taking one full cycle of the power-line frequency any interference adds just as much as it subtracts. The slow measurement rate can be good and bad—bad if you need an instantaneous value or perhaps a whole series of values, as for a digital oscilloscope, but good because the result automatically represents the value over the entire time period. The result is a highly noise-immune conversion for a low cost.

**Sigma Delta Converters.** *Newest technology. Serial conversion output. Used for new very-high resolution conversion.* This conversion technique is perhaps the most difficult to explain. It starts with a front end (mod-

---

[48]The "dual-slope" integrator can be built quite inexpensively and doesn't require any precision components (except the voltage reference). Twelve-bit conversion is quite common, with the ICs costing under $10.

ulator) that frequently compares the unknown voltage to the average of the last binary outputs—a sort of floating average. If the unknown is greater, the output is a 1, otherwise a 0. In a sense, you have a tracking converter that is always comparing the unknown with the last reading (here the average of the last n comparisons). Following the modulator is the filter—a circuit block that processes the bit stream to produce, say, a 16-bit binary number representing the current voltage. The process is called decimation and the filter is called a comb filter. To go further you have to go into DSP.

The main advantage of sigma delta converters is their ability to very rapidly provide very accurate values with inexpensive parts. They have become the sampling devices of choice for DSP applications.

**Flash Converters.** A much faster conversion technique is called *flash conversion*. It is fast because there is just a single step. The basic configuration is shown below. The unknown voltage is fed simultaneously to an entire array of comparators while the reference, divided by a resistor string into all the possible voltage steps, is fed into the other side of the comparators. As quickly as comparators can react, a voltage in the middle of the range will cause the lower half of the comparators to indicate "unknown higher than me" while the upper half report "unknown lower than me." The comparator outputs go into a digital logic array that transforms, say, 64 inputs into a 6-bit binary code. The conversion time is limited by the propagation delays of the comparators and the digital logic. Conversion rates of 10 MHz to 100 MHz are not uncommon, but the prices are correspondingly high.

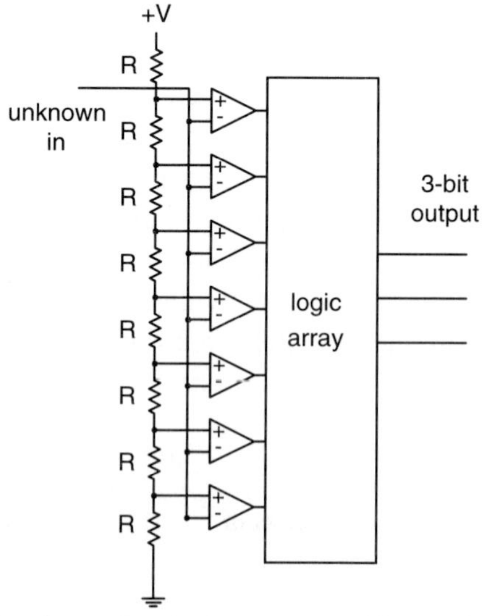


**Flash A-D Converter: Functional Diagram**

The logic for a flash converter makes a good exercise in digital design—particularly since the number of cascaded logic stages multiplied by the propagation delay of each stage determines the response time. You get a quadratically increasing amount of circuitry as the resolution increases, so even 4- and 6-bit converters are common. The cost for more resolution goes up rapidly from 4 to 6 to 8 bits. Since these converters are used most commonly with digitizing video,[49] higher resolutions are uncommon.

**A-D Converter—Voltage to Frequency.** *Frequency directly proportional to incoming voltage. Twelve-bit linear range (LM 331).* Ideal for input to microcontroller, the National Semiconductor LM 331 is a common voltage to frequency-converter chips. The frequency of an oscillator (pulse generator) changes proportional to the unknown voltage. The internal technique is surely an integrating one, so, like the earlier integrating converters, it is highly noise immune. Your microcontroller can either measure the time between two pulses or count pulses for a fixed period of time to determine the unknown voltage. This conversion method can give quite high resolution while being inexpensive. From the perspective of the microcontroller, it is very economical because it requires only one input pin. With a crystal oscillator, measuring time is something a micro does very well. See the algorithms section for more on measuring pulses. The single digital output also allows simple optical isolation for noise suppression, breaking ground loops, and computer protection.

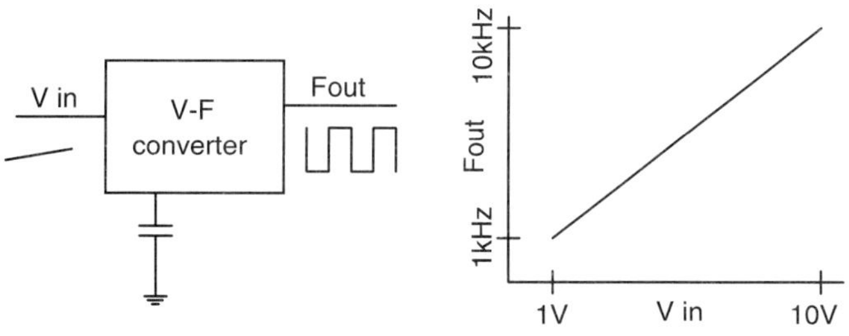


**Voltage to Frequency Converter**

[49]People have no problem with the smoothness of images having 256 shades of gray—usually 64 shades are enough. Most of the video displays boast at least 24 bit color—8 bits for red, green, and blue respectively. For extremely critical publishing applications there is 30-bit color—10 bits per color. I suspect this can be perceived only by comparison in the region near white.

**FREQUENCY IS BEST:** The conversion of an analog voltage to a frequency or pulse width is one of the most efficient choices you can make. Measurement can be as accurate as the crystal oscillator and hardware is usually minimized.

**Increasing A-D Resolution.** What can you do if you have a need for more resolution that your A-D converter supplies? Of course, you might buy a higher-resolution converter, but suppose you are using a microcontroller with an on-chip converter or you are limited due to cost constraints. Here are several suggestions:

**Limit the Input Range**. Make the incoming signal as large as possible to fill the range of the converter. Suppose your converter takes an input of 0 to 5 V, but your actual signal comes from a temperature sense chip and varies from 2.5 to 4 V for 0° to 150° C. If you feed the signal *directly* into the converter (suppose 8-bit or 256 levels), the resolution is 5 v/256 or about 20 mV. That gives a resolution of 75 over the temperature range (that worked out well for no prior planning!) or a resolution of 2° C. If, instead, you put an op-amp circuit such as the one shown below in front of the converter, the signal goes from 0 to 5 V over the stated temperature range. Then the resolution is 150/256 or about 0.6° C.[50]

**Oversampling.** If you can take *multiple* readings you can increase your accuracy by introducing what is called *dither*. You either vary the reference voltage of the converter slightly or introduce a small fluctuation in the incoming signal. Particularly if this fluctuation is a triangle wave, the signal will be above and below the comparator threshold. The closer the undithered signal is to the threshold, the more often the added dither will push it above the threshold and give the higher reading. Ideally, if you took sixteen readings and found that ten of them took the higher reading (say, $10110011_2$) and the other six took the lower (say, $10110010_2$), then you could say that the higher resolution reading should be $10110010\ 1010_2$—

[50]Does it bother you that it doesn't come out an even whole degree? You should realize that nothing is perfect in an analog world, and the higher the resolution, the better, usually. Particularly if there is some drift or software-calibration in the process, a non-integer resolution is probably better. If you are already calibrating, there will not be a sudden obvious shift around the even (.00) points.

2 bits above the mean of the two readings. If half the readings went both ways, then you could say that the high-resolution reading was $10110010\,1000_2$. I should add a disclaimer here that the analog world is not ideal and the A-D converters do not always give a perfect reading—the lsb is always in question. This technique is best used when you need *resolution* rather than *accuracy*.

## ELECTRICAL MEASUREMENTS

Voltage measurements have already been discussed under *Analog Inputs*. The assumption was that you have a reasonably steady voltage perhaps representing an already converted physical quantity. Get whatever it is you are measuring into a steady voltage and you can convert it to digital information for the micro.

While some of the quantities that follow are steady, there are also ones that are AC signals or vary in other ways.

### Resistance

Suppose the thing you want to measure comes out a resistance.[51] You have at least two ways to convert a resistance into a voltage. First, you can add a resistor to get a resistor divider. With a fixed voltage across the resistor series string, as the unknown resistor changes value, the output voltage changes as well. Choose a resistor of about the maximum value you expect for the unknown, so the output voltage of the divider ranges down from 50 percent of the driving voltage.

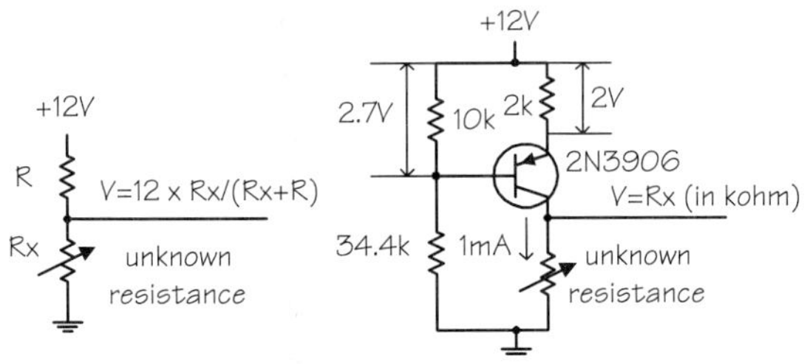


**Resistance to Voltage**

[51]The most common physical quantities measured as resistance are strain—bending of material that compresses or stretches the resistive material of a strain gauge—and devices such as thermistors that change resistance with temperature.

Second, you can drive a known current through the resistor and measure the voltage across it. Constant current sources are easily obtained with a transistor—by maintaining a constant voltage across the emitter resistor, you maintain a constant emitter current and, effectively, a constant collector current.

Finally, you can put the resistor in the controlling circuit of an oscillator and observe the change in frequency.

**Accuracy and Calibration.** In all these techniques, the question of accuracy and stability is significant. While some resistance measurements may not need high accuracy, the accurate ones are particularly sensitive to temperature drift. Many resistors change with temperature and the production tolerance is a significant part of the specification—20 percent, 10 percent, 5 percent, 1 percent—the cost varies accordingly. The best thing you can do in a system where the precise value is important is to include self-calibration features. For example, you can eliminate concern about accuracy of all sorts of elements if you can electronically switch between the unknown resistance and a *very stable* known resistance. Given a well-matched analog switch pair all on one chip, the known resistor can be measured and then the unknown can be measured relative to the known. Given a microcontroller, it is a simple matter to correct the reading mathematically. If there is both a gain and an offset shift, then measurement of two known resistors can give the information needed to calibrate.

This all assumes that the measurement is linear over the range. Most of the time that is true. If not, then you have a much bigger problem. If there are known logarithmic characteristics, there are analog devices that will give an output voltage proportional to the logarithm of the input (and vice versa). Anything else is a swamp. If you can define a piece-wise linear model, then something can be done electronically with diodes and various resistors, but this is not the stuff from which stable accurate electronic measurements arise.[52] For example, suppose you want to measure temperature with a thermistor—a predictable but nonlinear device. If you can establish one value at one temperature, you can establish all the rest of the scale. In the end, however, you make a lot of trouble for yourself, and had better be sure that cost or environmental requirements are forcing this—a thermocouple or solid-state temperature sensor have the advantage of being linear.

---

[52]The *Algorithms* chapter goes into an example of linear interpolation off a look-up table to convert non-linear data.

## Current

**DC Current.** As discussed earlier, most converters measure *voltage*. DC *current* can be measured by observing the voltage across a known resistor. Ohms law says that $E = I \times R$, so the voltage ($E$) is the product of the current ($I$) times the resistance ($R$) or, $I = E / R$. If you insert a low-resistance precision resistors in a current flow you can often get a small voltage without significantly upsetting the circuit delivering the current. You can quite easily get 1 Ω and 0.1 Ω resistors rated 10 W at 1-percent accuracy.

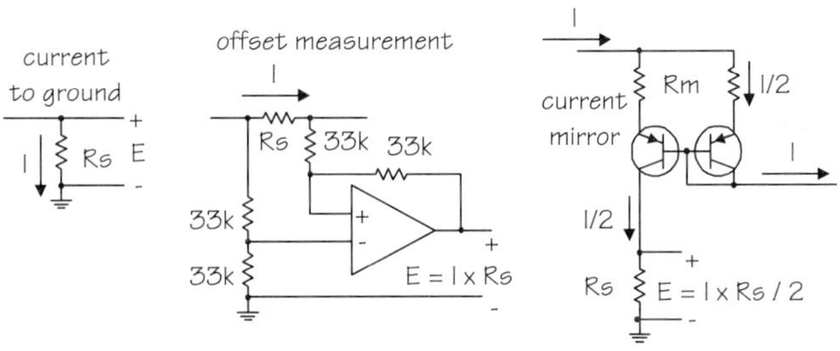


**Current Measurements**

**Current Offset From Ground.** If it is easy to put a resistor in the current path just where the current returns to ground, the voltage across the resistor can be easily fed into a ground-referenced A-D input. If the current can't be measured at ground potential, things get a bit more difficult. The resistor must be somewhere else in the current path, and it is necessary to measure the voltage across the resistor—the differential voltage.

Some A-Ds can take differential inputs as long as the two input voltages are within the power supply range of the A-D. Typically that means the potential on either side of the resistor can't get outside of ±15 V. If it does, it is necessary to do something like run resistor dividers from the high-voltage points to the inputs to the A-D. The second circuit above uses an op-amp in a differential configuration to give a single-sided voltage proportional to the current. The 33-k resistors are about as high as you should get to avoid inaccuracies due to op-amp drift. The circuit shown can measure current offset by about 30 V. Assuming the voltage difference across the current sensing resistor is quite small, the attenuation by the divider reduces the difference even more and adds inaccuracies due to mismatching in the dividers. It is possible to trim the dividers with pots to match and to give zero difference when the current sense resistor is shorted out, but you have to be careful about the thermal characteristics of the resistors.

**UTILIZE THE A-D'S CAPABILITY:** Differential inputs can allow you to measure a voltage away from ground with no extra circuitry—you don't always have to ground the negative input.

Another way to sense current when it is referenced to a voltage far away from the A-D is with a circuit called a current mirror. The base-emitter voltage across the transistor in the current path is matched across the mirror transistor that sends collector current down to the current sense resistor at ground potential. It is a good idea to add small emitter resistors to even things up in case the transistors are mismatched or drift due to temperature. This is particularly good where the current to be measured is being drawn from a high-voltage supply.

There are other ways to get current measurements. We will talk about AC next, but for DC it is possible to use hall-effect sensors that sense the steady magnetic field produced by the constant current through a wire. The principles are beyond the scope of this book, but the result is a signal directly related to the current in the wire, but totally isolated from the voltage of the wire.

Finally, there are ways to obtain the isolation optically. If the current to be measured is converted in an LED to a light level transmitted from the high side to a photo-diode near ground, the output of the photodiode can be converted to an easy-to-measure voltage. Commercial modules exist that do all of these, providing an isolated analog signal transfer. Some are related to the brightness of the light—perhaps in a two-direction feedback path—and some actually do the A-D conversion at the input side and transmit digital signals by light, avoiding some of the nonlinearity and drift of the optical intensity transmission. In any case, these are really potted subsystems and may be unnecessarily precise and too expensive for many low cost applications.

**AC Current.** Fortunately the laws of electromagnetic interaction make transformers quite appropriate for isolating signals with an average value of zero. There are devices called *current* transformers, but the principles are the same for them all. Ideally the secondary voltage is proportional to the turns ratio for an *open* secondary and the current is inversely proportional to the turns ration for a *shorted* secondary. The measurement of the secondary involves letting a little current flow in the secondary for voltage, or dropping a little voltage in a sense resistor for current. The only difference is the choice of windings—a current transformer probably has fewer turns made of heavier wire. If anything, you might take a normal step-down power supply transformer and run it in reverse to get a current transformer. A bigger voltage could result which is easier to measure without disturbing the primary current as much.[53]

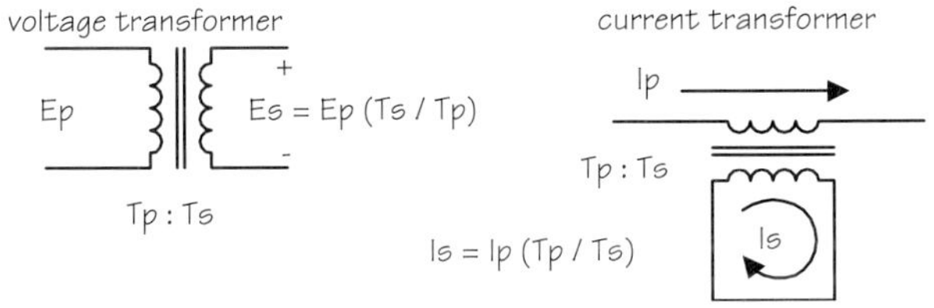


**Current and Voltage Transformers**

Once you have a voltage proportional to instantaneous current, you are faced with the problem of determining RMS. If the waveform is an ideal sine wave, the mathematics says the RMS is 0.707 of the peak voltage. The easiest thing to do is to take the peak value with a peak detector and multiply accordingly.

The other alternative is to take the square root of the average of the square, which is exactly what the term means. If you are processing in analog, you can feed the current into both inputs of a multiplier, feed that into a low-pass filter, and feed the result into a square-root circuit (another possible configuration for a multiplier). I suspect there are ICs that combine all these functions as well as ones that actually measure the heating value of the signal (the RMS) by actually applying its energy to a resistor and compensating by reducing the additional heating to maintain the resistor at a constant temperature.

Finally, if you want the RMS current, you can take frequent samples of the waveform and do the math digitally. Assuming you can get an analog reading every 50 μsec, you can 320 samples of a single 60-Hz cycle. It might be a stretch to carry out an integer-squaring operation and a summation in 50 μsec on an 8051—if that is a problem, you can either use a faster clock or perform the calculations in non-real-time. To check the idea, I wrote a program to do just that.

---

[53]You can't have a perfect open on the secondary if you're going to measure it any more than you can have a perfect short and still develop a voltage across a resistor to have a measurable signal. The general principle is that it is impossible to measure anything without disturbing it somewhat, but you can usually get "close enough." An amplifier across the small resistor gives a voltage proportional to current while the secondary of the voltage transformer gives a voltage proportional to the primary voltage.

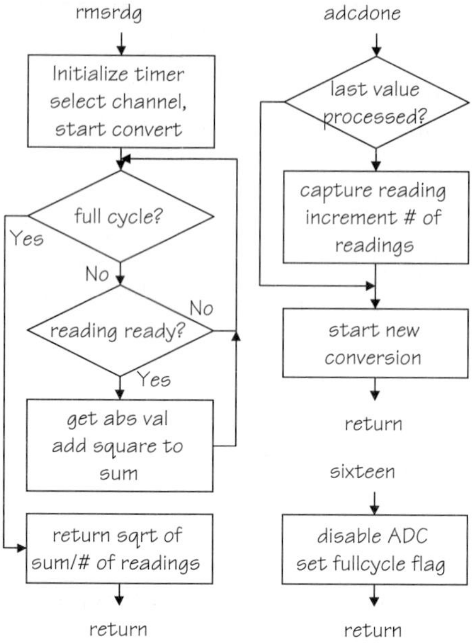


**RMS Version-1 Flowchart**

The programs that follow use an 80C552 to find the RMS of a reading for one cycle of a 60 Hz. Both versions use *timer 0* to end the sampling process and set a flag when 16.6 msec have elapsed so the sampling represents equally spaced samples over one cycle. Both programs use only 8 bits of the result, even though 10 bits are available.

The first version takes full advantage of the hardware by using the internal ADC with the *End Of Convert* interrupt (*ADCI*). The conversions go at the maximum rate of the converter if the math processing can keep up. Unfortunately, while the conversion takes 50 machine cycles (and the processing in the interrupt takes another 15 cycles), the squaring and summing loop takes 60 cycles so the conversion is done before the math![54] The fastest, then, would be 75 µsec (at 12 MHz) but the program here would ignore every other conversion for a rate of 100 µsec or 166 samples per cycle.

[54]This was determined by a quick run of that part of the loop with DS51. The bind is that no matter how fast you make the processor clock, the conversion will always be faster since it is tied to the same clock! Incidentally, the final square root and return took 2,889 cycles including integer division, conversion of the integer result to float and converting the square root back to an integer. That was with some dummy numbers inserted (sqrt (5000/7))—when it was sqrt (0/0) it only took 374 cycles.

**rms.c**

```
#include<reg552.h>
#include<math.h>
#define uint unsigned
#define uchar unsigned char
data uchar reading;
data uint noofreadings;
data uint current;
bit newval,converting;

void display(uint newval){
 uchar x;
 x=newval; /*dummy stub for now*/
}

void adcdone (void) interrupt 10 using 1{
 if(newval==0){/*discard if not ready to process*/
  reading=ADCH; /*capture upper 8-bits of value*/
  newval=1;
  noofreadings++;
 }
 ADCON|=0x04; /*start newconversion*/
}

void sixteen (void) interrupt 1 using 1{
 IP0=0; /*disable all interupts*/
 converting=0; /*end of one cycle of 60Hz*/
}

uint rmsrdg (channel){
 data uint sum;
 data uchar temp;
 ADCON=channel &0x07; /*select channel*/
 newval=0;
 converting=1;
 IP0=0x42; /*enable timer0 and ADC interrupts*/
 ADCON|=0x04; /*start new conversion*/
 TH0=16666/256;/*16.6 mSec@12MHz*/
 TL0=16666%256;
 TR0=1; /*start 16mSec timer*/
 while (converting){
  if (newval){
   if (reading<0x80)temp=-reading;
    temp=reading&0x7f;
    newval=0; /*allow next reading to be saved*/
```

```
      sum=sum+((uint)temp*temp);
    }
  }
 return sqrt(sum/noofreadings);
}


void main(void){
  TMOD=1; /*timer0 in 16-bit mode*/
  for(;;){
    current=rmsrdg(0);
    display(current);
  }
}
```

Because the math takes more machine cycles than the conversion, a better way is this second approach which abandons the *end of convert* interrupt and just paces things by the math—restart the converter each time just before the processing. The loop here takes sixty-five cycles,[55] so the conversion is always done before the math finishes. A dummy cycle is inserted at the start of *rmsrdg* give time for the first conversion to finish before the real math begins.

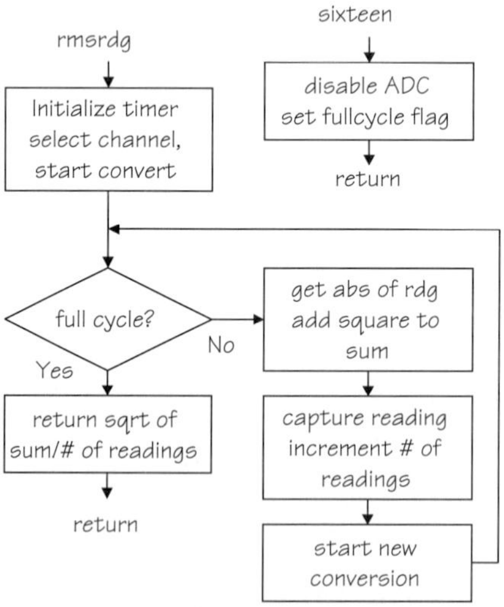


**RMS.c Version-2 Flowchart**

[55]Determined by running DS51 and looking at the cycle count before and after one time around the loop.

**RMS2.c**

```
#include<reg552.h>
#include<math.h>
#define uint unsigned
#define uchar unsigned char
data uint current;
bit converting;

void display(uint newval){
 uchar x;
 x=newval; /*dummy stub for now*/
}

void sixteen (void) interrupt 1 using 1{
 IP0=0; /*disable all interupts*/
 converting=0; /*end of one cycle of 60Hz*/
}

uint rmsrdg (uchar channel){
 data uint noofreadings=-1;/*1st is dummy*/
 data uchar reading=0;
 data uint sum=0;
 ADCON=channel&0x07; /*select channel*/
 converting=1;
 IP0=0x02; /*enable timer0 interrupts*/
 ADCON|=0x04; /*start new conversion*/
 TH0=16666/256;/*16.6 mSec@12MHz*/
 TL0=16666%256;
 TR0=1; /*start 16mSec timer*/
 while (converting){
  if (reading<0x80)reading=-reading;
  reading=reading&0x7f;
  noofreadings++;
  sum=sum+((uint)reading*reading);
  reading=ADCH; /*capture upper 8-bits of value*/
  ADCON|=0x04; /*start new conversion*/
 }
 return sqrt(sum/noofreadings);
}

void main(void){
 TMOD=1; /*timer0 in 16-bit mode*/
 for(;;){
  current=rmsrdg(0);
```

```
      display(current);
    }
  }
```

## AC Power

Fifty years ago, power metering was done with power meters where the needle deflection was related to the cross-product of voltage (converted to current) in one winding by the current in another winding. It was a clever use of magnetic principles and is difficult to surpass for efficiency. It was very simple because it produced a deflection directly related to the average of the product of voltage and current.

The simple approach of multiplying the *average* current by the *average* voltage will not give the true power for nonresistive loads like motors. In the worst case of a capacitor or inductor, the real power can be considerably less. Of course the electric utility is quite concerned about this (called the power factor), but that is beyond the scope of this book. If you had a course in AC power, you would know that the RMS (root mean square) value of a sine wave is about 0.707 of the peak, and that the 110 VAC power in the United States would have a peak-to-peak voltage of over 300 V. Most inexpensive analog meters actually measure the peak voltage and show it on a scale reduced by a factor of 0.707—the meter doesn't read true RMS at all, but assumptions a sinusoidal wave shape. Since the power company hasn't started supplying square waves, it is a tolerable approximation if heavy loads aren't messing up the shape.

**True RMS Metering.** If you need true root mean square (RMS) readings in the presence of noise or distorted sine waves, there are analog devices to make such measurements. Some true RMS devices go as far as heating a signal resistor by thermal conductance from a nearby resistive element, applying the unknown signal to the signal resistor, and then measuring the reduction in power in the resistive element necessary to restore the signal resistor to its original temperature. This is a true heat-value measurement.

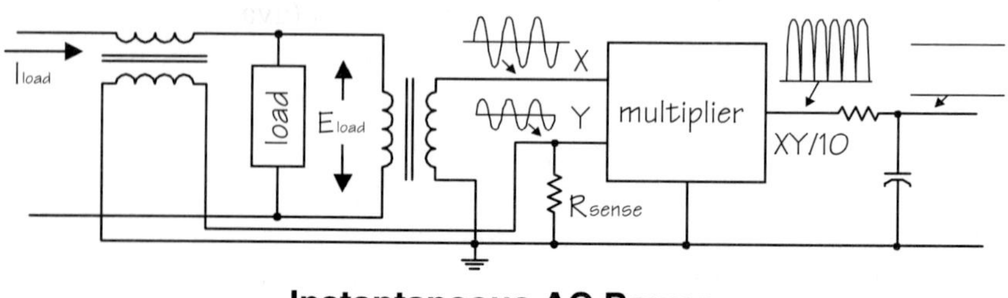


**Instantaneous AC Power**

A second type of integrated device (or circuit as shown above) makes instantaneous multiplications of the input by itself (the "square" of "root mean square") and feeds that into a square-root circuit. The average of the result can be fed to a simple analog meter or converted to a digital reading.[56]

For high voltage (like 110 or 220 VAC), it is good to have isolation and step the signals down to levels that normal ICs handle. Having obtained the voltage and current signals, feed them into an analog multiplier and out comes a signal proportional to the product—hopefully a sort of sine-squared waveform. A little filtering (integrating) and you have the average of the instantaneous power. Multiplier chips can handle frequencies far above those of 50- or 60-Hz power. Take the nice smooth (DC) voltage and feed it into an A-D converter and you have your power reading. It is efficient use of computing power because there is nothing to do except pick up a single voltage reading occasionally.

**Internal Multiplication.** If it can be done in analog, it should be done better digitally (says the modern digital designer). With two A-D converters and a digital signal processing (DSP) chip, this processing can be done reasonably well in real time. Can you do it directly in a micro? As mentioned before, you can't smooth the voltage and current signals *first* because the product must be instantaneous to give true power. The 10-bit converter on the 80C552, for example, completes a conversion in fifty machine cycles or about 50 μsec. Unless you have a sample-and-hold circuit, the voltage and current will be measured at slightly different times, but if toggling between the two is good enough, you can get eighty-three samples of each in a full cycle of 60 Hz. That's not too bad. You could multiply the current and voltage readings directly together in software or, if you're a stickler for details, multiply one by the average of the two neighboring values of the other. Is that good enough? It depends on why you need the reading. Unless there are some severe spikes and high frequency fluctuations, it should be more than enough!

Of course, integer multiplying 10 bits by 10 bits risks overflow and, in floating point, probably can't be done in real-time (every 100 μsec), but you could reduce the resolution and do $8 \times 8$ multiplies (available in hardware). Remember that the processor is yours to use during the time the A-D is converting. If there isn't much else to do, you might manage the math nicely. That way you've kept the processor employed and avoided some costly hardware.

---

[56]Note that the average of the instantaneous RMS values is not necessarily the same as the RMS of the average.

Failing in managing the math in real-time, you could easily capture a full cycle and process it later. Presumably that would be in a situation where the power does not change radically between successive cycles.

These techniques may not be appropriate where you need high accuracy or where there is lots of other processor work to perform, but they deserve serious consideration because they represent real hardware efficiency. Don't run away from the approach just because it looks complicated.

## Zero-crossing Detection

For many power-drive applications, it is necessary to know when the voltage (or current) drops to zero. Specifically if you decide to drive a phase-angle firing triac directly from the timing in the micro, it becomes critical to know when to initiate the delay that determines the output brightness or power level. The zero-crossing detection is included in many of the opto-isolator drivers discussed in the output section, but that provides no signal to the microcontroller. Also, if you are deriving a time base from the power line, you need to count the cycles.

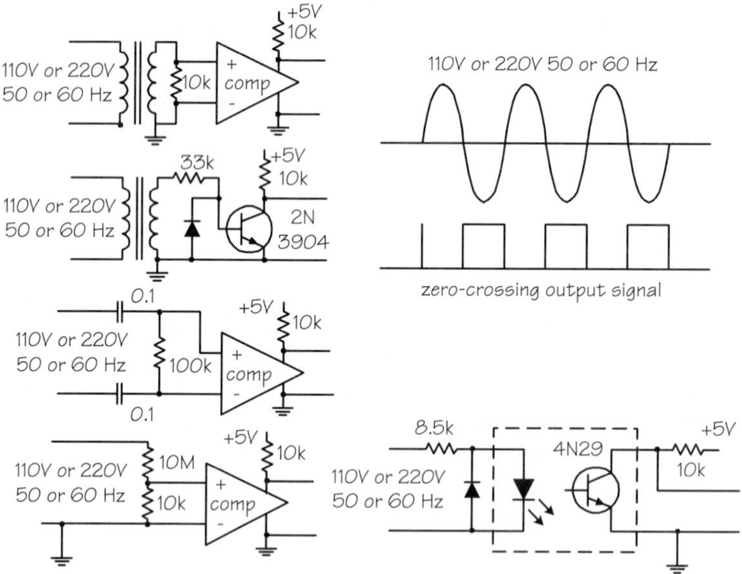


**Zero-crossing Detect Circuits**

One of the simplest ways to get the zero crossing is to take the signal out of a secondary winding of a power transformer. The awkward part is the need for an unused winding because the power supply winding will probably have a large capacitor and bridge attached. A close look at the waveform is in order if you are going to tap a working winding.

You can use a *separate* secondary winding to drive into a comparator or a transistor. By using the open-loop gain, a very sharp transition is possible when the current reverses. The only necessary step is to limit the current so the device is not damaged.

Another way is to capacitively couple from the primary side. That is not so good because the capacitors pass high frequency better than lower frequencies and any noise on the power line could come roaring through. If you want to maintain isolation, you should bring a capacitor from the neutral side as well as the hot side. If isolation is unnecessary, you can run a simple resistive divider. In general, it is poor design to give up isolation unless your product is totally insulated from the user (such as double-insulated drills and hot-chassis TVs).

Perhaps the best way to get a zero-crossing signal is with an opto-isolator with a large series resistor and a protection diode. Then for half a cycle the diode will conduct and send light to the detector. The other half-cycle it will be off. That can give a nice square wave at the output, safely isolated from the primary and the 0.7-V diode drop will be insignificant compared to the 340-V swing of the power line. The biggest drawback is the power consumption—since the diode needs at least 10 mA to conduct well (assume that it is limited to 20 mA at the peak of 170 volts) a resistor of about 8.5 k ohm is needed with a power dissipation of about 1 W—a 2-W resistor must be used. Also, the circuit comes on softly at the crossing. Power efficiency suggests that a watt is a lot of power just to find a zero crossing.

**POWER-GRID AS PRECISION CLOCK:** Because of the presence of synchronous-motor clocks, the power-grid is carefully adjusted to keep clocks accurate. If you want a calendar that is very accurate over long times, counting zero-crossings is a very low-cost way to do it.

## PHYSICAL MEASUREMENTS

### Flow

For many high-flow-volume applications, commercial flow sensors often use some sort of impeller, turbine, or paddle wheel. The rotating shaft comes out through a liquid-resistant bearing to either a small generator that produces a voltage proportional to flow, or to a optical mark shaft encoder to produce a pulse stream related to flow. If that is what you need, I suggest you buy it pre-built because the commercial mechanical designs have been refined over generations.

Another type of flow sense involves inducing (ultrasonic) vibration in the flow stream and detecting the doppler frequency shift at a detector down the line. The faster the fluid flows toward the detector, the higher the apparent frequency.

Since I emphasize efficiency, let me refer back to the project of *Volume I,* Chapter 14. All that I needed to detect was the presence or absence of flow—not the actual rate. After much looking, it was found that a simple steel ball would suffice. Captivated in one parallel branch of the plastic pipe, gravity would bring it to the bottom of the branch. When there is flow, the steel ball moves up to the top of the branch. By winding a coil around the pipe at the top, the steel ball changes the inductance of the coil. By putting the coil in the feedback loop of a Schmidt trigger oscillator, the frequency changes measurably when the steel ball becomes the "core" of the coil. This frequency could be measured directly in the micro, requiring only a single input pin! See the section on V-F converters for more details of the software options.

### Temperature

Normal ambient temperature ranges can be measured with solid-state devices, thermistors, or thermocouples. The ***silicon devices*** capitalize on the temperature effect in silicon junctions that most analog designs fight to eliminate. The so-called 0.7 junction voltage actually changes with temperature. I don't know the internal details, but companies such as National Semiconductor have for many years made devices such as the LM34 and LM335 that give a voltage linearly related to temperature. The LM335, for example, puts out 10 mV per degree Kelvin or just below 3 V for room temperature (I'm sure they don't work down near absolute zero!).

***Thermistors*** are not linear and may not be highly repeatable from device to device. They are devices that change resistance with temperature, but it is exponential rather than linear. For an extremely low-cost temperature measurement requiring only "cold" and "hot" or very coarse values, a thermistor can be the feedback of such an oscillator.

***Thermocouples*** are junctions where two dissimilar metals touch. I don't know the physics of it, but a very small voltage is generated at the junction that changes linearly with temperature. The main thing a thermocouple has going for it is the ability to sense over a far wider range of temperatures. The wire is relatively inexpensive and can sense temperatures as high as inside engines or industrial processes using molten materials. There are several types of thermocouples depending on temperature range and resolution. I believe your best approach is to locate a commercial thermocouple-interface chip and take the output into an A-D converter.

Let's return to the silicon temperature sensing devices such as the LM335. They provide a *voltage* related linearly to temperature—a sort of temperature-sensitive Zener diode. If you position such a device at some distance from the micro, however, the resistance of the wire enters in. In order to cancel out the effect, you can drive a known *current* through the wire, read the voltage, and then drive a different known current and get the new voltage. You can then factor out the resistance and know the voltage at the sensor. Graphically, it consists of putting the two points and tracing the corresponding line back to the axis.[57] See Chapter 7 for details.

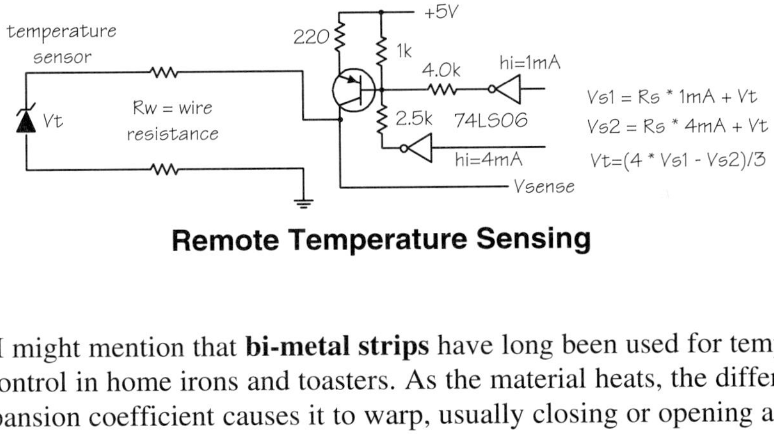


**Remote Temperature Sensing**

I might mention that **bi-metal strips** have long been used for temperature control in home irons and toasters. As the material heats, the difference in expansion coefficient causes it to warp, usually closing or opening a contact to turn the heating element on or off. Perhaps you could measure temperature by measuring the position of the strip end—use one of the many proximity/position techniques described earlier.

[57]Using this idea, a remote two-wire thermostat system for each room can use a potentiometer in series with a temperature sensor. The pot is used to set the *desired* temperature and the sensor reports the *actual* temperature over the simple two-wire connection. The two-point measurement of the previous paragraph now supplies the voltage (temp reading) and the slope or resistance (the pot setting of the desired temperature). All this is easily done with two known currents driven down the wire.

## Pressure

Pressure sensing is virtually always done with commercial parts. In addition to the devices that attach strain gauges to diaphragms and such, solid-state sensors offers an etched cantilevered part with deposited resistors that work as strain sensors. They measure the deflection of the silicon overhang. They are not terribly expensive and some will measure absolute pressure by comparison with an evacuated reference.

The only lower-cost way I can imagine for getting a value for atmospheric pressure is to design some sort of optical sensor of the mercury column in a conventional barometer (or a shaft sensor off an aneroid barometer). I suppose a linear CCD sensor could work or else some totally enclosed chamber where the amount of light transmitted varied linearly with the column height.

## Acceleration/Tilt

There is a strong correlation between tilt and acceleration since the earth's gravity is acceleration in itself.

There have recently come on the market accelerometers that are quite low cost—an entire system for under $50 including microcontroller and display. I believe they are targeted for the automotive market—measure the performance of your car: *0 to 60 in 6 seconds!* I suspect the sensors are solid-state—some sort of etched, suspended silicon with either strain sensors or some sort of magnetic suspension where the force needed relates to the acceleration. Most of the commercial accelerometers probably put out an analog voltage related to acceleration. Basic calculus says that by integrating acceleration you can get the velocity, and by integrating again you can get position.

I have at times proposed a very low cost tilt/acceleration sensor for use in car alarms made out of two potentiometers and a weight. The hardware would require some assembly and would be undamped, but it could do for some very simple applications. Perhaps the weight hanging down could be immersed in oil to provide the damping. Again, its primary use could be in a tilt sensor.

A very low cost, rugged tilt sensor can be made from a curved tube with a drop of mercury inside. If the object tilts, the mercury makes (or breaks) the shorting of the two electrodes poking through the glass.

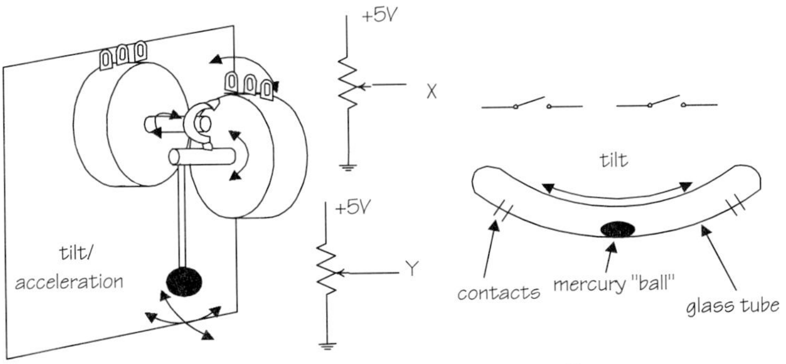


**Low-cost "Accelerometers"**

Another idea is to levitate a ferrous object by a magnetic field with optical sensors to determine if the object is remaining stationary. By measuring the current necessary to keep the ball in place, the change in drive could measure the acceleration. The two-pot affair just discussed could be used also with two coils added to keep the hanging ball fixed. Again, the two pots could be the position sensors and the field necessary to keep the ball in one place would reflect the acceleration.

### Humidity

I have not a lot of experience with humidity sensing. Numerous student projects here at Purdue have used a commercial humidity sensor whose capacitance changes with humidity. I believe they cost about $20 and were accurate only to a few percent. The other low-cost option is to attach a sensor to a "home weather station" meter that produce a rotation of a needle related to humidity. I don't know the principle but logic suggests it is based on a combination of two materials, at least one of which changes dimension with humidity, much like bi-metal strips bend with changing temperature. I have a vague recollection of something based on horsehair—very empirical measurements.

The only other method I know of is the traditional wet- and dry-bulb temperature measurements. Some sort of evaporative wick attached to one thermometer reduces the temperature due to the energy absorbed in converting the liquid to vapor. The more humid the air, the slower the evaporation and the less the temperature differential. The temperature sensing is easy to do electronically, but the need to supply water would make it very awkward for a measurement. The conversion of temperature differential to humidity is easily done with tables available from the weather bureau, and interpolation from tables is quite easy and discussed in the algorithms chapter.

## OPTICAL INPUTS

### Light

Light is now commonly detected with silicon photodiodes and phototransistors.[58] It turns out that photons can provide base current to a transistor in much the same you might supply it over a wire.[59] Silicon based detectors are naturally most sensitive to red and near-infra red (IR) rather than to blue or green light.[60]

If you are concerned about linearity—the output voltage linearly proportional to the input light intensity—then a photo***diode*** is the device of choice. Not only is its output linear over quite a wide range of input light intensity, it is very fast. If you can modulate your light source at 1 MHz or more, you need a photodiode to detect it.[61]

If, instead, you want to get as much output as possible for a small amount of illumination, the photo***transistor*** or photo***darlington*** is preferred. Such devices don't have the linearity or the speed (rise times of 3 to 30 μsec), but they can give a directly usable signal without the additional amplifier that is almost mandatory with a photodiode.

The light from fluorescent bulbs comes on at a 120-Hz rate that can confuse peak-detecting algorithms. Incandescent bulb light is virtually steady, but any time fluorescent light is also picked up there will be a variation. If the light you want to measure can be chopped much faster than 120 Hz, you can use a synchronous detector. On the other hand, if you want to detect a steady light that is not under your control, either use a "valley detector" or sample the light frequently and take the smallest reading over an 8 to 10 msec interval as "the" value.

---

[58] There are certainly other light-sensitive materials in use. Historic materials for such things as camera light meters are cadmium selenide and cadmium sulfide. I am not sure if they changed resistance based on the impinging light or if they actually produced current—in series with a resistor, the result is similar. The entire field of photoelectric power generation continues to involve research and has yielded devices practical for certain applications.

[59] In fact, there are some sensitive circuits that have to be covered to be sure that ambient light doesn't disturb them. At one point, a company marketed a black/white-only imaging camera that used as its sensor a conventional RAM chip with a clear cover that charged up from the light hitting it.

[60] Devices based on cadmium selenide and cadmium sulfide are sensitive to light in a more balanced way, but they are much slower and can't have any gain built in. If you aren't concerned about the reduced sensitivity to blue or if you are picking up light from a red or IR LED, the silicon devices are best.

[61] I'm not sure of all the developments in the area of fiber optics and telephone transmission, but there has been lots of development work there and you should investigate more if you need very high-speed light signaling.

A fascinating device (the TSL230 from Texas Instruments) combines a photodiode with circuitry to provide a frequency out related to light intensity. With a photo*diode*, it is linear over six decades of intensity (and has programming inputs to extend that over eight decades—100,000,000 to 1 light-intensity variation). As discussed in the section on A-D, it is extremely handy to go into a microcontroller with a frequency signal because it requires only one pin and can be handled with one of the internal timers. The detector costs only a few dollars, yet it can be extremely simple to interface and can work over a very wide range. Because it integrates to produce the pulse train (frequency), it also is very noise immune. It is, of course, most sensitive in the red and IR, but for *low-cost* intensity measuring systems, it is a major breakthrough.

### Color Detection

Detecting color *accurately* (colorimetry) is outside the scope of this book, but there are some applications where you need some type of sorting or rough distinguishing of dissimilar colors. One technique is to have three (or more) identical detectors with different colored filters in front of them. The red filter blocks all but the red light, the blue blocks all but the blue, and the green all but the green. By looking at the relative signals from the three detectors, a signature can be established. Of course, different shades of the same color may disturb the ratio. The big problem is that silicon detectors are very sensitive to IR and most of the gelatin filters don't block out the IR. The signals are blue + IR, green + IR, and red + IR. The differentiation may not be very good. An alternative is to go to cadmium selenide or cadmium sulfide detectors that respond much more evenly in the visible region and not so sensitive to IR.

Another approach is to use a single detector and illuminate the unknown object with three different colors of light—try Christmas tree bulbs for a start! This is a good idea if you are analyzing colors over an area using a monochrome CCD camera.[62] Another option for a monochrome camera to detect color is with a rotating set of filters in a color wheel.[63]

You need to seriously think about your application. Be sure you understand what you really need to detect before you jump in with a full-blown

---

[62]The CCD cameras are available surplus, at least, for under $100. I believe the tiny devices see increasing use in security systems.

[63]Color cameras are readily available, albeit not inexpensive. The recent introduction of the small "eye" cameras that interface to computers shows what new developments and aggressive marketing can do. For your moderate color sensing application you may find it more economical to use such a camera and add synchronizing circuitry to sample the video output at a specific time—perhaps sample-and-hold feeding A-D converter inputs.

colorimetry system. Perhaps a simple filter and the right lighting will save having to use an expensive system and add a lot of processing.

### Images/cameras

Entire images are another input. You can get video information with small CCD cameras for perhaps $100 for a complete system. Bringing that into a microcontroller is a challenge if the camera provides regular broadcast-rate video because the signal may be coming in at over 2 MHz—too fast for an 8051, generally. Frame grabbers provide a buffer between the video and the slower micro, capturing an entire image into a memory block for more leisurely processing by the micro at its convenience. This used to be very expensive, but now memory is much less expensive and higher levels of integration have changed things. Now a highly integrated device is available for about $200 that captures an entire color image and feeds it into the parallel port of a PC. You can easily simulate the parallel port interface with the micro if you want to capitalize on the low cost arising from the PC market volume and competition.

There are some CCD cameras that can be operated at much slower rates. It is easier to interface a slow scan, and this may be quite appropriate for applications that don't need a flicker-free display for humans. Unfortunately, they don't benefit from the high-volume demand of broadcast rate and resolution cameras. If you want to go back to basics, you can get both linear and area CCD array chips to build your own system. The linear arrays are used in scanners, so they are common now, and the area arrays are now a common part of surveillance cameras, so there could be some hope of building your own if the potential volume could justify the extra work. Linear arrays are particularly applicable for observing production lines where the thing to be seen can move by in the other dimension. Or you can add the second dimension with rotating mirrors as in many of the supermarket barcode systems.[64]

> **TRACK NEW DEVELOPMENTS:** As an engineer, you should be keeping an eye on new developments. Many approaches that were uneconomical yesterday may be very practical today.

[64]Camera input entails enough electronic technology distinct from the micro that you should hesitate to work up a system from scratch unless you really must for cost and ultimate production volume reasons. Always see if you can "go with the flow" by using or adapting devices that are readily available for high-volume markets. Let some one else fund the development costs wherever possible.

## IR Remotes

Suppose you have an application that takes input from a commercial infrared (IR) remote.[65] A hand-held IR remote works over a long distance because of the modulating (chopping) of the light. The sun, incandescent bulbs, and fluorescent bulbs are quite bright compared to what an ordinary LED can put out. The key is in chopping the light from the LED at about 40-kHz (much higher frequency then the "DC" up to 120 Hz emitted by the ambient sources). By filtering on the receive end, you can add all the gain you want until the very small 40 kHz signal is very easy to detect. Usually the signaling used is a simple on/off control of the 40 kHz. The received 40-kHz signal can be clamped or rectified and integrated to give an actual logic 1 when the 40-kHz signal is present. As long as the chopping frequency of the light is reasonably stable and no other nearby light sources are chopped at high frequency, distances of 25 to 50 feet are not unusual.

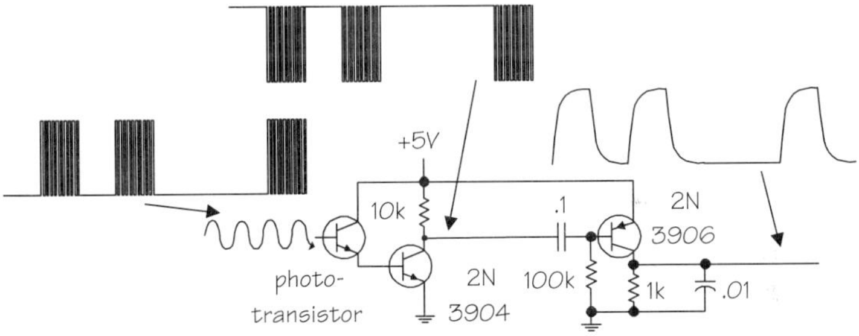


**IR Remote Detection Circuit[66]**

The IR signals, once the 40 kHz is out of the picture, have been designed from pre-microcontroller days to be easy to decode cheaply. They were designed to be *self-synchronizing*, meaning that they sent some pulses at the start to get an oscillator in sync with the signal and lined up with the

[65]Your project might be an extender for a conventional remote (convert the incoming signal to RF in another room, receive it near the stereo, and convert it back to IR to go into the stereo's IR receiver), or it might be an application where you capitalize on the mass-produced, readily-available transmitters and use them for your own system. Remotes have long been used for VCRs and television, but you might be spreading the wave to microwave ovens or ceiling fans! Historically, IR transmissions standard have varied among manufacturers so you have to "set up" a universal remote to work with a particular receiving device.

[66]Note that this circuit has not been tried an an actual remote and the capacitor values might need adjusting. Also, the final stage is not suitable to go into TTL—add another transistor.

on/off frequency applied to the 40 kHz.[67] At 40 kHz, the LED is on for about 12 μsec and off for the same amount of time (that is about the fastest an ordinary phototransistor can respond). The on/off switching of the pulse train may be at a 1-kHz rate, so a "1" may consist of a "burst" of 20 or less pulses of light. That is enough to work with easily.

The variation consists of the way information is sent. Most codes keep something at a regular frequency (typically a 1 to 5-msec period) in order to keep a wandering oscillator synchronized. Then the information in the signal may consist of a *second burst* later in the cycle that is either early or late (slipped—call it *pulse position encoding*). Alternately, the signal information may be in the *presence or absence* of the second pulse (*missing pulse encoding*). Finally, the signal information may relate to the *duration* of the synchronizing pulse (*pulse-width encoding*). One edge comes on time, but the other edge may be early or late depending on whether a 1 or 0 is being sent at that time.

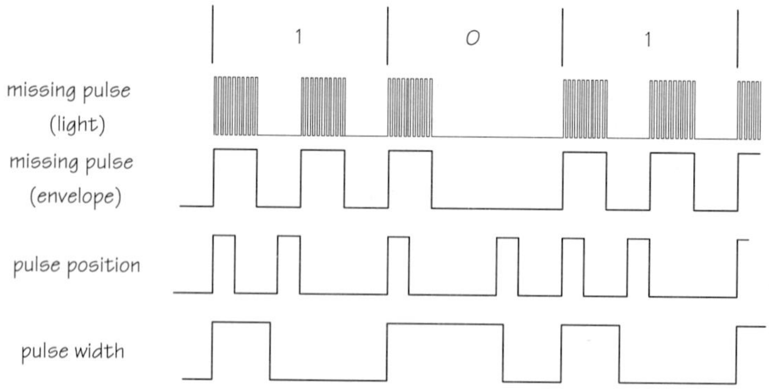


**IR Encoding Patterns**

The signal from the remote doesn't have to convey much information—8 bits is more than enough to represent any key on the remote. The bits don't have to be sent quickly, either, because this is a human interface where a 1/4-sec delay is not regarded as serious—how can the user tell when the key really clicked?[68] Most systems require at least two signal

---

[67]Incidentally, there is an entire field of study relating to information storage that applies here. The encoding on magnetic disks, for example went from FM to $M^2FM$ and beyond to store 1s and 0s with the fewest transitions of the signal while retaining the synchronizing transitions. If designers were starting from scratch now, they might shift much of the burden to the electronics and greatly increase the speed of the remote transmission. Then again, the amount of information is so small, they might not change.

[68]The design becomes much more demanding if you are setting up a remote communication between two electronic devices such as optically sending digitized music to a remote loudspeaker or optically communicating across a room.

blocks in agreement before they respond to the transmission. With a simple (ASCII-like) transmission, a block of 8 bits at 1 kHz would take 10 msec. Given a second transmission after a respectable delay, two transmissions could be completed in perhaps 30 msec. Slow it down and it might take 100 msec—still fast enough to satisfy any human user while retaining good noise immunity.

The discussion that follows might go in the algorithm chapter, but here I keep it with the input signals. Probably the different timings would make phase-locked loops and such-like difficult or costly for more than one type of input (quite unlike a detector for, say, a specific Zenith or RCA set, for example), and you should be thoroughly in favor of software over hardware for efficiency by now. The better solution is to sample the output of the integrator which is either a logic 1 (when the burst is coming in) or logic 0 (when no light is arriving). There will certainly be some lag between the light pulse coming in and the integrator reaching a logic 1, but presumably there will be a delay in going back to logic 0 also.

The most obvious approach is to take regular samples of the signal and store them in an array for later processing. If the 40-kHz burst is rectified and filtered, a sample every 100 μsec would catch all the bursts and give a general idea of the timing. That would be 10 samples per 1-msec bit interval, so *pulse-position* encoding *missing-pulse* encoding or coarse *pulse-width* encoding could all be detected. Fortunately, the IR encoding schemes have a lot of latitude to allow for slow, sloppy integrating and frequency drift with temperature. It is a bit inefficient to fill a byte array with single-bit information, but it's easier to process as separate array elements. The 8051 family supports bit variables, but not bit arrays. You could pack the bits into a byte and then store them, but since you want to quickly process them anyway, it is probably better to leave them one to a byte. Since the frequency could drift and the phase may not be synchronized with the sample times, you will have to allow for a bit of jitter in the pulses relative to the samples. An algorithm could walk through the array storing the array index each time a new string of 1's is found. Then go through taking the differences from the previous indices to get the time between pulses. If it is a pulse-position code, the sum of adjacent times should be nearly constant since the synchronizing bursts come at a regular time. If it is a missing-pulse code, then smaller numbers should add up to approximately the value of the bigger numbers because the time between sync pulses is constant whether or not a signal pulse is present. If these don't work, then a re-computation of the raw samples where the 0's start should give a pulse-width decoding scheme. There are even more sophisticated encoding schemes that have long existed for information storage of tapes, floppy discs, and hard discs. These are beyond the scope of this book, of course.

If you want to be more efficient with the processor and you can provide an interrupt line, you might run the interrupt off the signal from the integrator into the interrupt. Then you wouldn't have to watch the input line, and yet you could capture the *time* of each edge. Rather than a string of samples, you would immediately have the times of the edges and avoid one of the processing steps. Even better, you could have the interrupt compute the time from the last edge and make a decision about the significance of the new burst. If it has been a long time, you might be getting the start of a new bit sequence. This way you can ignore the input line until an interrupt occurs and process only when new data arrives. Save, say, the most recent twenty time differences, and you should be able to decipher a block of data as soon as the next one starts. Then the decision of when to recognize a completed block can depend on the delay to the next block and the software doesn't have to go on sampling forever, wondering if there is a full block yet. Having the last twenty time differences stored in a FIFO buffer doesn't take up much RAM and avoids lots of processing. The differences can be handled as in the previous paragraph. The actual differences should be stored as fairly coarse numbers, not with the 1-µsec resolution of the micro's clock—don't eat up RAM and add processing time by storing 16-bit numbers.

## MAGNETIC INPUTS

This is an area where the background knowledge is furthest from ordinary electronics—magnetics seems to have declined in general interest and is mostly limited to specialists in motor and transformer design. Incidentally, the mathematics can rapidly go beyond the background of any but engineers. There are several very useful input applications that rely on magnetic fields.

### Pick-up Coils

It is a basic law of physics that changing magnetic fields induce current in wires. That is the principle of electric generators. But it is possible to apply the same principles to other uses. One method of X-Y position sensing ("digitizing") described earlier involves inducing current spikes through wires in a flat grid and picking up the induced field with a small coil at the position to be measured.

A *permanent* magnet on a moving object might pass by a coil and, even though the magnetic field is not changing, the field changes relative to the coil. Now you have a noncontact pickup for your cruise-control from an automobile's drive shaft.

The magnitude of the signal from a coil is likely to be very small (unless many turns are involved or the field is very strong), so amplification, rectification, and some sort of comparator are likely necessary.

Another use of pick-up coils is in sensing the timing of automobile ignition (and thus engine speed). By wrapping a coil around one of the spark-plug wires (usually with a ferrite core rather than actually winding the coil around the spark-plug wire), a circuit is obtained which will have an induced current each time the spark plug fires due to the surge in current through the spark wire.

### Magnetically-driven Switch

Glass-enclosed "reed" switches are frequently used as input devices. The most common use is in the security systems for windows where a switch is attached to the frame and a permanent magnet is attached to the sash. As long as the magnet is in place, the switch is closed and the circuit is complete. If a burglar slides the window open, the switch opens and the alarm sounds.

The same thing is an option for drive-shaft sensing for tachometers or add-in cruise controls. Attach a magnet to the rotating shaft and mount the switch close enough to have it close once each revolution.

### Changing Inductance

If you have ever noticed demand-triggered traffic lights, you have seen an example of changing inductance. Buried in the pavement is a loop that makes up the inductive component of an oscillator. When a vehicle passes near the large loop, the ferrous material presents a shift from an air core to an iron core, changing the inductance, and thus the oscillator frequency.

In the same way, I described in the *Volume I* (Chapter 14) a flow-sensing input where a steel ball was trapped inside a short length of plastic pipe with a coil wound around one end. As the water flowed, it moved the ball out of the coil, changing the inductance and thus the frequency of a simple oscillator. That change of inductance (frequency) is sensed by the microcontroller. The frequency shift is not critical in this application since the microcontroller determines the no-flow and full-flow values at calibration time. A frequency of a few kHz is adequate to be resolved easily by the controller and can be determined by the capacitor choice and the number of turns to the coil.

### Hall-effect Sensors

While all the earlier magnetic methods rely on changing fields, a somewhat obscure magnetic/electric interaction, called the Hall effect, has been put to use to produce a sensor that gives a voltage proportional to the magnetic field (rather than the *change* in field). Such devices give a voltage directly proportional to the magnetic field passing through the device in a certain orientation. A pair of devices can be combined to produce an electronic compass detecting the orientation of the earth's magnetic field. Virtually all the other magnetic methods above can be adapted to the Hall-effect devices, usually at a somewhat higher cost. They have the advantage that the signal is not frequency-dependent so, with a drive-shaft sensor, for example, the speed could be tracked down to a full stop.

## SOUND ANALYSIS

Suppose you want to recognize sounds—not speech *per se*, but a telephone ringing or other predictable sounds from a limited set of possible sounds. One project I saw planned recently was a system for a deaf person who wanted to be notified if the front doorbell rang, the telephone, rang, the microwave oven timer beeped, and so on. This would not be a demanding analysis job, but there are several ways to do it.

### V-F and Peak Detector

One approach is to feed the sound from a microphone and amplifier into a frequency to voltage (F-V) converter along with a peak detector. When a loud sound is detected, the voltage out of the F-V is measured to give a two-number "signature" for the sound. It looks good in theory, but it overlooks the way the F-V operates. The input must be a *digital* input that causes transitions at each edge. Feeding an analog (audio) signal directly in can confuse the converter. The amplitude may be too small for a weak sound.

The problem with recognition by raw F-V conversion is that the transitions can be multiplied in the presence of higher harmonics. F-V is fine when there is only the fundamental, but when there are significant higher frequency harmonics or noise, the conversion can be fouled.

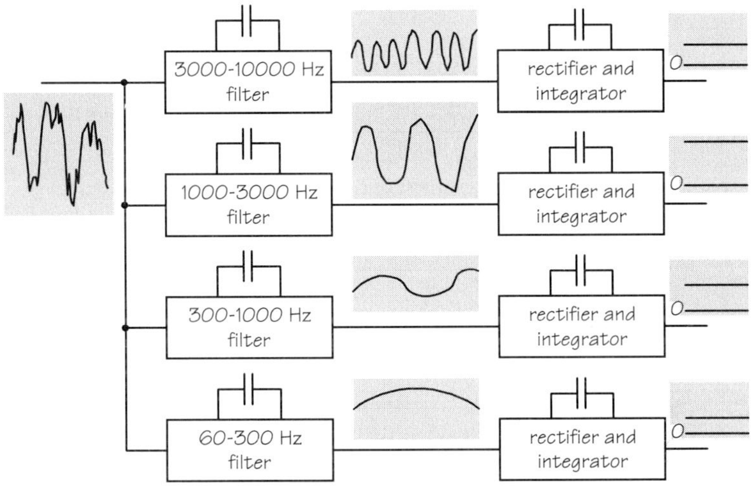


**Band Energy Detection**

A second option is to use a bank of active bandpass filters with rectifiers and lossy integrators. The filters might have bands between 100 and 500 Hz, 500 to 1 kHz, 1 to 3 kHz, and 3 to 6 kHz. The design of active filters is beyond the scope of this book, but they can be done with standard configurations and the components can be chosen in a "cookbook" fashion. For this application the actual sharpness of the rolloff (how well the filters reject frequencies outside their band) is not really important so fairly simple designs are permitted.

Having gotten rid of the "squiggles" of noise and higher frequencies, voltages can be obtained by rectifying and integrating the waves out of each filter. Again, precision is not important so a simple integrator is possible with a diode and a passive R-C filter, but it might be better to have an ideal rectifier so low-level signals can still be detected and won't be lost in the 0.7-V diode drop. With a bit of careful design, the rectifier and integrator can be combined in a single op-amp, so four bands can be handled with a single quad op-amp.

## Switched-capacitance Filter

An entire bank of filters is a bit much in the hardware category if the incoming signals are slow enough or simple enough. If you want a spectrum analyzer or signal detector for a sound that holds reasonably steady, a single bandpass filter that can be moved around is useful. That is essentially what a switched capacitance filter can become. With the ability to change the clock frequency, you can change the pass band. And changing a clock frequency is quite easy in a microcontroller. To analyze the sound levels over a band, you can start low and work up in frequency without the confusion of harmonics to separate. The output of the filter could be rectified and integrated as with the previous example, or the signal could be fed directly into an A-D and processed in the micro. If the micro's converter can sample, say, every 20 μsec, then it can do a decent job for frequencies somewhere up to 25 kHz—particularly when aliasing is not a problem due to the preceding filter. (*Aliasing* is thinking that you have a much lower frequency because the samples come just about on successive peaks). The point here is that you can watch the signal well up into the audio range without any special hardware. As the pass band is moved up, you obtain a measure of the maximum amplitudes over the frequency range. As long as the signal stays constant, you have a spectral analysis.

For the sound recognition system (telephone, doorbell, and so on), a bit of intelligent programming should be able to determine some algorithms to distinguish different sounds. Perhaps you stay at one frequency band until a signal is detected. Measure the energy and quickly shift to a second band to distinguish the different sounds that might have come in. You establish a signature for the sound. Perhaps a measure of the duration or repetition of the sound is significant, such as with a ringing telephone. Remember that the filter takes a bit of time to settle, but nothing like human response times. It should be possible to pack a few different samples in the time a telephone completes one ring or a microwave announces it is finished.

### Direct Sampling

If you want to go for a minimum system, it should be possible to do signal processing directly in the microcontroller. Ideally, this would be done with a DSP chip, but, if the constraints are not too great, it can be done in the micro. That way the extra chip is avoided and the micro is available for other activities as well. DSP algorithms are beyond the scope of this book (meaning I don't know much about them), but the process can be done on a burst sample of raw incoming sound taken in by the A-D. The analysis doesn't have to be continuous and real-time if the sampling can be started by the detection of an unusually large acoustic signal. The processing for frequencies can be like adding samples 1, 3, 5 and subtracting samples 2, 4, 6 to detect a signal at half the sampling frequency. Likewise 1, 5, 9 minus 3, 7, 11 would detect at one-quarter the frequency. The idea is that signals at much different frequencies average out to zero. For much more about algorithms, you would have to explore DSP.

### Pattern Matching

There is no law that a sound has to be detected by breaking it down into frequency components. It is quite feasible to store a waveform and process it for a pattern match—correlation. Take perhaps 100 samples of the raw audio at 250-μsec intervals over ¼-sec. Then take a later set of 200 samples at the same interval. Now walk through the 100-sample pattern laid against the first 100 samples, squaring the difference at each sample to get a number for the degree of fit. Next take the 100-sample pattern against samples 2 through 101, getting a new number. Go through all 100 times, sliding the pattern over by 1 each time. Then look through the 100 numbers representing the fit. Find the smallest one. If it is below some threshold, you can decide that you have a match. Since each measure of fit takes 100 subtractions, 100 multiplications, and 100 additions, the whole process takes 10,000 of each, which, even with 1-byte math is easily 1/10-sec and with integer math is probably more like 2 sec. You can see why DSP chips are designed to do these repetitive operations much more efficiently. I put these simple algorithms in only to get you thinking of the crude and low-cost ways you can do things when precision or speed are not important.

## SPEECH RECOGNITION

Speech recognition is much more complicated than speech synthesis. The production of speech can be with one particular voice and pronunciation, but recognition, ideally, would detect words even with variations in pronunciation, pitch, and rate of speech. Ideally, recognition would be speaker independent, handle an unlimited vocabulary, and work for connected words and phrases. In reality, the present systems are good for only limited vocabularies and isolated words. If the vocabulary is very limited—perhaps the numbers and a few commands—then it is possible to be reasonably speaker independent for isolated words. Simple chips are available from electronics stores that can control toy cars by voice—"left, right, stop, go"—but the choices become much less when larger vocabularies are involved. There are large software packages that run on PCs to handle large vocabularies as well as smaller systems that handle limited vocabularies, but they all are somewhat *speaker-dependent*, meaning that they have to be trained by having the speaker say words that are to be recognized several times until the system has recorded and averaged the features of the particular speaker. The day may come when the computing power necessary for *speaker-independent* recognition makes it possible to include highly effective speech recognition in small systems, but for now the affordable systems for small applications are much more limited.

### Do You Really Need Speech Recognition?

Efficient applications make the best use of the available technology. Before embarking on recognition you should decide how important such a difficult feature is to the user. If there are significant background noises, the system could require a headset microphone and additional noise-canceling circuitry. Is the user willing to put up with that? How important is a verbal input? Could a set of buttons handle all the user inputs? Could a joystick and a screen navigate through inputs, or could a menu selection on an alphanumeric display suffice? IR TV remotes have become an acceptable input for the typical "couch potato" and could be easily expanded to more work-related settings. Historically, speech recognition has been used for data entry in low-light situations, for situations where the user's hands are fully occupied with other tasks such as inspection or very delicate operations, and for the handicapped. The high cost has kept it out of the more imaginative applications such as dictating a report directly to a computer, but that may change.

The cost of computing is dropping steadily and there is a potentially huge market on the horizon as personal computers get smaller and smaller. Laptop computers have keyboards about as small as human fingers can reliably use. The palmtop computers require a stylus to use. The only viable alternative currently is the "pen computer" that allows the user to write on a small area and applies handwriting recognition to decipher words. The next step may have to be voice input, since a microphone can be quite small—perhaps with voice synthesized output to save display space! To be publicly acceptable, such a system would have to accept ordinary speech from a variety of users and apply rather complex rules of grammar and context in deciphering the input.[69]

### Do-It-Yourself

There are recognition applications where neither large vocabulary nor connected speech is necessary. For example, there are robotic toys which

---

[69]Currently there are several speech-recognition systems on the market for PCs that do moderately well, but they require at least 32 Mbytes of RAM and do only passably with a 200-MHz computer—not something to incorporate with an 8051! While the algorithms are probably very complex, the future *may* see solutions boiled down to a very elaborate standard chip using DSP and specialized firmware. What might have been primitive research on a supercomputer a few years ago may become a custom IC a few years from now.

can recognize three words—*go, stop,* and *turn.* Likewise, hands-free data collection systems can recognize numbers and a few commands from any speaker and in the presence of significant background noise. The key is in the limited vocabulary.

In the first case, for example, *go* is much shorter than *stop* or *turn. Stop* begins with the high-frequency *sss* sound. *Turn* ends with a lower-frequency *nnn* sound. A very simple algorithm could distinguish the three words. (Of course an *Oh!* would pass for a *go* and *Tim* would pass for *turn*!).

Historically the analysis of speech has relied on the energy distribution within frequency bands over time corresponding to the phonemes that make up speech. That is still the best way for limited-vocabulary recognition with conventional microcontrollers. If you time-normalize a discrete utterance and mark when the energy within a frequency band exceeds a threshold, you get a *signature* of the word. By storing a library of the average signatures for the expected words, best-fit pattern matching can return the most likely word if the fit exceeds a set threshold.[70]

In a simpler manner, you can take acoustic inputs for simpler tasks as in clapping to control the intensity of a lamp. In such a case, the system would high-pass filter or analyze the sound envelope to detect short, sharp changes in loudness. The analysis would have to rely on the duration of a clap as opposed to a shout, but it could not rely on the frequency of the sound because a clap can vary in pitch (try it!). Once you have brought the digitized sound or envelope into the microcontroller, the processing should be relatively simple, but a lot of experimentation may be necessary to block out other sounds. Perhaps the triggering requires two claps separated by limited range of times.

> **DO ONLY WHAT IS NEEDED:** Make sure you are not complicating a job by requiring more capability than is necessary. Speech recognition is a particularly good area to apply this rule.

## INPUTS FROM THE HANDICAPPED

A large variety of devices for the handicapped and the development efforts have been motivated by compassion out of proportion to the economic market. It is encouraging to think that there are those who are motivated by care

[70]Actually, if speakers say a word in distinctly different ways (dialect variations), *two* signatures can return the same word as in a Bostonian saying *caaah* for *car.*

for others. The input devices vary with the type of disability. There are very high-tech ones to do with sensing eye position.[71] There are position sensors—X-Y and joystick as with the electric wheelchairs. For the severely paralyzed, I understand there are inputs operated by the mouth and even puff sensors. In most cases, the inputs are related to the ones already discussed, but often the processing has to be much more intelligent to filter out extraneous signals as you may have observed with someone having poor motor control. This is a field where all the microcontroller "intelligence" you can add is worthwhile.

## POWER SUPPLIES

This isn't precisely an input category or an output category, but it doesn't rate a chapter of its own. Here I will talk about supplies that involve no microcontroller to function—they provide power to the micro. In the output section, I have a section on the use of microcontrollers as a part of the supply itself.

### Typical Linear Supply

For almost all applications you want transformer isolation. Also, for any low voltage output from 110-V power lines, you ought to step down the voltage to improve efficiency.[72] The circuit below is full-wave rectified, meaning that both half-cycles contribute to the output. The energy storage capacitor acts as a filter to "filter out" the 120 Hz of the rectified signal, leaving a DC value that can be dropped down by the linear regulator.[73]

**Transformer.** Be sure you have enough current capability. The internal resistance of the windings reduces the output voltage as the current drain increases—or, if you prefer, lower current gives higher output voltage than at the rated current. Overcurrent can cause heating problems in trans-

[71]Also used by the military in directing armament—a new dimension of point-and-shoot!

[72]The only exception to this is a switching power supply that drops the voltage. By drawing current only briefly, it can maintain high efficiency. The temptation to avoid isolation is there but should be resisted except for insulated applications.

[73]Twenty-five years ago power supply regularors were a much different proposition. From tube-based regulation and inductive filters, it moved to the 723 IC from Fairchild and then to the three-terminal regulators. The latter were a marvelous improvement because, in addition to short-circuit protection, they included integral thermal shut-down. The entire design process became dramatically simpler.

former (even a fire if it is not well-designed). The higher rated the transformer, the less the output changes with current. On the other hand, some of the transformers have enough internal resistance to be self-protected from short circuits—particularly the wall-mount transformers.

The secondary voltage needed relates to the size energy-storage capacitor—higher voltage, smaller capacitor. I will go into the options as I discuss the capacitor.

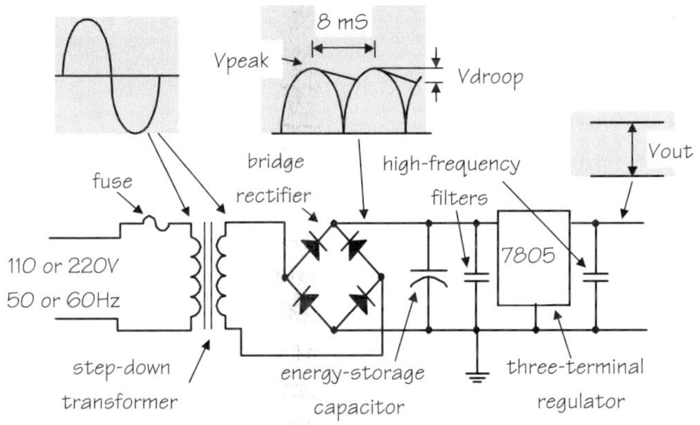


**Typical Single-Supply with Linear Regulator**

**Fuse.** Even with all the protection of the regulator, you definitely should have a fuse on the primary side to protect from a transformer short circuit. In choosing a fuse value, consider the output power. Here the design calls for 1A out of the 5-V output so, a 10-V secondary and 1 ampere implies a normal primary current (RMS) of $10/110 \times 1\text{A} = 91$ mA. So use very small fuse—say 1/4-A (250 mA)—well above normal current, but less than short circuit current.

**Energy-storage Capacitor.** You need to always have enough voltage at the capacitor for regulator head room (1.5 V over higher than the output voltage). If you look at the waveform, you can see that the voltage droops between charging cycles. It is the capacitor size and current drain that determines droop (lowest voltage at regulator input). You can trade voltage against capacitor size. I suggest about an 8-V RMS secondary winding for 5-V supply.[74]

[74]Your best bet is a secondary of 8 to 10-V. *Do not* use an ordinary 6.3-V secondary (commonly available) because the voltage will be too low and you will have to use an inordinately large storage capacitor to avoid ripple.

The reasoning in picking the capacitor is the peak secondary voltage is 13.6 V (about 1.7 times the RMS value of the secondary). The pair of bridge rectifier diodes drop about 1.4, so the peak voltage at the rectifier output is about 12.2 V. From the 12.2-V peak subtract the 5 V out, as well as the 1.5-V headroom (the minimum voltage lost in the regulator) and you have 5.7 volts of allowable droop. Now, assume you are supplying constant current at the output so you have a linear droop by the equation: $\Delta V = I \times \Delta t/C$. Solving for the capacitance: $C = I \times \Delta t/\Delta V$. In the specific case where we want 1 A of current and full-wave rectify (@60 Hz one half-cycle is 8.3 msec) the equation becomes: 1 A × 8 msec/5.7 V = 1400 μF.

Next we need to be sure the regulator handles the power dissipation. Since power = Current × voltage drop, here 1 A × (12.2 − 5)V = 7.2 W. Look up free-air power rating or max junction temp and °C/watt to determine if heat-sink is required.

A closer check of the data book may indicate that, in addition to the 0.1-μF ceramic capacitors that reduce any high-frequency effects, you may need 10 μF at output also. So here is the result:

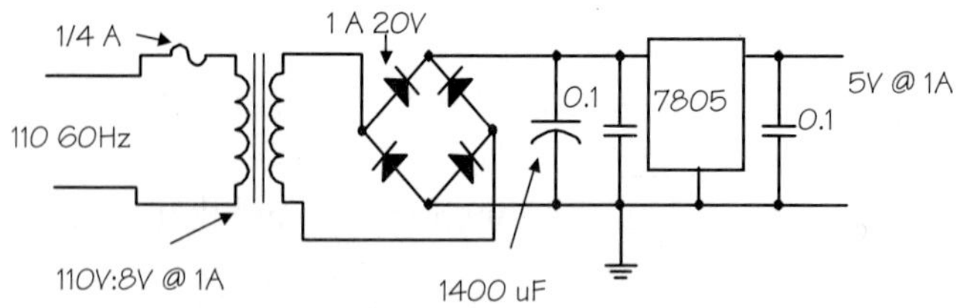


**Basic Linear Power Supply**

A second example shows a power supply with three output voltages. I suggest that you purchase a specifically-designed transformer for this with dual secondary windings.[75] The only difference from the previous example is the obtaining of the plus and minus voltages with a single bridge and a center-tapped transformer winding. Since it is generally the case that you need much more +5-V current than ±12-V current, it is inefficient to take the raw 5-V source from the +12-V regulated output or the +12-V raw source. Again, note the trade-off of voltage and efficiency with capacitor size. The 20-V secondary (10-V each half) rectifies to 16.3-V peak or a headroom of only 2.8-V. As a result, the capacitors must be a bit larger, but

[75]You can find them in *Digikey* catalogs as well as direct from my favorite, *Signal Transformer*.

the efficiency is higher—less heat is dissipated in the 12-V regulators than, say, if a 24-V secondary were used with smaller capacitors.

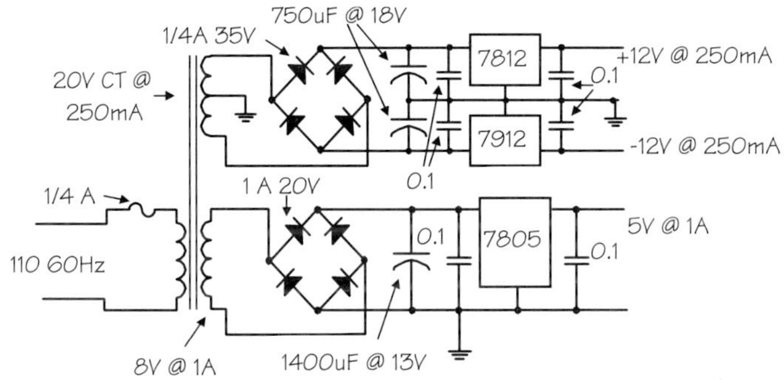


**Typical Triple Output Power Supply**

## Switching Power Supplies

There is a lot more to this field than I can cover here. The key advantage is that a switching power supply works at much higher frequencies and therefore requires less iron in any transformers, and therefore weighs less. In addition, power devices in switching regulators operate outside the linear region—they are either on or off and therefore dissipate less power, resulting in higher efficiency. For complete supplies operating off 50 or 60 Hz, it is common to directly rectify the incoming power—no isolation and no step-down. Obviously this non-isolated portion must be kept from being touched since the 190-V peak can be hazardous. Then the rectified (and filtered) coarse supply is chopped by high-voltage transistors to produce a square wave in the 10s to 100s of kHz range. This signal can be simply filtered as shown in the picture, or it can be put through an isolation transformer. The transformer now can be much smaller and lighter—stepping down at 100 kHz requires far less material. If the supply is to be isolated, then the feedback must be isolated as well. This can be most easily done with an opto-isolator that transfers the *too-high/too-low* signal in digital form to the control circuitry. The feedback comparator and modulator reside on the secondary side while the transistor drive circuitry is on the primary side. The sorts of control techniques tried have varied you can use pulse-width variation or fixed-width, varying frequency pulses to drive the transistor(s) on.

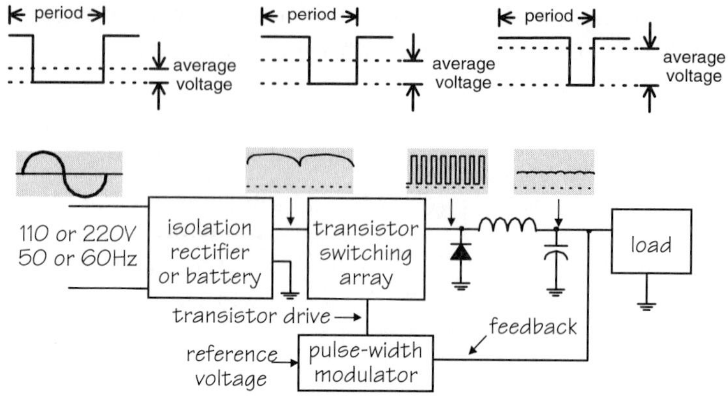


**Pulse-width Modulation**

In addition to entire power supplies, it is increasingly common to use switching techniques for regulation, even with the transformation being done at 50 or 60 Hz. In addition, the efficiency is particularly attractive for battery-powered devices where there is already a steady DC voltage, but the value needs to be changed. The regulators that drop the voltage (as with a 9-V battery supplying 5-V logic) are called ***buck regulators*** while the regulators that step the voltage up are called ***boost regulators*** (as with a microcontroller running off a pair of 1.5-V AA cells). Some configurations can do either as needed, so a 6-V battery supplying 5 V could be adjusted down to 5 V until the battery drops below 5 V and the regulator changes to raising the voltage until the last bit of energy is extracted from the battery.

A number of companies manufacture the ICs for such regulation and the specific configurations vary. The biggest problem, in my experience is obtaining the inductors when needed—it is not just the inductance that is needed but also the current capability for the desired frequency. Locating such devices is not always easy when you are building one unit.[76]

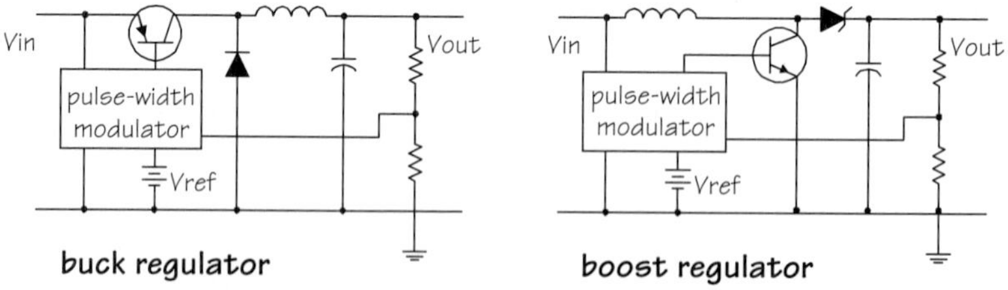


**Switching Regulator Configurations**

[76]Maxim, at least, has some low-power regulators that require no external inductors. I had a student whose humidity-sensitive fan control project included a direct-off-the-AC line regulator with no inductor that supplied the very-low power for the two or three chips involved. It had to be insulated from the user, of course.

As mentioned, there are two regulator configurations (sometimes supported by the same IC). The buck regulator dumps energy into the inductor by switching the input side to the incoming high voltage. When the incoming connection is opened, the diode clamps off the negative-going spike that occurs when the current through the inductor would be stopped. The output capacitor holds the charge that passes through the inductor—if you will, it filters the pulse train that comes through the inductor. The resistor divider serves as the sensor for feedback to control when the transistor is on or off.

The boost regulator uses the flyback principle to boost the voltage. By switching the inductor, output-to-ground current is started through it. When the transistor switch is opened, the output voltage flies up because of the inductance and therefore dumps charge forward through the Zener diode into the output capacitor. Again, the feedback controls the duty cycle and thus the output voltage.

**BE REALISTIC:** In picking a regulator for a power supply, weigh power efficiency against design complexity. A linear regulator for a low current or a wall-driven system may be much more practical despite the small amount of energy wasted.

## CONCLUSIONS

That completes the input device section. Any of the really complicated inputs comes with electronics to reduce the final result to digital, analog, or a serial stream. I would have liked to have more examples with very specific devices, but they change all the time and finding the exotic new devices is what applied engineering is all about.

# 5

# Outputs

## TTL AND MICROCONTROLLER CAPABILITIES

Since TTL *outputs* feed multiple TTL inputs down the line, they are only good at current *sinking.*[1] With *standard* TTL, the basic logic device *outputs* sink up to 16 mA for a fan-out of 10 (the basic *LS* logic devices sink 8 mA and the normal 8051 ports sink 1.6 mA).[2] The same standard TTL outputs can source only 250 µA (100 µA for LS devices and 60 µA for 8051 ports). Thus, if you must drive significant current, try to arrange your circuits to use current *sinking.*

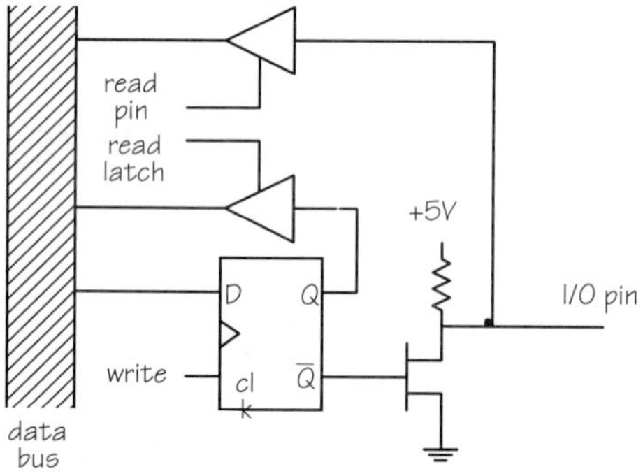


**Basic 8051 Internal Port Pin**

Internal ports of the 8051 are unusual in that they are only *partly* bi-directional (the term is *quasi*-bi-directional). You can see the hardware in

[1]Remember that you must *sink* about 1.6 mA for a logic 0 into a standard TTL device (0.4 mA for an LS device and 0.05 mA for a CMOS microcontroller) while you *source* almost no current for a logic 1.

[2]The specifications for the 8051 ports indicate that its ports can sink 1.6 mA with a voltage of no more than 0.45. It also indicates an absolute *maximum* of 15 mA for any one port pin, but makes no indication of the voltage at that point. It would seem that there is much more than 1.6 mA available, but tying a port high would surely burn it out.

the schematic for one pin here.[3] There is no special disabling or setting of port direction.[4] Also, when using P0 as an output port you must add a pull-up resistor.

The numbers for CMOS 8051 outputs are: <0.45 V when sinking 1.6 mA (3.2 mA for P0) and >2.4 V when sourcing 60 μA. Additionally, no one pin should sink more than 15 mA; an entire port should never sink more than 26 mA total; and the total sunk by all outputs should not exceed 67 mA. The implication is that you can sink more than 1.6 mA if you give up the 0.45-V limit.

**TTL OUTPUTS** *sink* current very well; they *source* poorly

## LEDs

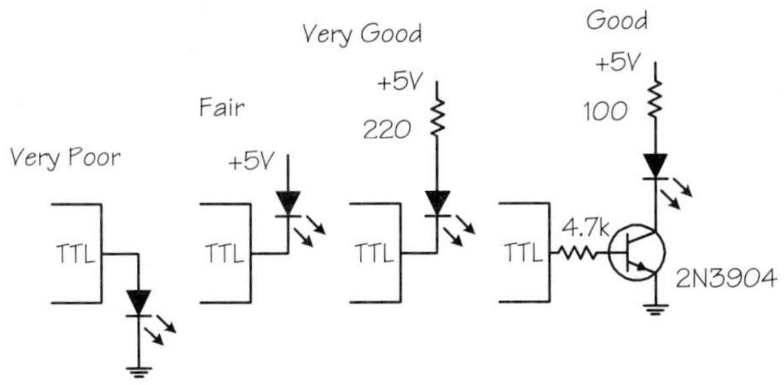


**LED Drive Circuits**

Light-emitting diodes (unlike silicon junctions with a forward voltage drop of about 0.7 V) have a forward voltage drop from 1.7 to 2.5 volts.[5] Most are rated for forward currents up to 20 mA except that they may be pulsed at higher current for a short time in a multiplexed arrangement.

[3]There are other functions for some of the port pins. Suffice it to say, if you are using a pin for any other functions you should not use it as a port.

[4]You can always use a port for output, but be sure to remember that an input must have a 1 output last written to it if it was ever used as an output (see start of Chapter 4).

[5]The lower voltages are for the ones with gallium arsenide phosphide—the IR ones and some of the green to yellow to red ones. There are also gallium phosphide LEDs that are red and drop about 2.2 V. I have never investigated the much newer blue ones or the very expensive laser diodes.

The first circuit on the previous page, *Very Poor,* is very common and is about the worst possible. The TTL output won't source above 1 mA at best and the LED will appear quite dim. Students use it because they insist on having a logic high turn on the LED.

The second circuit, *Fair*, relies on standard TTL gates going into current limiting at about 16 mA. It is tempting to rely on the current limiting to take effect before the LED fails, but you have a "war" between the TTL output trying to reach 0 V and the LED conducting heavily at about 2 V—the extra 3 V has to be dropped in the TTL circuitry or else the LED would burn out. That is marginal design—the current limiting is not specified to be consistent from device to device, and it causes high power dissipation in the TTL device. The former can cause variation in brightness and the latter can reduce the reliability of the product.

The third circuit, *Very Good*, adds a resistor to limit the current. Calculate the resistor by assuming about half the 5-V supply across it and no more than 20 mA through it. You may be able to sink 20 mA into standard TTL, but only 16 mA is guaranteed without the logic low rising above 0.9 V. If you assume a red gallium phosphide diode (2.2 V) and the TTL only pulling down to 0.9 V, you have 1.9 V across the resistor, so the 220-Ω resistor shown would limit current to 8.6 mA. On the other hand, if you assume a gallium arsenide phosphide diode (about 1.7 V) so the current would then be about 10.9 mA. You should never use a resistor below 100 Ω in any case.

If you are bothered by the negative logic or need more "on" current, *Good* is an option. The NPN transistor inverts the signal so the presence of a logic high turns on the LED. With standard and LS TTL families, the high-level output current is 400 μA at at least 2.7 V.[6] Assuming the transistor has a $\beta$ of at least 100, there is enough base drive for the transistor to sink 40 mA—more than enough for one LED.

The high-voltage open-collector devices (hex buffers or hex inverters) can sink more current. The standard family 7406, '07, '16, and '17 can all sink 40 mA. An alternative is to use one of the buffer/driver chips such as the 74LS28 (quad NOR), 'LS37 (quad NAND), or 'LS40 (dual NAND), which sink 24 mA. Also, there are 74128 (quad NOR) line drivers that sink

---

[6]Note that coming from a microcontroller port, you have basically no current (60 μA from a small internal pull-up or none at all), so you must add a significant external pull-up resistor—but not one so small that the port cannot pull the resistor low. Driving an LED direct high from a port is very poor and even going into an NPN is not a good idea.

48 mA and *source* 42 mA.[7] In these design choices the number of lights needed to drive comes into play. While transistors may be more readily available and interchangeable, they force you to assemble more components on a circuit board. Board real estate consumption with transistors might actually be less and discrete components might make PC trace routing easier, but the balance will tip quickly to driver chips if there are more than two or three LEDs to drive.

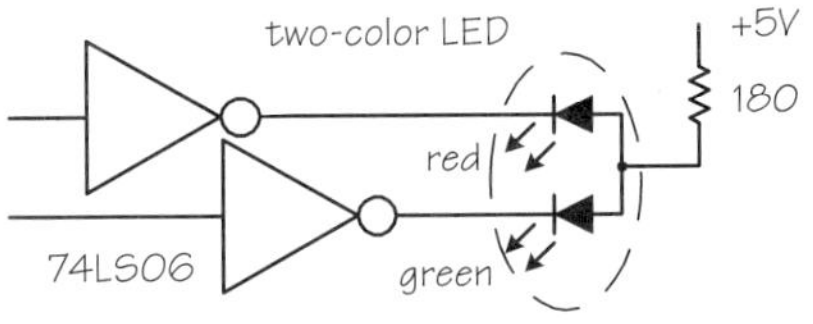


**Two-color LED**

A single-package two-color LED (actually two separate devices in one package) reduces the cost of mounting hardware.[8] Otherwise, two separate diodes might be less expensive. In the case, here the diodes came connected common-anode so, assuming only one is to be lit at a time, a single resistor can be used. The usefulness of this arrangement as a status indicator is that it can provide far more than two pieces of information. Obviously, red can indicate trouble and green can assure of proper operation, but by introducing flashing at different rates it is possible to indicate urgent trouble by flashing rapidly or a need for routine service by flashing slowly. Rapid or slow flashing of the green could indicate different stages of a process while assuring the observer that things are OK. Remember when you are setting out the details of a new design that there are many low-cost ways to communicate instead of requiring an alphanumeric display module.

---

[7]I should mention that there are newer chips that are designed specifically to drive LEDs, but most of them come into play with seven-segment displays and other more elaborate devices. Check availability and price before you commit a design to one.

[8]There are now multicolor diodes on the market.

## SEVEN-SEGMENT (NUMERIC) LEDs

Seven-segment LEDs are packages holding seven (or eight including a decimal point) LEDs arranged often with plastic light guides to have single emitters show as bars, making up a figure-eight. These LEDs are always connected together on one end–either common cathode or common anode.[9] While there are some circuits that drive the anodes, if you have only a few digits, the best choice is to use common anode LEDs and drive them with 7447 TTL Drivers. 7447s are 4-bit BCD to 7-segment encoders that sink 10- to 20-mA per segment and supply a low for the *on* segments. If you have two digits to drive, you can drive them individually with separate 7447s and an 8-bit port.

### Multiplexing

Since microcontroller ports are usually limited, *multiplexing* is a good option. For display multiplexing,[10] as shown next, the decoder/driver for the seven segments goes to all eight of the digits and the hardware is shared for the digits—each one gets the use of the decoder for one-eighth of the time. LEDs can be driven at much higher currents as long as the *on* time is short and is a small percentage of the total time. Ordinary 20-mA LEDs can perhaps handle over 160 mA as long as the on time is no more than 1 msec out of 10 msec. The LED will be much brighter during the "on" time, so the average (perceived) brightness comes out the same. You must drive the LED at least sixty times a second so the observer won't perceive flicker. For driving up to eight (or sixteen) digits from one port, you can tie a single 7447 to the cathodes of all the digits and tie a 1-in-8 (74138) (or a 1-in-16) decoder to PNP transistors that drive the individual digit selects (common anodes). As long as the transistors can handle the maximum sevenfold segment current (when all the segments are lit for an 8), the situation is only limited by the drive capability of the 7447s.

---

[9]Unfortunately, I never learned diodes that way, so I have to work backwards from the anode of a CRT being the most positive voltage to conclude that current flow can go from anode to cathode but is blocked in reverse meaning that the tip of the diode's arrow is the cathode—talk about a mental block!

[10]In communication, time multiplexing is the process of sharing a wire to send different signals. There are also means of frequency-division multiplexing, but I stray from the topic at hand.

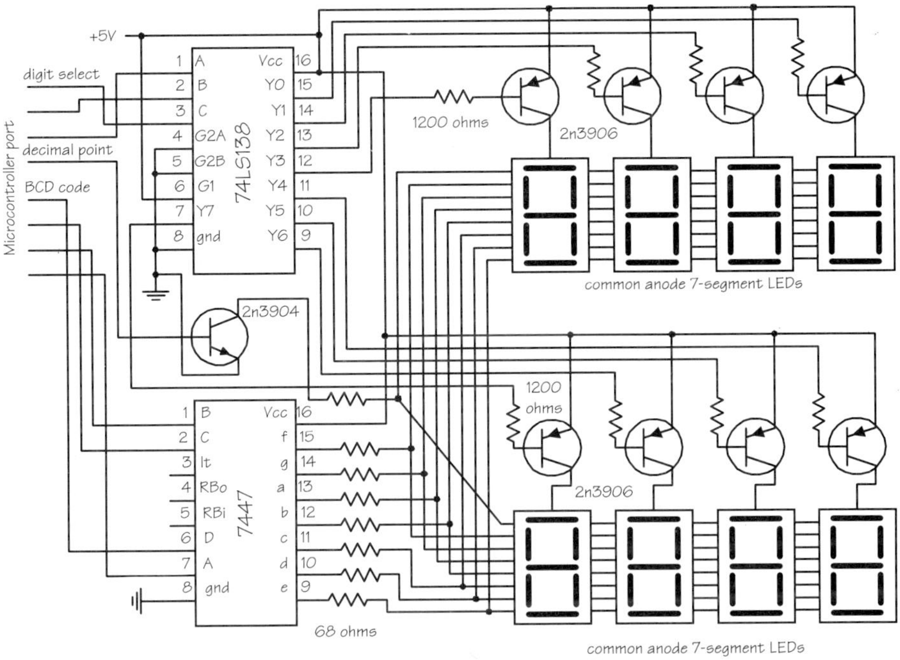


**8-digit Multiplexed 7-Segment LEDs**

For the example above, look at the partial schematic on the next page showing a single digit. First look at the digit to the left. The 7447's specification of 40 mA per line is reflected in the 68-Ω resistors to the segments [(5 V – 2.2 V)/40 mA = 70 Ω)]. The base resistors to the PNP transistors are sized to reflect 40 mA per segment or 280 mA maximum per digit (4.3 V/280 mA = 1536 Ω). Ideally you should have eight (or sixteen) times the 20 mA of current to each segment, but directly from a 7447 encoder chip that is impossible. In the example above, each digit is driven with only twice its steady-state current for one-eighth the time, so the perceived brightness would be about one-fourth of the maximum.

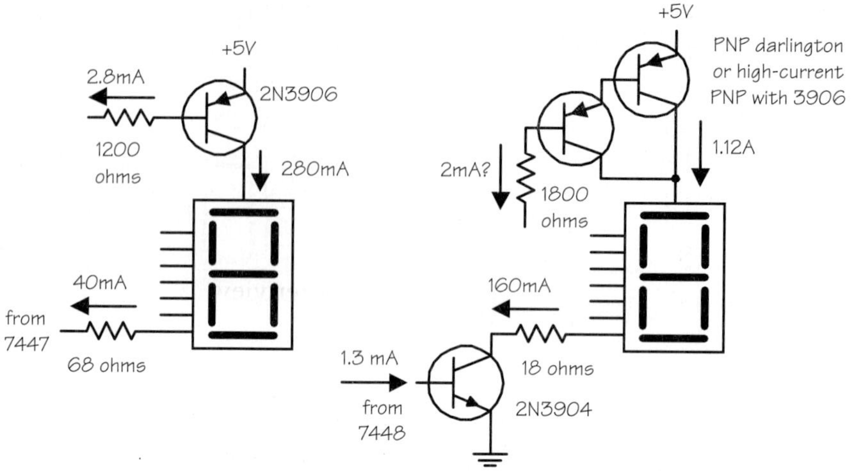


**Multiplexed Digit Driver Schemes**

Another option for common anode displays is shown above on the right where 7448's (positive-true) outputs go to NPN transistors that boost the drive to the 160 mA needed for maximum brightness with a one- in eight-duty cycle.[11] If the segment current is to be increased, the digit drivers must also be boosted to handle the very high current. The partial schematic assumes a Darlington pair with an overall β of 1000.

Since this problem isn't new, it is appropriate to point out that there are high-current drivers designed just for these applications. The real design decision you face is the trade-off of the number of parts against the cost of the devices. The designs shown here use very common parts. A careful check of what is readily available is in order, since there have been chips with great promise that have gone in and out of production in a few years.[12]

Having said all those discouraging words, I want to show the use of the MAX 7221 LED display driver chip. It is essentially all the previous circuitry in one 24-pin chip! The three-wire serial interface works well with the shift-register mode of the 8051 family. The serial port can transfer 8 bits

[11]The 7448s supply the reverse logic (supposedly for driving common cathode digits, but they cannot source the necessary current for LED displays). To drive common-cathode segments the best option may be to drive PNP transistors from a 7447.

[12]Some of my students who are much more proficient on the web have located 8-digit multiplexing chips from Maxim (introduced next) and four-character (alphanumeric) 5 × 7 LED arrays with drivers built in (from Siemiens). The latter seemed very inexpensive, while the former cost something less than $10 in small quantities, I believe.

of data to the external device in an automatic shifting mode, supplying the shift clock in synchronism with the data bits. If you want to cascade two chips, you have to make sure that the shifting includes 2 bytes before issuing the load command. You will need the data sheet from MAXIM to program the 7221 because there are a set of registers in each chip for the mode (it can drive decoded seven-segment data or any pattern of sixty-four LEDs), the brightness (sixteen levels), and the number of digits to scan (it can short-cycle for less than eight digits). Of course, there are also the registers for the individual digit data! Needless to say, this is better viewed as a programmable subsystem than as a simple driver chip.

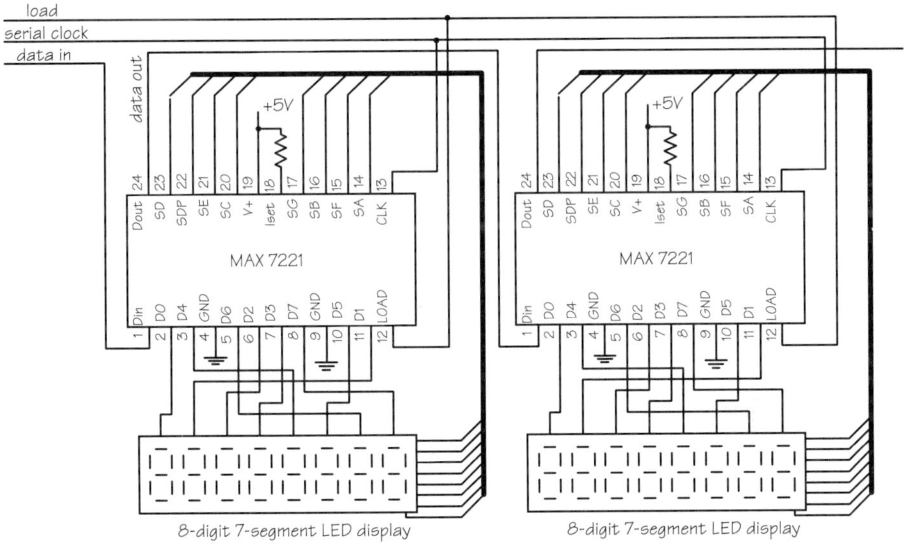


**Multi-digit LED Display Using Integrated Driver Chip**

In the area of integrated LED displays, watch for new developments—many are saving space and some may also be less expensive.

## NUMERIC LCDs

Liquid crystal displays (LCD) are made of glass plates with a liquid sandwiched in-between. Transparent electrodes are deposited on the inside of the glass plates and a voltage across the liquid changes its state. The excitation can switch it from reflective to nonreflective in the common ones where the display changes from gray to black. There are also displays that utilize light from behind that switch from transmissive to opaque. The voltage to switch is well under 5 V and the excitation is usually a square wave—often, I believe, with one electrode switching in phase with the other to keep the display *off* and out of phase to turn the display *on*.

The simplest displays use seven-segment digits for *large* several-digit numeric displays. The decoding is much like that for LEDs. I believe there are driver chips for such applications although it should be straightforward to drive a few digits from a microcontroller.

## ALPHANUMERIC LCDs

| Pin | Symbol | Level | Function |
|---|---|---|---|
| 1 | $V_{ss}$ | 0 V | ground |
| 2 | $V_{cc}$ | +5 V | Power Supply |
| 3 | $V_{ee}$ | | for LCD Drive |
| 4 | RS | H: Data I/O; L: Command I/O | Register Select Signal |
| 5 | R/W | H: Data Read—Module->MPU<br>L: Data Write—MPU->Module | Read Write Signal |
| 6 | ENA | H, H->L(No Pull-up Resistor) | Enable Signal |

| Pin | Symbol | Level | Pin | Symbol | Level |
|---|---|---|---|---|---|
| 7 | DB0 | H/L | 11 | DB4 | H/L |
| 8 | DB1 | H/L | 12 | DB5 | H/L |
| 9 | DB2 | H/L | 13 | DB6 | H/L |
| 10 | DB3 | H/L | 14 | DB7 | H/L |

**LCD Module Pinout**

I am ignorant of LCD *numeric* displays because alphanumeric ones are inexpensive and much more powerful. You can get small modules with 1 × 8, 1 × 16, 2 × 20, 4 × 20, 2 × 40, and 4 × 40 characters for prices from \$10 to \$80.[13] The characters are formed on the LCD screen by arrays of 5 × 7 dots.[14] Most of the *alphanumeric* LCD modules use the same interface

---

[13]You can get exceptionally good prices for modules from overstock/salvage/surplus dealers that make very good sense for one-of-a-kind projects.

[14]There are many more options with an array than with seven-segment or even fourteen-segment formats. The display drivers include RAM space for eight user-defined characters so you can have any pattern you want within those thirty-five dots. You can even generate a cumbersome graphing function but, if that is your goal, a full matrix display would be better.

chips and having the same basic command set. Since LCDs and keypads are a part of perhaps 50 percent of the student projects I see, I will go into more detail.[15]

The examples use a SEIKO M4032 with two rows of forty characters, but it is typical of virtually all the alphanumeric (non-graphical) ones. I show the software with *two* drivers—one with connections from *ports* of an 87C751 and the other with *memory-mapped* connections off the expansion bus.

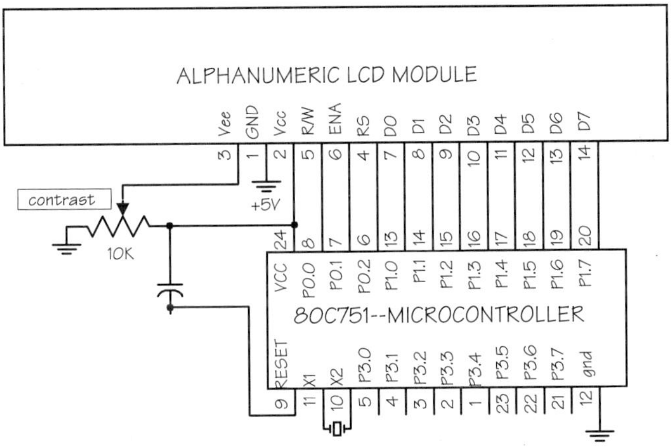


**Port-driven LCD Display Module**

The basic writing sequence is as follows:

1. Initially set the *ENA* line and the *RS low* (for commands) or *RS high* (for sending display characters). Assume *R/W* is wired high or left high by software.
2. Write *ENA high* while holding *RS* and *R/W* unchanged.
3. Send the desired data to the eight data lines (*DB0-DB7*).
4. Return *ENA low* while maintaining *ENA* and *R/W* unchanged.

Using I/O ports, you can do these four steps as separate output instructions with no delays between steps. This sequence may seem unnecessarily long, but the specification calls for *RS* and *R/W* to be present for 140 nsec *before* start of *ENA* and the data present must *lead* the *end* of *ENA* by 320 nsec.

[15]Some of this material is repeated form *Volume I,* pages 158–163 and pages 45–47.

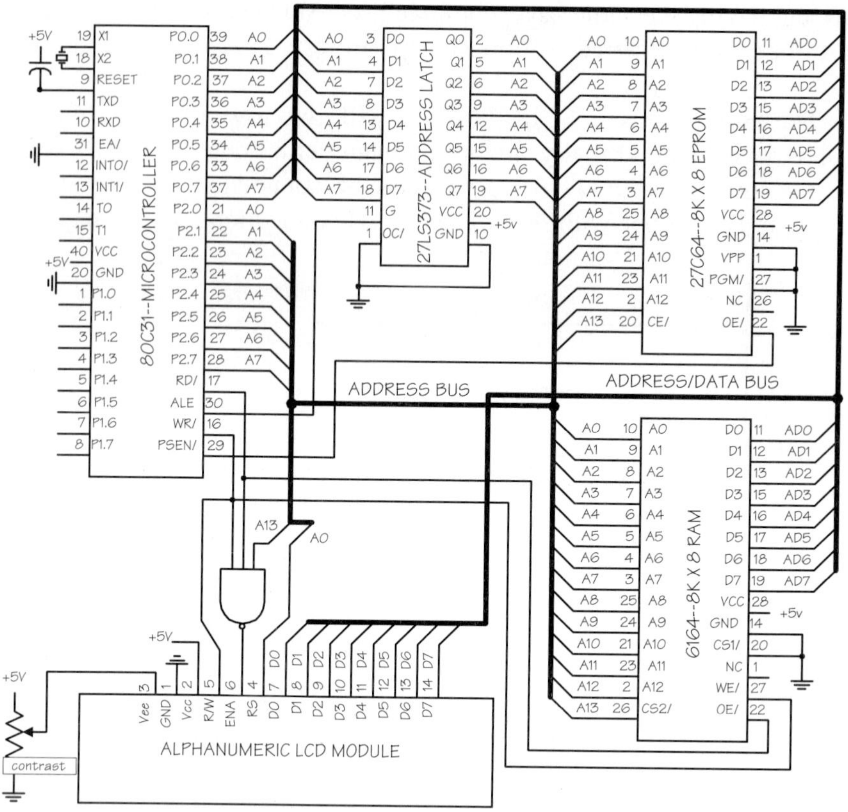


## LCD Module Directly Addressed on Expansion Bus

The ***direct*** connection of the LCD to the expansion bus is shown next. It presents an interface problem as seen on the timing diagram on the next page. The alphanumeric LCD modules require a 450 nsec write time, which *should* rule out interfacing them directly to the bus of a 12-MHz 8051. Decoding the *ENA* from the *RD* or *WR* of the microcontroller (and address decoding) produces a pulse technically too fast for the specifications of these display devices,[16] but many projects using these devices have wired them directly to an 8051 and had no problems over normal temperature ranges.

[16] A 12-MHz crystal produces a *RD/* or *WR/* of 400 nsec. You get 275 nsec at 16 MHz gives 275 nsec. Twenty-four MHz gives 150 nsec. If you use a 10-MHz crystal you have 500 nsec and the typical 11.059 MHz gives 443 nsec. Perhaps that is the key—since 11.059 is needed for 9600-bd serial communication, the speed is close—443 minimum versus 450 minimum for the display. I have never determined if the student success was because the manufacturers of the display drivers were conservative in their specifications, if manufacturers have improved chip performance without advertising, or if the specifications apply to an extreme temperature range that the student projects avoid. Needless to say, if you are designing for production, you will not let your job security hinge on unsupported assumptions!

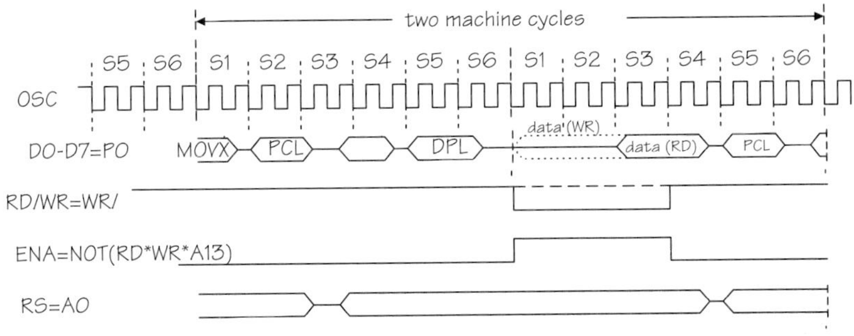


**Timing for Direct-connect LCD**

The module initialization sequence used in the program examples that follow is best understood from the next table taken from the data sheets. Among the LCD instructions are ones to do a status read-back from the module (the busy flag as well as the current cursor address) and to read back the data from the display RAM or the character-generator RAM. If you do *not* read back the status, it is wise to allow at least 40 μsec after most commands and display writes to be sure the instruction will be accepted. The display *may* be ready sooner, but without polling the status (busy flag) you cannot be sure. Unless you poll the busy flag, an arbitrary long delay is the easiest solution. (No *damage* will occur to the display by writing too quickly, but the results could be erratic.) Even with 100-μsec delays between writes, an entire forty-character display is filled in 8 msec, so the writing speed is not slow to an observer. Two of the commands—cursor home and clear display—take longer and should have 1.7 msec to execute as shown in the last column of the next table.

**LCD Commands**

| Instruction | RS | RW | Code 7 6 5 4 3 2 1 0 | Description | Execution1 Time(max) |
|---|---|---|---|---|---|
| Clear Display | 0 | 0 | 0 0 0 0 0 0 0 1 | Clear display; return cursor to home position (Address 0) | 1.64 msec |
| Cursor At Home | 0 | 0 | 0 0 0 0 0 0 1 * | Return cursor to home position (Address 0); return display to original position (no change to DDRAM) | 1.64 msec |

| | | | | | |
|---|---|---|---|---|---|
| Entry Mode Set | 0 | 0 | 0 0 0 0 0 1 I/D S | Set cursor move direction; specify whether to shift display (during data write & read) | 40 μsec |
| Display On/Off Control | 0 | 0 | 0 0 0 0 1 D C B | Set ON/OFF of: all display(D) cursor ON/OFF (C), blink cursor (B) | 40 μsec |
| Cursor/ Display Shift[1] | 0 | 0 | 0 0 0 1 S/C R/L * * | Move cursor; shift display without changing DDRAM contents | 40 μsec |
| Function Set | 0 | 0 | 0 0 1 DL N F * * | Set interface data length(DL) # display lines(L) character font(F). | 40 μsec |
| CGRAM Address[2] | 0 | 0 | 0 1 CG Address (6 bits) | Set CGRAM address, CGRAM data is sent or received after this setting | 40 μsec |
| DDRAM Address[3] | 0 | 0 | 1 DDRAM Address (7 bits) | Set DDRAM address, DDRAM data is sent or received after this setting | 40 μsec |
| Busy Flag/ Address Read | 0 | 1 | BF Address counter[4] | Read Busy flag(BF), read address counter contents | 0 μsec |

[1]Assuming internal (to module) clock, fcp is 250 kHz. When frequency changes, execution time also changes. A useful mode, not used in the example, is to have the display shift left automatically as you write characters. This gives a scrolling effect that is quite useful for showing long messages on a short display. Position the cursor at the right edge of the display and as each character is written, all the previous characters will move one space to the left until they *fall off* the left edge. Be careful to check the relationship between the multiple rows since they may all move and the one row will probably not flow into the next row. (For $1 \times 16$ displays the *one* row of sixteen characters is actually *two* blocks of eight *separated* by unused addresses!) Row two begins at address 64 ($40_{16}$). If selective erasing is desired, remember that a blank is $20_{16}$—not 0. With the addressable cursor you can have a static display and update individual numbers if you wish.

[2]The character-generator RAM allows up to eight characters of your own (codes 0 through 7). Write 8 bytes in succession beginning at CG address 0, 8, 16 . . . for your own patterns of $5 \times 7$ dots. Address 0, for example holds the top 5-bit row of character 0, address 1 holds the next row down, etc. Once you have filled up 8 bytes of CG RAM, display that character by sending it's code (0–7) to the display just like you display an "A" by sending a 41h. This is useful if you want non-ASCII symbols such as the up and down arrow (see example in Chapter 7).

[3]The cursor can be moved directly. For the $2 \times 20$ LCD, row two begins at address $64_{10}$ ($40_{16}$). If you want selective erasing, remember that a blank is $20_{16}$—not 0. With the addressable cursor you can keep the display static and only update individual numbers.

[4]Address Counter is used for both DDRAM and CGRAM addressing—be sure it is set where desired before sending data.

[5]Although I have not reproduced the entire character set, it turns out that all the *displayable* ASCII codes map directly while the bottom eight *(control)* codes are replaced by the custom characters programmed by the user.

| | | | | | |
|---|---|---|---|---|---|
| CGRAM/ DDRAM Write | 1 | 0 | Written Data (8 bits)[5] | Write data into DDRAM or CGRAM | 40 μsec |
| CGRAM/ DDRAM Read | 1 | 1 | Read Data (8 bits) | Read data from DDRAM or CGRAM | 40 μsec |

I/D = 0: Decrement ; 1: Increment
S = 1: display shift with new char; 0: no shift
S/C = 0: Cursor moves; 1: Display shifts
R/L = 1: Shift to the right; 0: Shift to the left
DL (data length) = 1: 8-bit; 0: 4-bit

N(number of lines) = 1: 2 lines; 0: 1 line
F(format) = 1: 5 × 10 dots; 0: 5 × 7 dots
BF = 0: New Instruction OK; 1: Busy
CGRAM: Custom Character Storage RAM

Well-written software will isolate the interface in a driver routine so, if the hardware changes, there is only one place needing the software change. The C program below shows both a driver routine for *port* I/O interfacing and one for *memory-mapped* interface. The use of drivers here is perhaps unnecessary, but it indicates portable code where you just change the drivers when the hardware changes.

**Drivers for LCD Connected to 87C751 Ports**

```
/*put this in initialize()*/
sbit ena=0x81;
sbit rs=0x82;
sbit rd_wr=0x80;
rd_wr=1; /*all writes to LCD*/

void lcddatwr(uchar dbyte){
 rs=1;
 ena=1;
 P1=dbyte;
 ena=0;
}

void lcdcmdwr(uchar dbyte){
 rs=0;
 ena=1;
 P1=dbyte;
 ena=0;
}
```

**Drivers for Memory-mapped LCD**

```
/*connections of direct-driven LCD */
 uchar xdata *LCDcmd;
 uchar xdata *LCDdat;
[17] /*put next two lines in initialize()*/
 LCDcmd=0x2000;
 LCDdat=0x2001;

 void lcddatwr(uchar dbyte){
  *LCDdat=dbyte;
 }

 void lcdcmdwr(uchar dbyte){
  *LCDcmd=dbyte;
 }
```

**LCD Driver *Use* Program**

```
#include <reg51.h>
#define forever for(;;)
#define uchar unsigned char
#define uint unsigned
void lcddatwr(uchar b);
void lcdcmdwr(uchar b);
[18] void display(uchar startloc,uchar *s){
 lcdcmdwr(0x80 | startloc); msec(1);
 while(*s)lcddatwr(*s++); msec(1);
}

[19] void numdsp(uchar startloc, uint number,bit dpt){
 lcdcmdwr(0x80 | startloc); msec(1);
 uint m = 10000;
 bit ldgzero = 0;
 do{
  if((((number/m)>0)||(m == 1))||((m == 100)
                    &&(dpt == 1))) ldgzero = 1;
```

[17]Keil software has the absolute access macro *XBYTE* so the use of the pointers could be replaced by *XBYTE[0x2000]=dbyte* in the command function and by *XBYTE[0x2001}=dbyte* in the data function.

[18]This function looks simple despite its ability to handle messages from *code* or *xdata* because, by default, a generic (3-byte) pointer is used and the third byte notifies the function which memory space is to be used.

[19]This function is useful, but it is too complex to absorb quickly. It takes an unsigned integer and converts it to a decimal display on the LCD. It suppresses leading zeroes but maintains alignment by substituting blank spaces.

```
    if (ldgzero == 1) lcddatwr(number/m+'0');
    else lcddatwr(' '); msec(1);
    number %= m;
    if ((dpt == 1) && (m == 100)) {
     lcddatwr('.'); msec(1);
    }
[20]  }while((m /= 10)>0);
 }

 void msec(uint x){
  uchar j;
  while (x-- > 0){
   for (j=0;j<125;j++);
  }
 }

 void initialize(void){
  msec(15); /*set up LCD modes*/
  lcdcmdwr(0x30); msec(4);
  lcdcmdwr(0x30); msec(1);
  lcdcmdwr(0x30); msec(1);
  lcdcmdwr(0x38); msec(1); /*1/16 duty?*/
  lcdcmdwr(0x0e); msec(1); /* dsp on*/
  lcdcmdwr(0x01); msec(2);
  lcdcmdwr(0x06)I; msec(1);
 }

 void main(void){
  uint x=0;
  initialize();
[21]  display(0,"test");
[22]  display(64,"time");
  forever{
[23]   numdsp(69,x++,0);
   msec(1000);
  }
 }
```

[20]I do not know if you prefer to put the *while* on the next line—presumably the *do* will alert you to look for the *while*.

[21]This line sends top row display information.

[22]This moves to the second row of the 2 × 40 display.

[23]This sends the value of *x* for display. Since the count goes up every 1000 msec, it is a seconds count to the accuracy of the delay loop. Other methods are much better for obtaining accurate times, but the goal here is to show the display.

### Other Alphanumeric Displays

Less common are displays that utilize vacuum fluorescence (typically a blue-green color) and ones that use gas plasma (typically an orange "neon" color). Both technologies require much higher voltages to function but the configurations are similar to that of LCDs. If you want such displays but are going to work from the raw display level, you are into electronic design that is outside the scope of this book. I should add that some students recently located some very inexpensive LED modules from Siemens that interface much like LCDs with the same parallel interface and dot-matrix LEDs.

From a microcontroller interface perspective, the many modules that have been developed recently utilizing other display technologies mimic the pervasive LCD alphanumeric modules. You can switch display color and brightness by changing modules and little else, as testified by the project from Chapter 1 that switched to a vacuum fluorescent module. All that changed was the power and the price.

## GRAPHIC DISPLAYS

Graphic displays are ones that display both characters and anything else—graphs, pictures—whatever you can do with dots. Technically you can go on to gray-scale (not just on or off) and color displays as the entire PC field has done, but somewhere you have gone beyond reasonable microcontroller applications.[24]

### LCD Arrays

There are small dots-everywhere monochrome LCD displays that are a step up from the $5 \times 7$ dot or $5 \times 10$ dot alphanumeric ones. Some of the newer *fish locators*, for example, have an LCD matrix display that shows numeric information as well as a profile of the bottom. I have no details of the interface except to have heard that there are integrated drivers so the every bit on the module can be addressed as a bit in an array of bytes. Once the byte is written, I believe it continues to be refreshed by the driver chip(s) until overwritten.

---

[24]If I were faced with such a project today I would look closely at the low-end video cards remaining from bygone PC systems. In fact, rather than utilizing chip-level designs, I would seriously consider mimicking the PC's ISA bus and installing a commercial card. The ISA bus *maximum* timing is too fast for the standard 8051, but judicious control of wait-state signals could probably get the job done. If you really must have a video display with maximum performance, you have no business using an 8051! Send the data to a laptop instead.

Also in the "I don't know" category are the cast-off laptop computer displays. The oldest, most primitive ones are still high-class for a microcontroller. I suspect you would have a serious electronic design job to adapt one, but I have no knowledge of anyone doing it.

### CRT (Video) Drivers

As mentioned in an earlier footnote, CRT drivers are best bought from a PC application. You can't hope to match the volume cost reductions and development efforts that have gone into such cards. The enormous resources going into designing smarter faster devices makes a small, low-budget development effort difficult to justify.[25] Besides, good engineers are inherently lazy and avoid "reinventing the wheel." If complex displays are vital to your project, I encourage you to move to the PC environment where all sorts of powerful hardware and software are already available for a very low cost.[26]

## DC MOTORS

When you opt for a DC motor, the voltage rating is crucial to any interface. Some very small DC motors run off 5 V but even they probably require too much current to be driven directly from TTL or a microcontroller. Assume, instead, you have chosen a 12-V motor. Assume in the first circuit in the next figure you used only a *single* NPN driver. Virtually any transistor can handle 12 V and the main design issues are the $\beta$ and current. Suppose you have a small motor needing 200 mA. Even a 2N3904 is rated for that, but with a $\beta$ (current gain) of perhaps fifty at that current, the micro or TTL device must *source* 4 mA to the transistor. Reviewing the start of this chapter,

---

[25]The magic of backwards compatibility means that the huge new driver chips still support the simple, basic, write-to-a-bit modes. They even support black and white (non-gray) modes if you want them. The performance wars are fought in the realm of the shortcuts—if a new chip can generate complex shapes on its display with only terse commands from the host microprocessor, then the slow rate of communication need not slow down the display changing. Where a 20 × 64 character display on the original PC could be totally rewritten in a few commands, rotating and shifting a high-resolution, full-color display needs all the help it can get.

[26]I am reminded of a quote about "a dwarf seeing farther than a giant by standing on the giant's shoulders" (and several others take-offs about being trampled on by giants, etc.).

you can see that is impossible. Also, if the 2N3904 comes out of saturation (doesn't stay fully on) due to insufficient base current, the voltage drop across it and current through it will cause significant power dissipation and the transistor may overheat.

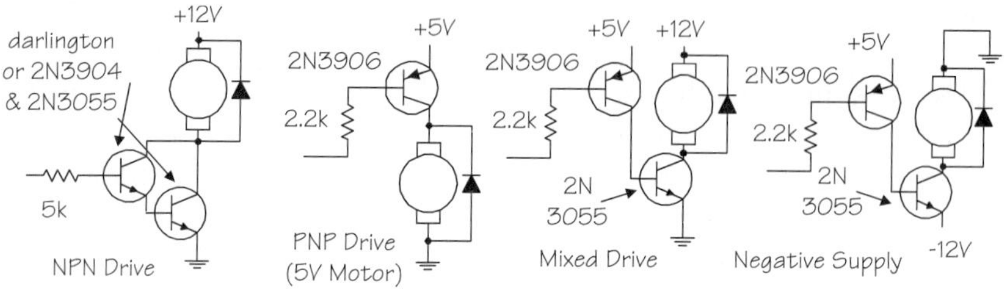


**One-direction DC Motor Drives**

One answer is a Darlington transistor. With a β of 1000, say, the TTL device need only source 0.2 mA into 1.4 V. That is quite feasible, but, of course, Darlingtons are not as readily available (at least order a few spares!).[27]

A much better alternative is to go first into a PNP transistor, but with a single device you can't use the second circuit with anything higher than 5 V. There are high-voltage open-collector TTL drivers you could put between the microcontroller port and the PNP transistor.[28] The last two circuits show drives with a power NPN driving the motor and a PNP doing the level shifting and current gain. With a port sinking at least 1.6 mA, assuming a β of 100 for the 2N3906 and a β of 20 for the 2N3055, the circuit could drive 3.2 A.

---

[27]Another very good option is a field-effect transistor—some of them can drive high current and, as FETs they take virtually no current. They may be a bit more expensive, but they reduce the circuitry dramatically. The challenge comes with driving bi-directionally with a pair—the drive on has to go higher than the power supply, hence the development of *high-side drivers*.

[28]The TTL 7406 or 7407 (hex inverters and hex buffers) are rated for 15 V and the 7416 and 7417 are rated for 30 V. Less attractive, you could put an NPN transistor in-between to level shift up, but you have the same problem as the first circuit with low source current.

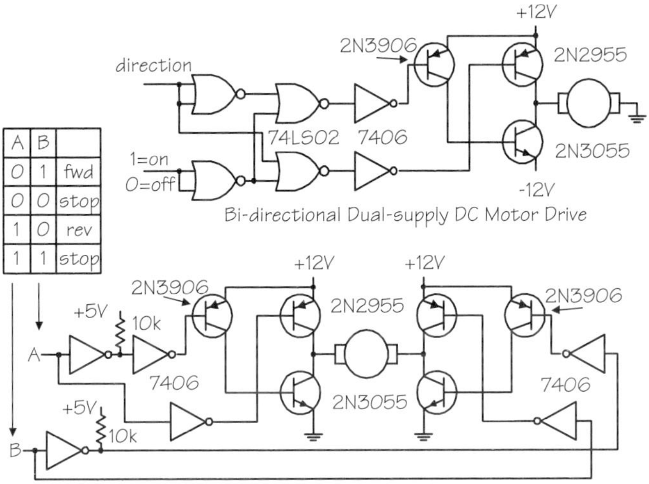

Bi-directional Dual-supply DC Motor Drive

Bi-directional Single-supply DC Motor Drive (H-bridge)

**Bi-directional Drives**

Often it is desired to drive a DC motor in *both* directions. There are a number of options. Of course you could use a two-pole relay to simply reverse the connections to the motor, but if you wish to stay all solid-state, you will either need a dual power supply or an ***H-bridge***. The first circuit above shows a dual supply drive. It is easiest to use a 7406 to do the level shifting and then the 2N3906 to shift down for the negative side. The 74LS02 is not strictly necessary, but it protects from turning on both transistors at the same time and gives you a separate direction input. You could go directly into the 7406 from the microcontroller if you could guarantee that it never put out a high on both lines at the same time.[29]

If you want to use a single supply you can use an H-bridge as shown in the second circuit. If both inputs are high or low, the net voltage across the motor is zero. If one input is high and the other low, the motor turns one way, while the reverse conditions cause it to turn the other way. The inverters at the front ensure not driving both transistors at the same time (except for a brief time of crossover—hopefully not long enough to damage toe transistors). They could have been 7404s, but this way the drive is all done by the single hex 7406 package.

---

[29]Sincc an inactive port would come up floating high, you could go to a 7407 (non-inverting buffer) so the floating high would deactivate both drivers. Then you would have to ensure never sending simultaneous lows.

There are *integrated* H-drivers with several-ampere ratings available from several manufacturers including Allegro (formerly Sprague).[30]

Depending on the repetition rate and the DC motor inductance, it is possible to vary the speed of a motor by varying the duty cycle of the drive. Such a signal can control speed with very little hardware—you can chop the on/off input or there may be a separate input provided on the integrated drivers—and then use of only one port pin for speed control. In many of the higher-performance 8051 family members, there are pulse-width modulated (PWM) outputs which will run settable duty cycle and repetition rates without the continual involvement of your software. PWM counters derive their separate *on* and *off* times from two internal special-function registers and may also have a pre-scaler to set the overall frequency. The hardware PWM functions can run at a higher frequency and with higher resolution than a strictly software-driven PWM output.

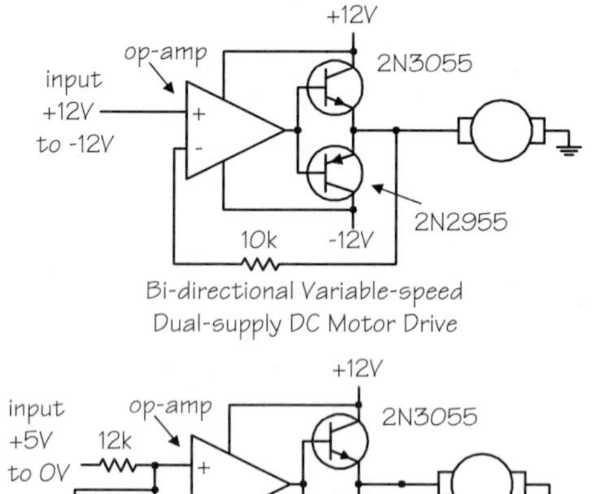


**Analog Variable-speed Drives**

If you want *variable speed* and have a D-A converter then you would swap the output transistors (NPN on top and PNP on the bottom—a push-pull configuration) with an op-amp and feedback from the output or even re-

[30]At one point, my students were plagued by a lack of availability in small quantities from distributors. Today that problem seems to have abated, but you ought to do all the cost analysis before you go either discrete or integrated. One may be more readily available and less expensive but the other is simpler, perhaps more flexible, and takes less board space.

place the whole thing with a *power* op-amp, but power op-amps can be expensive and hard to get.

The schematics on the previous page show analog drives with either a ±12 V input or a single-ended 5 V input (here mapped to a ±1 V across the 1Ω sense resistor or a range of ±1A to the motor).[31]

## RELAYS AND SOLENOIDS

From an output standpoint, relays and solenoids can be treated just the same as one-direction DC motors—they consist of inductive loads driven from the microcontroller. Your choices with relays relate to the contact current rating and the coil-driving voltage and current. Very small relays, called "reed relays" have a single glass-enclosed contact that is normally open and is closed by the field from a coil wound around the glass. These are not rated for high current (only a few 100 mA), but they are quite small and can be easily circuit-board mounted. There are also relays in dual inline (DIP) packages that can go into standard IC sockets and operate with 5-V activation. The big drawback of 5-V operation, though, is that they take perhaps 60 mA to activate—less efficient and too high a current to drive directly from ordinary TTL. If you have 12 V available, it is much better to use 12-V relays because the currents are more reasonable and the actual power consumption is less. Take your circuits from the figure on page 258. Consider driving the relay coil off a filtered, *unregulated* voltage if your coil drop-out specifications allow it. If your relay will be usually closed, consider a normally closed relay to save energy.

**Use 12-V relays:** Since 5-V relays require too much current to drive direct, if you have a 12-V supply, use it to drive lower-current 12-V relays or even run them off the unregulated supply for the 5V.

Solenoids behave much the same. You may want to investigate bi-directional and rotary solenoids for some applications. A very common ap-

[31]If you have had a full course in circuit design, you can compute the resistors to get the desired effect from the 0- to 5-V range. I am not precisely correct here but have chosen standard values as near as possible. If you used this circuit, you would have to either add a trim pot or a self-calibration stage to determine what D-A code gives zero current. Ideally 2.5 V or a code of $10000000_2$ would be a dead stop, but the tolerances and drift could make that value be a slight creep one way or the other. Hopefully, the error would be so small a current the motor would not overcome bearing friction.

plication of solenoids is for valves that control fluids or air. Pneumatic/hydraulic control is an entire field in the mechanical engineering domain. Most of these solenoids operate on 110-V AC, so they will be covered in the section on AC interfacing.

## STEPPER MOTOR

Common stepper motors have a permanent-magnet rotor and a pair of windings that produce multiple poles around the stator (not just a simple, single north-south). In most cases there are six magnetic poles giving twenty-four steps per revolution. The two windings are oriented to produce fields 15° from each other. By activating first one winding and then the other, first one way and then the opposite way, it is possible to produce fields at every 15° increment so that by the time the first winding is activated the same way again, the rotor has moved one-sixth of the way around 60°. Thus it takes six repeats of the four transitions to complete a full revolution. If only one winding is activated at a time it is called ***wave drive***.

While wave drive, just outlined, does move the rotor, there is an advantage of higher torque and faster possible speeds if you activate *two* windings at a time.[32] The basic sequence is *windings 1 and 2 forward,* then *winding 1 reverse with winding 2 forward,* then *windings 1 and 2 reverse,* then *winding 1 forward with winding 2 reverse,* and then back to the initial condition. This is called ***full stepping.*** It does require more power since current is driven through two windings at any one time.[33]

***Half-stepping*** is a third choice: a field orientation between the normal twenty-four positions is produced by alternately activating one and then two windings. The result is forty-eight possible positions for the motor on this peripheral board.

The table on the next page shows all the sequence options.

---

[32]A stepper motor is not damaged by stepping too fast—it just stops. The rotor can't keep up with the magnetic *detent* as it moves around, so it stops and *buzzes* as the field rapidly alternates between making it move forward and then backwards.

[33]Stepper-motors windings are usually driven all the time so the more current the more power. If you wanted, you could deactivate when the motor was stopped, but then you give up one of the main benefits—the high ***holding torque***.

| | bipolar | | unipolar | | | |
|---|---|---|---|---|---|---|
| | *winding 1* | *winding 2* | *winding 1A* | *winding 1B* | *winding 2A* | *winding 2B* |
| wave | fwd | off | on | off | off | off |
| | off | fwd | off | off | on | off |
| | rev | off | off | on | off | off |
| | off | rev | off | off | off | on |
| full step | fwd | fwd | on | off | on | off |
| | rev | fwd | off | on | on | off |
| | rev | rev | off | on | off | on |
| | fwd | rev | on | off | off | on |
| half-step | fwd | fwd | on | off | on | off |
| | off | fwd | off | off | on | off |
| | rev | fwd | off | on | on | off |
| | rev | off | off | on | off | off |
| | rev | rev | off | on | off | on |
| | off | rev | off | off | off | on |
| | fwd | rev | on | off | off | on |
| | fwd | off | on | off | off | off |

**Stepper Motor Activation Sequences**

Incidentally, there is another drive method called ***microstepping*** where the windings are driven by varying (analog) currents to produce a net field *anywhere* in-between the normal poles desired.

In choosing a stepper motor, you should keep in mind that there are two common types of windings. *Bipolar* motors have only two windings (four wires) and must (at different times) have current driven both ways through the coils. Like reversible DC motors, reversing current requires either two supply voltages or an H-bridge (actually *two* H-bridges). While H bridges are even available in pairs in single chips, such drivers have been fairly costly. See the schematic on the next page.

Somewhat less efficient in utilization of copper, but much easier to interface, are *unipolar* motors. The two windings each have a center tap (six leads total) and current can flow from the center (at, say, the positive voltage) to one half or the other of the winding depending on which transistor is sinking the current.[34]

[34]Do not overlook smaller motors with gear reductions since many applications (stepper *or* DC motor) do not need *both* high speed *and* high torque. Many low-cost motors are available from electronic salvage houses if your project is not intended for volume production.

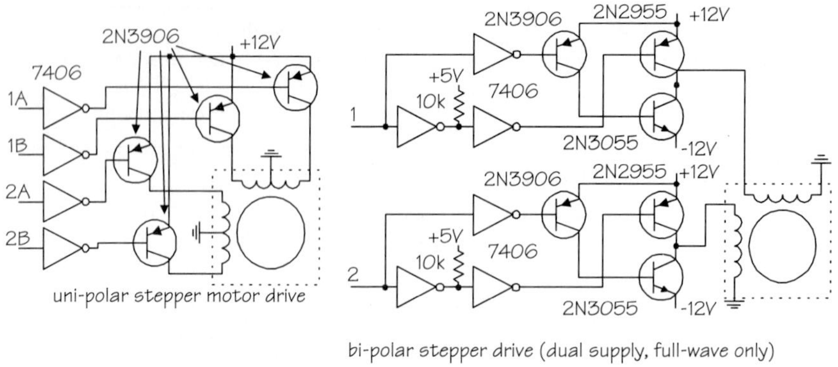


**Unipolar and Bipolar Drive Circuitry**

Stepper motors drive circuitry is much like that for DC motors except there are several windings. You can get a higher stepping rate by using a higher-than-rated voltage with current limiting resistors; because the higher voltage overcomes the winding inductance more quickly. For example, if you have a 12-V 60-Ω motor, you can safely drive it at 24 V if you add a 60-Ω resistor in series with the winding. Just remember that this approach consumes a lot of power—in the example, you need greater than 2.5-W resistors and you are using 5 extra watts for a full-wave drive. I recall a high-current drive system that you could have heated your lunch on! There are also switching-type current-sense drivers that can supply the very-high voltage needed during the step transition without wasting power during the holding.

With or without the quicker field changes due to higher voltages, stepper motors reach stepping rates faster than their specified starting rate if they are *ramped* to overcome rotor inertia (see later discussion).

If you can't tie up 4 bits of a port and commit the microcontroller to watching the windings, you might use a ***programmable logic array*** (*PAL, PLD,* or *GAL* depending on the manufacturer and various trademarks).[35] You can program it to generate the individual winding drive signals from *step* and *direction* inputs using a counter and decoding logic. The equations below describe registers configured as a 3-bit (divide by 8) counter and the

[35]The file here was for a 16R4 PAL (PALASM software) done many years ago. The PAL was a fusible-link device that was burned once. Now the same equations could configure a GAL that is re-programmable. Fortunately, the software tools still accept the old format information.

logic to convert those counter values to an appropriate sequence of motor coil activations to give steps in either direction.[36]

**PLD Equations**

```
TITLE                    EET STEPPER
PATTERN                  1-MMI
REVISION                 0-1
AUTHOR                   T. W. SCHULTZ
COMPANY                  PURDUE
DATE              APRIL 25,1986
CHIP              STEPPER PAL16R4

clk dir iw ix iy set e f gnd /ena z y d c b a x w vcc

EQUATIONS:
/c:=c*/set
/b:=b*c*/dir/set +b*/c*/dir/set +/b*c*dir*/set
                                   +/b*/c*/dir*/set
/a:=a*b*c*/dir*/set +a*/b*c*/dir*/set +/a*/b*c*/set
                    +/a*b*dir*/set +/a*/c*/dir*/set
/w=/e*b +/e*/f*c*/b+e*f*a +e*f*c*b +e*/f*/iw
/x=/e*/b +/e*/f*c*b +e*f*/a +e*f*c*b +e*/f*/ix
/y=/e*/f*/b +/e*/c*b +f*c*/b +e*f*/b*a +e*f*b*/a
                                          +e*/f*/iy
/z=/e*/f*/b +/e*/c*b +f*c*/b +e*f*/b*a +e*f*c*/b
                                   +e*f*b*a +e*/f*/iz
```

[36]Depending on two additional inputs (the *mode* switches), the chip is programmed to produce either a standard four-phase (full-step) drive (01), an eight-phase (half-step) drive (11), a drive activating only one winding at a time (wave drive) (00), or a condition where four inputs go directly to the outputs (10). In the last case, the user controls the drives in software and sends the phase information via 4 bits of a parallel port. An eight-phase counter requires the wave and full-step drives to cycle through the sequence twice for one full counter sequence.

| | half step | full step | wave drive | direct drive |
|---|---|---|---|---|
| *e f (switch)* | *1 1* | *0 1* | *0 0* | *1 0* |
| a b c (counter) | w x y z | w x y z | w x y z | w x y z |
| 0 0 0 | 1 0 1 0 | 1 0 1 0 | 1 0 0 0 | Iw Ix Iy Iz |
| 0 0 1 | 1 0 0 0 | 1 0 0 1 | 0 0 0 1 | Iw Ix Iy Iz |
| 0 1 0 | 1 0 0 1 | 0 1 0 1 | 0 1 0 0 | Iw Ix Iy Iz |
| 0 1 1 | 0 0 0 1 | 0 1 1 0 | 0 0 1 0 | Iw Ix Iy Iz |
| 1 0 0 | 0 1 0 1 | 1 0 1 0 | 1 0 0 0 | Iw Ix Iy Iz |
| 1 0 1 | 0 1 0 0 | 1 0 0 1 | 0 0 0 1 | Iw Ix Iy Iz |
| 1 1 0 | 0 1 1 0 | 0 1 0 1 | 0 1 0 0 | Iw Ix Iy Iz |
| 1 1 1 | 0 0 1 0 | 0 1 1 0 | 0 0 1 0 | Iw Ix Iy Iz |

**PLD Truth Table**

Several techniques are possible for more elaborate stepper control. A particular small stepper motor used in the labs at Purdue steps at a 250 steps/sec rate from a cold start, but it is possible to get up to at least 500 steps/sec by ***ramping***. Ramping is the technique of starting with a relatively long delay between the first and second steps and then gradually shortening the delay to speed up the motor. Depending on the inertia of the system, the number of steps and the reduction in delay will vary. If it becomes too short too quickly the rotor will stop (which doesn't hurt the motor at all). With the given system, the ramping can be quite quick—one or two slower steps are sufficient to get to top speed.

Although probably not significant for tiny motors, it is possible to get settled at a new position more quickly. Normally you will see a certain *rubbery* action as the disc stops due to the under-damped characteristics, but by issuing a *reverse* step some time after the last step and then reissuing a forward step some time after that the reverse field provides dynamic braking.[37]

The program example on the next page shows both a home routine (assuming a single-mark pulse at one point in a revolution) and a back-and-forth driving of the motor to illustrate the use of the driver software. A delay is necessary to set the time between steps (and thus the speed), but those techniques, be they timers, schedulers, or software loops, have been discussed elsewhere.

[37]The actual timing of this depends on the system and the information for calculating such timing is beyond the scope of this book. I suggest that you run empirical tests with the actual hardware, but realize that worst-case designs for large production runs will have to correct for effects of heating and variations between systems.

Both examples show a driver using a PLD and a driver going direct to four unipolar connections. To use the latter driver, simply substitute it for the *pstep* routine—everything else remains unchanged. Obviously, it is possible to combine routines and reduce the passing of parameters, but this illustrates techniques of modular programming that bury details in driver routines and build up to higher level routines that use the software drivers.

**Stepper-Motor-Drive Example**

```
 #include<reg51.h>
 #define uchar unsigned char
 #define cw 1
 #define ccw 0
 sbit P1dir=0x91;
 sbit P1stp=0x90;
 sbit P1mk=0x97;
 uchar pattern[]= {1,5,4,6,2,0xa,8,9};

[38]void msecs(uchar time);

[39]void dstep(uchar dir){ /*use for direct drive*/
 static uchar phase;
 if (dir) phase=++phase&7;
 else phase=--phase&7;
 P1=(pattern[phase]);
 P1=(pattern[phase]|0x80);
}
 void pstep(uchar dir){ /*use for driving PLD*/
[40] if (dir==ccw) P1dir=1;
  else P1dir=0;
  P1stp=1;    /*toggle lsb to take one step*/
  P1stp=0;
}
```

[38]You must supply some sort of time delay routine or else set up a scheduler as in Chapter 3. In the latter case, the step routine would be part of the interrupt routine with a flag to stop stepping when the motor was to stop.

[39]For direct drive, I assume the phase bits are tied to *P1.0* through *P1.3*. Then the direction is not an output but rather a choice of which way to move in the pattern table. Here it shows a half-step mode. Alternately the table could be reduced to four elements and the phase pointer anded with 3 instead of 7.

[40]*P1* bit 1 is here a direction-set bit if the drive is through a PLD.

```
void step (uchar dir,uchar speed,uchar steps){
 uchar i;
 for(i=0;i<steps;i++){
  pstep(dir); /*use dstep for direct drive*/
  msecs(speed);
 }
}

[41]void home (void){
 while (P1mk==0) step(cw,8,1);
}

void main (void){
 home();
 while (1){ /*or for(;;) */
[42] step(cw,8,48);
  step(ccw,6,48);
 }
}
```

---

**Ramping Step Alternative**[43]

```
void step (uchar dir,uchar speed,uchar steps){
 uchar ramp[]={10,8,6,5};
 uchar I, rmp;
 if ((steps>8)&&(speed<8)){ /*enough to ramp*/
  rmp=1; steps-=8;
  for(i=0;i<4;i++){
   pstep(dir);
   msecs(ramp[i]);
  }
 }
 while (steps>0){
  pstep(dir);
```

[41]The assumption is that the msb of *P1* is tied to a mark detector that goes high at one narrow part of the total revolution. Finding that point gives a position reference.

[42]These arbitrary calls make the motor move one way slowly and then the other way more quickly. For a twenty four-step motor, given a half-step interface, this would be one full turn each way.

[43]Use this longer step function if you want to have ramping up to speed. It only ramps if the requested speed is high enough and the total travel is long enough. You could probably make a better algorithm to ramp all the time. You surely should experiment with the rate of increase since the dynamics of the system determine what is possible.

```
  msecs(speed);
 }
 if (rmp){
  for(i=3;i<0xff;i–){
   pstep(dir);
   msecs(ramp[i]);
  }
 }
}
```

**Assembly Stepper-drive Example(s)**

```
VARS SEGMENT DATA
RSEG VARS
  PPTR: DS 1

CSEG AT 00H ;COLD START JUMP
  LJMP STRT

STEPPING SEGMENT CODE
RSEG STEPPING

STRT:;MAIN PROGRAM
 LCALL HOME
RX1:MOV R2,#1 ;ARBITRARY BACK & FORTH MOTION
 MOV R3,#8
 MOV R4,#48
 LCALL STEP
 MOV R2,#0
 MOV R3,#6
 MOV R4,#48
 LCALL STEP
 SJMP RX1

HSTEP:;HALF STEP DRIVER
 MOV A,PPTR
 CJNE R2,#0,H1
 INC A
 SJMP H2
H1:DEC A
H2:ANL A,#7
 MOV PPTR,A
 MOV DPTR,#HTBL
 MOVC A,@A+DPTR
 MOV P1,A
```

```
 RET
HTBL: DB 5,1,9,8,0AH,2,6,4

PSTEP:;PLD DRIVER
 CJNE R2,#0,PX1
 SETB P1.1
 SJMP PX2
PX1:CLR P1.1
PX2:SETB P1.0
 CLR P1.0
 RET

STEP:;COUNTS DOWN & ADDS DELAY
 CJNE R4,#0,S1
 RET
S1:ACALL PSTEP
 LCALL MSECS
 DJNZ R4,S1
 RET

HOME:;BRING TO MARK HIGH
 MOV R2,#0
 MOV R3,#8
M1:MOV R4,#1
 JB P1.7,M2
 ACALL STEP
 JMP M1
M2:RET

MSECS: RET;DUMMY DELAY SUBROUTINE
END
```

## SERVO-MOTOR

Historically, ***synchros, resolvers,*** and ***servos*** dominated much of the world of controls. Essentially, they are a combination of a fixed set of two separate windings (resolver) or three interconnected windings (synchro) and a rotating coil. By activating the rotating coil with AC, the three (synchro) or two (resolver) windings pick up a field that varies with the orientation of the drive coil. It works the same way a transformer works—magnetic coupling. If the picked-up signals are relayed to a similar device with a similarly-activated rotating coil, the stationary field causes the moving part to line up the same way as the sending unit. Thus a user can orient the coil on the sending unit and the receiving unit will follow. With amplifiers, a small effort at a sender can translate into a large force at the receiver and the user can control distant parts of airplanes, for example, with light pressures in the cockpit.

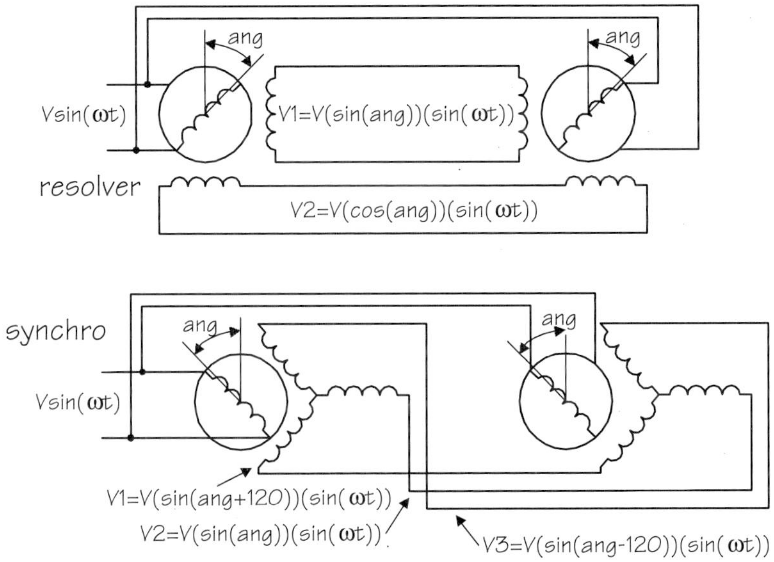


**Resolvers and Synchros**

These components, while designed to connect together with no electronic devices, can be used separately. You can measure the phase of the signal in from the two or three windings to get an input-angle system. Likewise, you can generate (in software or analog electronics) sinusoidal signals to drive a synchro/resolver. There are complex chips available to do both these functions, but this is an area in which I have no experience.

## AC DEVICES

AC motors have the advantage over DC motors in that they have no commutating brushes to wear out (or spark in explosive atmospheres). In addition, voltage shifting is easily accomplished by transformers.[44] Most high-

[44]The fight over AC versus DC power distribution was fought early in the century between Westinghouse and Edison.

power devices operate off AC, so interfacing is important for many industrial applications.

### Relays

These were discussed earlier, but most high-power ones operate off either 110 or 220 V—they can generally operate on AC or DC. Relays come with single-*normally open (NO)* contacts and with multiple combinations of *NO* and *normally closed (NC)* contacts. Except for the tiniest, lowest-power relays, the contacts are usually rated for at least an ampere at 110 V. Since many of the high-power relays require 110-V activation at significant current, from a microcontroller it is common to use a small DC-activated relay to switch the 110-V to the coil of the power relay.

The big drawback of mechanical relays is the deterioration of the electrical contact due to arcing as the contact opens—particularly with inductive loads which generate a high voltage as the current flow stops. Contacts are rated in terms of number of operations at full load—usually ten thousand to a million operations depending on the device and energy involved. If something is turned off and on once a day, that can be a life of over thirty years, but if it is a contact that cycles, say ten times a minute, that is a life of less than a day to three months! Now you see why certain applications call for solid-state switching using triacs and SCRs.

If there is something else that can control the current (perhaps if the relay is just reversing direction, but the on/off is by a separate switch), it is possible to extend the relay life by *dry-switching*—the contacts open only when the power is off so there is no arcing.

You may have encountered ***solid-state relays***. They are a combination of the circuits discussed next.

### Triacs and SCRs

If you are going to switch AC power frequently, the best choices are SCRs and triacs. They are devices that conduct by supplying a small gate current. Once they are conductive, no more gate current is needed. It is almost impossible to shut them off by anything you do at the gate—they shut off when the current through them drops to zero.[45] Because of this need to

---

[45]While SCRs and triacs are considered AC devices, they *can* switch DC if there is some way to bring the current back to zero. I believe there is a way to force them off by forcing a very high current in the gate, but I believe it is incompatible with the energy levels you derive from microcontrollers.

have no current to shut off, there can be up to an 8- to 10-msec delay before shutoff when the gate current is removed. In the same way, you can turn on a device somewhere in a cycle and then remove the gate drive and it will stay on. That is how *phase-angle firing* works to give a reduced brightness to an incandescent light bulb.

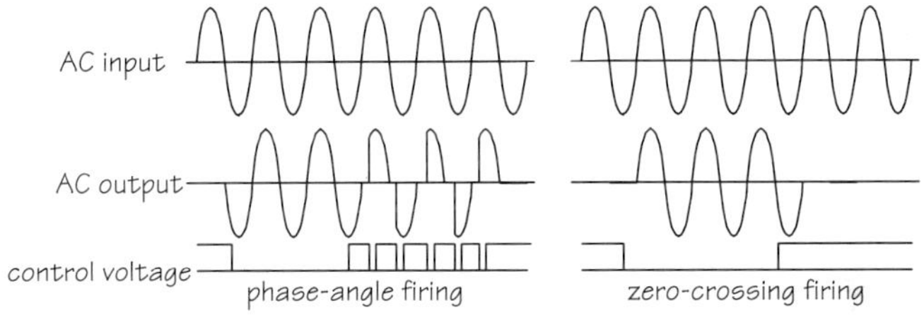


**Phase-angle and Zero-crossing Firing**

For most interfaces to 120-V AC from microcontrollers you should have isolation—either through a relay or an opto-isolator such as the MOC 3011 or 3030.[46] If you include an isolated driver, the triac is more expensive than a relay, but it should have a longer life.[47] Circuits for opto-isolated triac drivers are readily available in the manufacturer's data books.

For interface purposes, the input to the opto-isolator is the same as for an LED. They typically require from 10 to 30 mA of current to emit enough light internally to activate the triac driver on the output side. The MOC 3011 activates the output at once and can be used for phase-angle firing. The MOC 3030 includes a *zero-crossing detector* so the output is activated only when the input is driven and the output voltage is zero. The advantage of that is the elimination of in-rush current and the resulting high-frequency emissions.

[46]The exception is a circuit where the user has no contact with the circuitry, the power is derived directly from the AC line, and the cost of the electronics does not justify isolation. For example: a circuit that fits in an electric box with the sensors and the drive electronics all insulated from the user could be non-isolated.

[47]For safety reasons, with the hot-water circulating pump in Chapter 14 of *Volume I,* I decided to isolate rather than float the entire unit and drive the triac directly. Because the temperature sensor goes directly into the water pipe, improper polarization of the connection to 110 VAC could put high voltages in the water supply and water impurities could conduct them to a user.

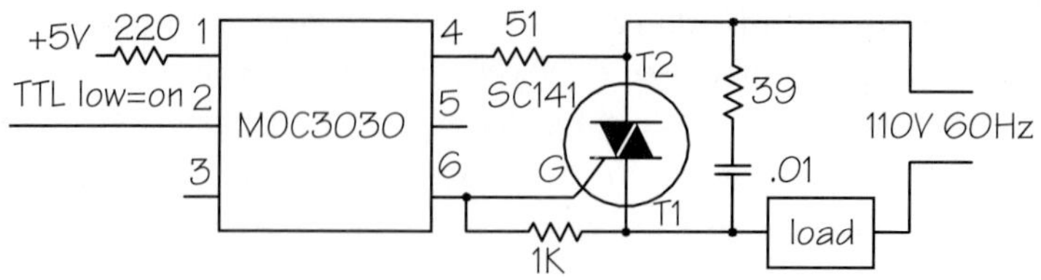


**Opto-isolated AC Drive**

I first tested the pump drive circuit of *Volume I,* Chapter 14 by driving a 50-W light bulb, grounding and opening the input to the opto-isolator driving the triac. Then I drove it with a simple program from the micro turning it on and off at a 1-sec rate. Only in the integration phase did I use an actual pump.

## AC POWER OUT (INVERTERS)

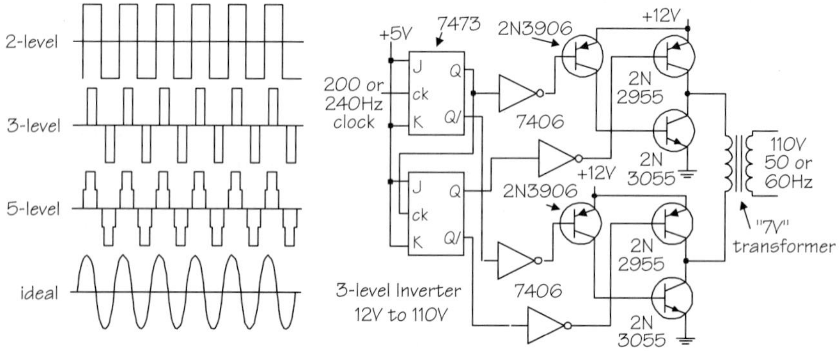


**AC Wave Synthesis**

Particularly for powering AC-operated devices when away from the normal power sources, it is useful to generate an equivalent power from batteries. This DC to AC conversion can be done with varying degrees of precision and smoothness. A simple square wave at 50 or 60 Hz can work for many devices, but it is more common to generate a three-level signal as shown in the schematic (on one way, on the other way, and off) to reduce the third harmonics. The circuit shown is a simple adaptation of the H-bridge shown earlier in this chapter with a pair of flip-flops to give the sequence of *fwd, off, rev, off.*[48] The oscillator should run at four times the desired frequency because there are four equal duration phases to the drive. With a transformer to boost the resulting 24-V peak-to-peak signal, most of the higher harmonics are filtered out. If the goal is 110-V RMS, since the 24-V peak-to-peak coming into the transformer represents about 7-V RMS, pick the latter as the "secondary" voltage of the transformer.

If you wish to come closer to an ideal sine wave, there are techniques using several more levels. The ultimate, not generally used because of the inefficiency, is to amplify a sine wave with a high-power, high-voltage amplifier. This requires very good heat sinks and high-power devices, but it is used particularly for test systems where it is desired to test products under low or high voltage conditions as well as varying frequencies. For efficient, low-cost systems the more coarse two- or three-level approximation is sufficient.

## D-A CONVERTERS

In order to get an analog voltage out from a digital computer, some sort of digital to analog (D-A) converter is needed. One type uses *a* ***resistor ladder*** network to add binary-weighted values from a reference (on the chip) to an output current. This can be converted to a voltage and scaled by an op amp. An 8-bit D-A converter could be scaled to put out 0 V when the input is all 0's and +10 V with all 1's input. Each bit change then produces a change of about 40 mV (10v/256 levels). For typical D-A converters the output voltage settles in about 1 µsec.[49] The final output is a function of the amplifier configuration. Some converters can run on only 5 V and produce a 0- to

---

[48]I suppose you could use four port pins on the micro to do this, but it seems a high price to pay unless you have extra pins. If it is important that the high voltage output not stop, it would make sense to use a separate oscillator for the 200 or 240 Hz. On the other hand, if frequency accuracy is important, the crystal oscillator of the micro is a very stable time base and toggling some port pins every 5 or 4.16 µsec is not a difficult programming job. Getting the odd time interval for producing 60 Hz might be a challenge if other things were required to run on 1-msec increments.

[49]This depends on the nature of the current-to-voltage op amp. The AD558, for example, has an internal amp to get this performance. Using the very inexpensive MC1408 with a 741 op amp gives a slew rate of 1V/µsec so a 10-V range could take 10 µsec to settle. The AD558 has an internal latch (transparently enabled on the lab board) which makes it suitable for direct interface to a microcomputer bus. It can also operate off only 5 V (0- to 2.56-V output range).

2.56-V output range, or with an offset of the amplifier, produce a ±10-V range. Remember that the input codes still go from $00000000_2$ to $11111111_2$—you do not get a sign-magnitude representation.

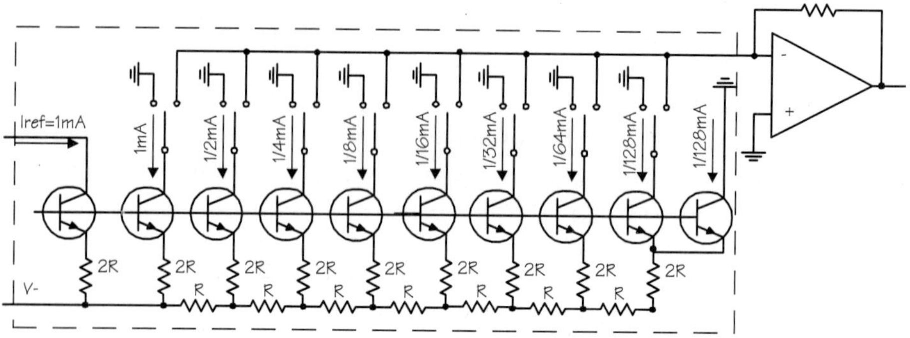


**R-2R Ladder Principles**

Aside from the resistive-ladder type of conversion, there are at least two other techniques, both of which involve a single-bit stream and an analog integration component. The first is *frequency-to-voltage* conversion where the output from the computer is a variable frequency pulse train. The higher frequency produces the higher voltage output. One configuration of the LM331 discussed in the last chapter puts out 10-V with a 10-kHz input, 1V with a 1-kHz input and progressively lower voltages with lower input frequencies.

If you only need slowly changing D-A outputs, the mere addition of a resistor and capacitor can deliver an analog voltage output by integrating the variable-width waveform. Add a simple resistor/capacitor filter attached to a pulse-width-modulated output. You may have PWM outputs directly in the microcontroller or you can use the real-time interrupt approach of Chapter 3. As the modulation approaches 100 percent, the voltage across the capacitor approaches 5 V. Turn the pulse on for almost no time and the capacitor voltage approaches zero. This is sloppy and will either be slow to change or will have pronounced ripple. Depending on the digital output, the pull high might not bring it to 5 V (add a pull-up resistor or an open-collector TTL device as a buffer).[50] You will have to trade ripple against maximum

[50]If you want the output voltage to be more accurate and you happen to have some A-D inputs to spare (there are often eight multiplexed inputs in A-D chips or on microcontrollers), you can use these extra A-Ds to monitor your "D-A" outputs and close the loop—making adjustments until the desired voltage is reached. In fact, if you run the A-D in a continuous mode with an interrupt, you can simply turn the output on or off depending on whether it is above or below the target value. Most A-Ds convert in 50 μsec or less, so you can update 2,000 times a second.

rate of change, but many applications have no need of the fast performance of a normal D-A. A crude approach may be most efficient for your particular application![51]

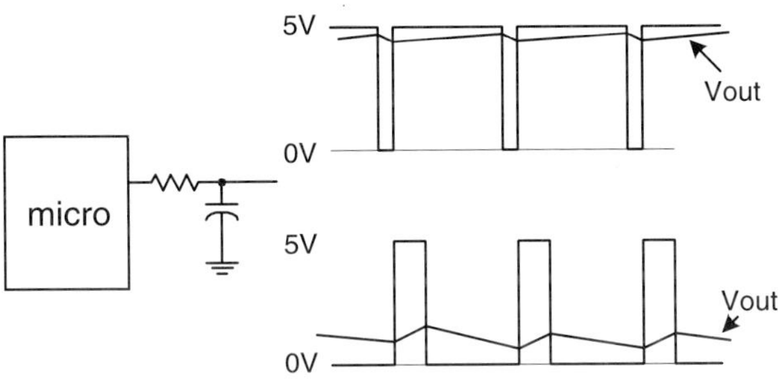


**D-A by PWM**

The second technique, commonly used for high-resolution audio output as in digital audio or CD players, is the reverse of *sigma delta* conversion. The analog output comes from a bit stream driving a one-bit converter feeding an integrator. A stream of 1's integrates to the highest output and a stream of 0's give the lowest output. An alternating stream of equal 1's and 0's would produce a mid-range value.

With both techniques, the rate of change is limited but there can be very real benefits to a system that is I/O-pin limited or aiming to minimize hardware costs. Perhaps the biggest drawback is the fact that a continuous signal is needed to maintain an output. Granted that is not too difficult with a real-time clock or separate timer/interrupt hardware, if high accuracy or rapid change is needed it certainly increases the processor load.

Since D-A is the output to the world in many cases, and since it is often desirable to output to many devices at the same time, there needs to be some way to hold the various voltages without tying up the controller. Obviously latching makes sense. The definition of "microprocessor compatible" is that a device has an internal latch and address decoding signals that fit with *WR/* and a *CS*. While it takes some extra hardware add ports, if the latch is internal to the D-A it can go directly on the bus.

But suppose you haven't expanded the bus for anything else or you have a 87C750 with no expansion capabilities. Do you have to give up the

[51]One application I envision is an analog display of a value. Suppose you have a system measuring power digitally, but the user wants a simple meter-type display of the result. Since humans can't absorb rapidly changing readings anyway, a PWM drive into a capacitor should be more than adequate (and perhaps less expensive than digital displays).

on-chip port pins to add A-D capability? Not necessarily. Two very interesting serial-interfaced devices are available from Maxim. On one device, the latch data can be shifted out in the shift-register mode while another device provides eight D-As all driven by an $I^2C$ compatible serial bus.[52] A large number of devices could be put on a bus that involved only two I/O pins of the micro. That is efficient!

## D-A Specifications

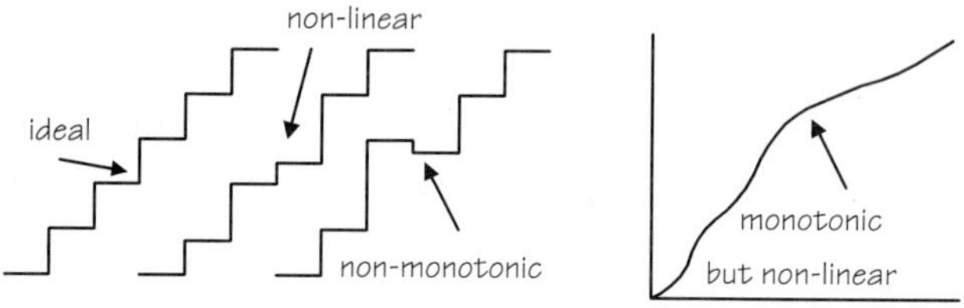


**Linearity and Monotonicity**

In specifying D-As, be aware of several terms. First, there is a difference between ***accuracy*** and ***resolution***. The former indicates how far away from the digital value the actual output voltage can be. If you get out 5.04 V when the code coming in is $10000000_2$ and full scale is 10 V, the *accuracy* is either 0.4 percent expressed as the percent of full scale or 0.8 percent expressed as percent of reading. Resolution is an expression of the number of bits involved—the number of steps or of discrete values that can be put out. As you can see from the previous diagram, there is a world of difference between resolution and accuracy.

An important feature in R-2R D-As is ***monotonicity***. As the code values increase, there comes a point where one resistor replaces the effect of all the lower ones. If the tolerance is poor, instead of going up, the output can go down. That is *non-monotonicity.* It may not matter if you are just outputting a single voltage, but if you are generating what is supposed to be a smoothly increasing wave, that is a problem.

***Settling time*** is another feature to specify. Specifications have become a sort of game where the way performance is expressed is carefully chosen to make the device look good. For example, with an external op amp, the current out of the D-A can settle quite quickly (well under 1 μsec) while an

---

[52]Serial modes are discussed in Chapter 11 of *Volume I* (pages 302–311). The MAX521 is priced at about $5 (1,000 piece).

inexpensive op amp will never keep up with that. Also, if you can configure for 2.5 V or 10 V, the performance at the latter will be slower. Make sure you compare devices in terms of the performance that matters to *your* application.

If your device includes a ***reference***, make sure you understand the stability involved. If the reference does not come with the device, if high temperature stability is important, you may have to add an expensive device.

## SWITCHED-CAPACITANCE FILTER

This area is not clearly output or input. If you are familiar with electrical theory, you know that a filter is a device to pass a range of frequencies.[53] If you fed a 1-V 1-kHz sine wave into a simple *high-pass* filter with a corner of 100 Hz, you would get out about a ½-V 1-kHz sine wave. A 1-V 100-kHz sine wave would come out almost unchanged while a 1-Hz wave would hardly come out at all. A *low-pass* filter with the same corner frequency would let the 1 Hz through almost unchanged while the 100-kHz would almost disappear. A band-pass filter would let through most of the 100-Hz sine wave but block the 1 Hz and 100-kHz. Most filters today are made up of op-amps, resistors, and capacitors. Generally the more stages, the sharper the rolloff.[54]

While most filters are fixed in frequency, a fascinating device that has been available for only a decade or two is the switched-capacitance filter. Overlooking the principles, the function of such a device is to provide a filter function based on charge transfer. Depending on the configuration you can have a high-pass, low-pass, or band-pass filter, but the attraction is that *the characteristics of the filter depend on the switching frequency*. If you supply a 10-kHz clock you may get a corner frequency of 300 Hz, but if you clock it at 100 kHz, you get a corner of 3 kHz. You have a device where you can dynamically change the filter characteristics. As will be mentioned next, if you want to produce a sine wave, you can filter a square wave (which is made up of, say, a 1-kHz fundamental and lesser amplitudes of 3 kHz, 5 kHz, 7 kHz and so on—the fundamental and the odd harmonics). If the rolloff is sharp enough,

---

[53]This is very basic information. There are many details that are not even hinted at here, and you could spend years in an electrical engineering program learning them.

[54]A simple R-C filter rolls off at 3 dB per decade, and is down 3 dB at the corner frequency *Decibels* are a measure of amplitude (actually power) and when referred to voltage –3 dB is 50 percent, –6 dB is 25 percent. A *decade* is a factor of ten, so a decade down from 100 Hz is 10 Hz. An *octave* is a factor of two as is used in music.

you can get a decent 1-kHz sine wave out. Then if you decide to make a 10 kHz sine wave, step up the square wave to 10-kHz and increase the clocking frequency to the filter to move the corner correspondingly.

## MUSIC

If you want fairly primitive tones—beeps, buzzes, or repetitive beeps—you can generate them quite easily directly from a port pin. If you get into more complex sounds, you *may* be able to handle the generation in the microcontroller, but you may find that it takes all its attention. Somewhere along the scale of complexity, it makes sense to add a chip specifically for that purpose.

### Direct Synthesis

First you can make a square wave just by toggling a port. If you want a 500-Hz tone you can toggle on every 1-msec interrupt. If you want a 1-kHz tone, you can speed up the interrupt to 500 μsec and make the rest of the scheduler activity divide between alternate interrupts. If you want a high pitch, say 5 kHz, you might go to a software delay loop in the main (background) program and have everything else happen during the interrupts. Unfortunately, the software loop will stop during the interrupt processing so you will not get a continuous tone! Perhaps you can turn the liability into a benefit and claim you intentionally chose the sound.

If you have a timer free and need a high-frequency sound, you can configure that timer to cause a *high-priority* interrupt and just toggle the bit in the interrupt as shown next. I revise my earlier opinion and say this is quite efficient—the total interrupt time is 5 μsec and it takes only 3 bytes of code plus the initialization of the SFRs. To generate a 10-kHz signal, the software eats up 10 percent of the processor time and never takes out more than 5 μsec at a time. No other software activity is needed because the timer is in the auto-reload mode. Any time you want to stop the sound just set *TR1* to 0 which stops the timer temporarily. If you want a pulsing sound, just use the regular scheduler interrupt to determine when to set or clear TR1. It is difficult to imagine something simpler.

---

**Assembly HF-Sound Generator**

```
SBIT P1SND=090H; USE P1.1 TO SPEAKER

CSEG AT 00H ;COLD START JUMP
 LJMP STRT
```

```
CSEG AT 1BH ;TIMER1 INTERRUPT
55CPL P1SND
  RETI

SOUND SEGMENT CODE
RSEG SOUND

STRT:;MAIN PROGRAM
 MOV TMOD,#20H ;T1-8-BIT RELOAD
 SETB PX1; T1 INTERRUPT HIGH PRIORITY
 MOV IE,#88H; UNMASK INTERRUPT
 MOV TH1,#50 ;50 USEC PER HALF CYCLE=10KHZ
 SETB TR1
M1:SJMP M1 ;ENDLESS LOOP
END
```

If you are using one of the 8051 family having PWM outputs, you can put out frequencies between 100 Hz and 20 to 30 kHz (depending on the crystal) with *no* software involvement or loss of processor time.[56] It is probably not sufficient grounds in itself to switch processors, but if you already have the capability available, use it.

In all of this square-wave output, there is one problem in driving a loudspeaker—the signal is offset around 2.5 V and when shut off does not hold at 2.5 V. Assuming you have an 8-Ω speaker and are driving with a ±2.5-V square wave, you could never deliver more than 0.78 W into the speaker. If it were a ±2.5-V sine wave because you filtered out the higher harmonics, you could only deliver 0.27 W. This is not a lot of power, but might well be sufficient for an alerting feature in a system. The circuits on the next page are best described as sloppy. The first one biases the speaker coil off center—check that it doesn't damage it. The second one is inefficient. The third one is perhaps best since it puts a smaller load on the port and drives the speaker low as well as high. All three come very close to the maximum current rating of the transistor(s) and the first two call for much more than the 1.6 mA specified rating on the port (but well less than the

[55]Since this interrupt routine occupies only 3 bytes, it can go directly at the vector address. The total routine takes 5 μsec in a 12-MHz system.

[56]A quick survey of Appendix 5 of *Volume I* suggests that PWM outputs are available on the x055, x167, x168, x267, x268, x552, x558, x562, x580, x598, x749, and x752. I have a vague sense that the family members with I²C bus interfaces might be able to put out a clock signal that would work, but it might be of little benefit.

15 mA *absolute* maximum). The 1-μF capacitor should be sufficient if you are producing a frequency of 1 kHz or above. These are not circuits for a home stereo system!

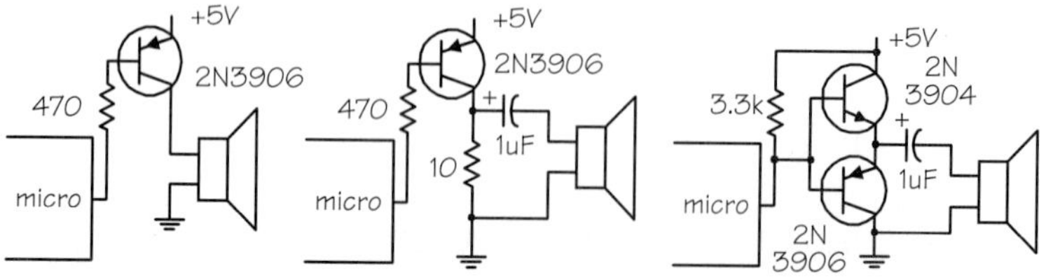


**Low-cost Sound Circuits**

## Programmable Sound Generator

If you need more elaborate sounds without continuous involvement of the microcontroller then I suggest a separate chip. Specifically, a very old chip (from the video game market), the AY-8910/8912 Programmable Sound Generator (PSG) can produce a variety of sounds under software control. It has sixteen internal registers that store data for the tone generator control, amplitude control, mixer control, and the envelope generator control.

Interfacing from the microcomputer can either be done with an 8-bit port and two additional control lines or directly from the expansion bus. The functions of the two control lines are as follows:

**PSG Control States**

| BDIR | BC1 | Function |
|---|---|---|
| 0 | 0 | Inactive |
| 0 | 1 | Read from PSG |
| 1 | 0 | Write to PSG (8-bit data goes to previously set register) |
| 1 | 1 | Latch register address (4 bits of data sets register pointer) |

Valid data to the PSG must be present by the time *BDIR* and *BC1* go to 0. Thus with port addressing the data or register number should be kept on the chip's data port at least until the control lines return to zero. In the same way, when reading *from* the PSG, the data should be captured before *BC1* is returned to zero. The program sequence to write *to* a register (driving control lines from port bits) is:[57]

[57]The chip can be interfaced directly to the external memory bus, as shown on the next page.

1. Put the register address ($00_{16}$ to $0F_{16}$) out the port.
2. Drive *BDIR* and *BC1* high
3. Drive *BDIR* and *BC1* low
4. Put the desired register contents out the port
5. Drive only *BDIR* high (*BC1* stays low)
6. Return *BDIR* low

The maximum allowable rates for these signals are much higher than you can achieve driving the control lines by port bits, so no delays need be put between these steps. If the *same* register is being written successively with different values (say a tone period or an amplitude slide), it is possible to use only steps 4 to 6). In general, it may be easier to just use a function that writes both address and data each time (or make an intelligent function that remembers the most recently addressed register and automatically omits the step).

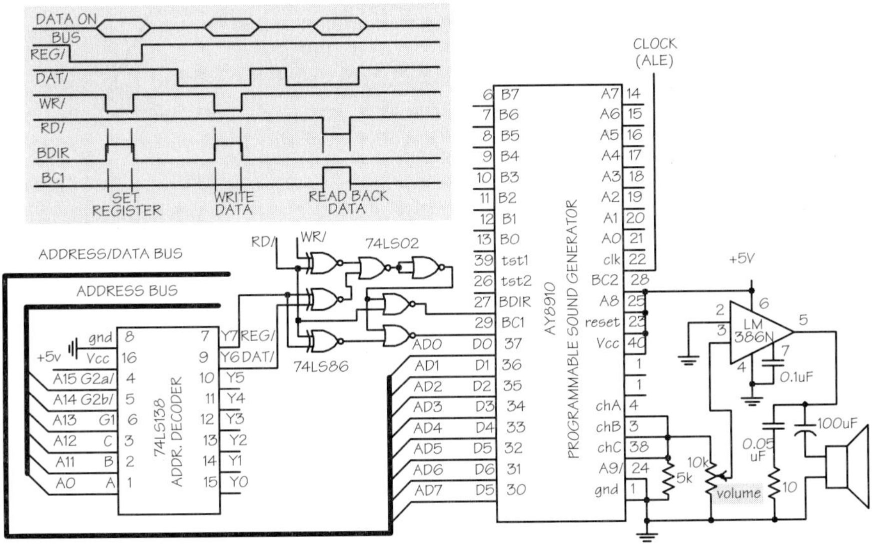


**Bus-Interfaced PSG**

Attaching the PSG directly to the memory expansion bus is not as simple as I expected. The signals are not directly compatible with the 8051's normal operation and a significant amount of logic is necessary. I dusted off my K-mapping and came up with a design above using two logic chips, but you might find it worthwhile to shift to a PLD since it involves only four inputs and two outputs. As usual, you face the board space versus parts cost dilemma—I'm sure the two standard SSI chips cost less than a PLD, but if you are already stocking glue logic for other uses or have additional logic to include in the single chip, the balance might shift. If you are a hobbyist with no PLD programmer and only stock TTL chips, building one unit, the balance shifts the other way!

Incidentally, with the decoding shown above, the register select address is $3801_{16}$ and the data (read or write) address is $3800_{16}$.

The PSG chip requires a clock signal. The maximum clock frequency into the chip is about 2 MHz.[58] With a 12-MHz (or less) crystal you can use *ALE* for the clock.[59] All frequencies and periods are scaled directly by the clock frequency so, if you use a different clock frequency, be sure to change the register contents proportionally to maintain the desired pitch.

The chip's actual output signal comes from three separate D-A converters (which are usually summed together in the following amplification stages). The quality of the sound depends on the speaker but programmable filters should be used as well to cut out the higher harmonics that result from the discrete time segments of the PSG outputs.[60]

There are three ***tone*** channels, A, B, and C, in the PSG (see the table on the next page). The 12-bit clock divide factor for each channel (coarse and fine) must be loaded into two registers. The channels can have a frequency range of fclk / $000000000001_2$ (110 kHz) to fclk / $111111111111_2$ (27 Hz).[61]

A single ***noise*** source, different from the tone sources, cycles through a pseudo-random noise sequence at a rate determined by a 5-bit period control. You can adjust it to get percussion effects as well as sounds for explosions and gun shots. It is controlled by the lower 5 bits of register 6. There is

---

[58]I believe the times in the tables that come later are with a 2 MHz clock which suggests an internal pre-scaler of sixteen or eighteen.

[59]This is a slightly sloppy clock since ALE is omitted during any *MOVX* or *MOVC* instructions but otherwise runs at one-sixth the clock frequence. If you have a 16-MHz crystal you get about a 2.7 MHz clock which is too high—you will have to add a flip-flop to divide it by 2.

[60]The chip does not synthesize a sine wave as far as I know. Although the gain is variable, it is the gain of a *square* wave that is involved. Three switched capacitance filter (such as MF10s) configured as low-pass filters could be driven by three programmable clock generators (such as in an 8253) to shift the corner frequencies to track the PSG programming. By cutting off all but the fundamental on each of the three tone channels you would have three *sinusoids* to sum into a final output. While the software might seem complex at first, the divide on the counters would relate directly to the tone period sent to the PSG.

[61]For music applications, once you have the 12-bit code for a note in the lowest octave, the nearest approximation to the note in the next octave is the first code shifted right one bit position (which reduces the divide by a factor of two and doubles the frequency).

only a single noise source. The actual period of the pseudo-random sequence is sixteen times the 5-bit value, times the clock period.

**Tone and Noise Registers (R0-R5)**

| Tone Channel | Upper Nibble Register Address | Lower Byte Register Address |
|---|---|---|
| A | R1 (lower 4 bits) | R0 |
| B | R3 (lower 4 bits) | R2 |
| C | R5 (lower 4 bits) | R4 |
| noise | — | R6 (lower 5 bits) |

The mixer control is a single register that turns on and off the A, B, and C channels for both tone and noise (as well as setting the direction of the unused I/O ports). Register 7 is programmed as shown next.

**Mixer Control Bits (R7)**[62]

| bit 7 | bit 6 | bit 5 | bit 4 | bit 3 | bit 2 | bit 1 | bit 0 |
|---|---|---|---|---|---|---|---|
| I/O pin B | I/O pin A | C noise | B noise | A noise | C tone | B tone | A tone |

The actual amplitude of channels A through C is determined by R8 through RA.[63] The lower 5 bits determine the amplitude. Bits 0 through 3 determine the amplitude if bit 4 is a 0. Otherwise, a 1 in bit 4 causes the envelope generator to come into play.

**Amplitude Registers (R8-RA)**

| A-channel amplitude | B-channel amplitude | C-channel amplitude |
|---|---|---|
| R8 | R9 | RA |

In addition to RD controlling the shape of the envelope (amplitude over time) described next, the *rate* at which the change in envelope takes place is determined by the envelope period registers:

---

[62]Note that a channel must *also* have amplitude (R8-RA) = 0 to be shut off.

[63]The *manufacturer's* data sheet, if you should happen to obtain them, list register numbers in octal.

**Envelope Period Controls (RB, RC)[64]**

| high byte | low byte | Envelope Period |
|---|---|---|
| RC | RB | 256 * (RC, RB) * ($T_{clock}$) |

The envelope generator varies the overall amplitude of the sum of the three channels. The pattern of variation is determined by 4 bits of RD in a way that is best described by the envelope drawings.

**Envelope Shape Controls (RD)**

| B3 continue | B2 attack | B1 alternate | B0 hold | Envelope Shape |
|---|---|---|---|---|
| 0 | 0 | 0 | 0 |  |
| 1 | 0 | 1 | 1 |  |
| 0 | 1 | 0 | 0 |  |
| 1 | 1 | 0 | 0 |  |
| 1 | 0 | 0 | 0 |  |
| 1 | 1 | 0 | 1 |  |
| 1 | 0 | 0 | 1 |  |
| 1 | 1 | 1 | 0 |  |
| 1 | 0 | 1 | 0 |  |
| 1 | 1 | 1 | 1 |  |

Two remaining registers are the on-chip ports—Port A (*RE*) and Port B (RF) which can be addressed in the same way as the internal registers. They are not as convenient as normal memory-mapped ports because the register must be set first, but they come at no extra cost!

The results of programming are easiest to determine by empirical ("playing around") methods. For example the "whistling bomb" example involves a program loop which successively lengthens the tone period after short delays to imitate a drop in pitch as the bomb falls and then switches to a non-repeating envelope of noise to give the explosion sound. You can easily create ambulance and air raid sirens the same way. Also, use of the attack, hold, and decay envelopes can give a rough approximation to piano or

[64]Note that the larger the number the slower the tone amplitude will change (actually the period of the envelope is 256 times (RC, RB) times the clock period).

drum sounds. Since the microcontroller can easily outpace the rate of change of amplitude of any musical instrument, considerable variety of sounds should be possible including the varying overtones (3rd and 5th harmonics at reduced amplitude) for a violin vs. a sax. Again, empirical methods make a good place to start. The effects of outlined in the next few tables are adapted from the manufacturer's data sheets and assume a 2 MHz clock.

Some of the more common sounds require only the use of noise and envelope generator (or processor control of channel envelope if other channels are using the envelope generator). Examples of this, which can be seen below, are gunshot and explosion. In both cases pure noise is used with a decaying envelope. In the examples shown, the only changes are in the length of the envelope as modified by the coarse tune register and in the noise period. Note that a significantly louder explosion can be obtained by using all three channels operating with the same parameters.

**Gunshot Sound Setup**

| Register | Load Value | Explanation |
|---|---|---|
| R6 | 0F | Set noise period to mid-value. |
| R7 | 07 | Enable noise only on channels A, B, C. |
| R8 | 10 | |
| R9 | 10 | Select full amplitude range under direct control of Envelope Generator. |
| RA | 10 | |
| RC | 10 | Set envelope period to 0.586 sec. |
| RD | 00 | Select envelope "decay," one cycle only. |
| all other registers | 00 | |

**Explosion Sound Setup**

| Register | Load Value | Explanation |
|---|---|---|
| R6 | 00 | Set noise period to max. value. |
| R7 | 07 | Enable noise only on channels A, B, C. |
| R8 | 10 | |
| R9 | 10 | Select full amplitude range under direct control of Envelope Generator. |
| RA | 10 | |
| RC | 38 | Set envelope period to 2.05 sec. |
| RD | 00 | Select envelope "decay," one cycle only. |
| all other registers | 00 | |

### Wolf Whistle Setup

| Register | Load Value | Explanation |
|---|---|---|
| R6 | 01 | Set noise period to max. value. |
| R7 | 2E | Enable tone on channel A, noise on channel B. |
| R8 | 0F | Select maximum amplitude on channel A. |
| R9 | 09 | Select lower amplitude on channel B. |
| all other registers | 00 | |
| R0 | 40<br>20 | Sweep effect for channel-A tone period via a processor loop with approximately 12 ms wait time between each step from 40 H to 20 H (0.572ms/1748Hz to 0.286ms/3496Hz). (Wait approximately 150 msec before continuing.) |
| R0 | 40<br>30 | A processor loop with approximately 25 msec wait time between each step from 40 H to 30 H (0.572 msec/1748 Hz to 0.429 msec/2331 Hz). |
| R0 | 30<br>68 | A processor loop with approximately 6 ms wait time between each step from 30 H to 68 H (0.429 msec/2331 Hz to 0.930 ms/1075 Hz). |
| R8 | 00 | Turn off channels A and B to end effect. |
| R9 | 00 | |

### Race Car Effect

| Register | Load Value | Explanation |
|---|---|---|
| R3 | 0F | Set channel-B tone period to 34.33 msec (29 Hz). |
| R7 | 3C | Enable tones only on channels A and B. |
| R8 | 0F | Select maximum amplitude on channel A. |
| R9 | 0A | Select lower amplitude on channel B. |
| all other registers | 00 | |
| R1/R0 | 0B/00<br>04/00 | Sweep effect for channel-A tone period via a processor loop with approximately 3 msec wait time between each step from 0B00 to 0400 (25.17 msec/39.7 Hz to 9.15 msec/109.3 Hz). *Decrement R1/R0 as a whole number, e.g., start at 0B00, then 0AFF, then 0AFE, etc. |
| R1/R0 | 09/00<br>03/00 | A processor loop with approximately 6 msec wait time between each step from 0900 to 0300 (20.6 msec/ 48.5 Hz to 6.87 msec/145.6 Hz). |
| R1/R0 | 06/00<br>01/00 | A processor loop with approximately 6 msec wait time between each step from 0600 to 0100 (13.73 msec/ 72.8 Hz to 2.29 msec/436.7 Hz). |
| R8<br>R9 | 00<br>00 | Turn off channels A and B to end effect. |

Because of the independent architecture of the PSG, many rather complex effects are possible without burdening the processor. For example, the wolf whistle effect shows two channels in use to add constant breath hissing noise to the three concentrated frequency sweeps of the whistle. Once the noise is put on the channel, the processor only need be concerned with the frequency sweep operation.

The laser, whistling bomb, wolf whistle, and race car sounds all utilize frequency sweeping effects. In all cases, they involve increasing or decreasing the values in the tone period registers with variable start, end, and time between frequency changes. For example, the sweep speed of the laser is much more rapid than the high-gear accelerate in the race car, yet both use the same computer routine with differing parameters.

Other easily achievable results include "Doppler" and noise sweep effects. The sweeping of the noise clocking register (R6) produces a Doppler effect which seems well suited for "space war" type games.

**Laser Sound Effect**

| Register | Load Value | Explanation |
|---|---|---|
| R7 | 3E | Enable tone only on channel A only. |
| R8 | 0F | Select maximum amplitude on channel. |
| all other registers | 00 | |
| R0 | 30<br>70 | A sweep effect for channel-A tone period via a processor loop with approximately 3 msec wait time between each step from 30 to 70 (0.429 msec/2330 Hz to 1.0msec/1000 Hz). |
| R8 | 00 | Turn off channel A to end sound effect. |

**Whistling Bomb Effect**

| Register | Load Value | Explanation |
|---|---|---|
| R7 | 3E | Enable tone only on channel A only. |
| R8 | 0F | Select maximum amplitude on channel A |
| R0 | 30<br>C0 | Sweep effect for channel-A tone period via a processor loop with approximately 25 msec wait time between each step from 30 to C0 (0.429 msec/2330 Hz to 1.72msec/582 Hz). |
| all other registers | 00 | |

## SPEECH

There are two fundamentally different ways of producing speech output from a computer. First is ***natural synthesis***—the speech derives from a recording of actual speech by person. You approach speech by analyzing the pressure waves in air (sound) coming from a speaker, and head in the direction of reproducing acoustic wave forms of an actual speaker. With waveform compression techniques you can now efficiently reproduce virtually *any* sound.

Second is ***artificial synthesis***—there is no real person, rather tone generators adjust their combined output to model the effects of the vocal tract. Rather than analyzing sound, the approach begins by analyzing the way the sound is created. You start with the impulses of the vocal cords and the resonances of the vocal tract, and develop circuits that can be controlled the same way as the vocal tract. Artificial synthesis may or may not be less "natural" sounding, but it is usually much more flexible in producing new words that have not been processed before.[65]

### Artificial Synthesis

Humans make sounds when impulses from the vocal cords start the air in the throat, mouth, and nasal cavity resonating. The frequency of the resonances depends on the size and shape of the cavities much like blowing across the end of a bottle will produce a different frequency depending on how much water is left in the bottle. These resonances result in peaking of the sound energies at different frequencies. For vowels these frequencies of resonance are called ***formants***.[66] Bursts of air at, say, 100 Hz initiate the higher-frequency resonances that decay away between bursts.[67]

---

[65]Historically artificial synthesis has had much more attention until the last decade. It dates to the time when memory was expensive and when a small amount of analog circuitry was generally more efficient than a much larger amount of digital circuitry. Recently the explosion in compression circuitry related to telephony has changed the equations. Inexpensive stand-alone solid-state answering machine chips can store and play back *minutes* of recorded sound and other compression techniques relate to reducing the bit-rate for digital telephone transmission.

[66]You change the shape of your vocal cavity unconsciously as you speak. For example, say "aaaah" and then say "ooooh" and compare the shape of your mouth.

[67]Not all sounds involve the vocal cords. ***Unvoiced*** sounds involve no closing of the vocal cords, but instead produce sound by the random vibration of the air passing out the vocal tract. Sounds including this vibration are called ***fricatives*** because they involve *friction* of the air. They can be with vocal-cord vibration, voiced fricatives, or without, unvoiced fricatives.

The rate at which the vocal cords produce pulses does not alter the resonant frequencies of the vocal tract—it is not the changing of the frequency of the vocal cords that makes speech—actually it has very little to do with the communication of speech.

***Phonemes*** are the name given to the fundamental sounds of speech. In English, at least, there is not always a one-to-one correspondence between written letters and speech sounds as can be seen from the table below. There are about forty basic sounds (***phonemes***) in English (quite different ones are possible in other language groups!).[68]

**Spelling Irregularities**

| | Same sound represented by different letters | Different sounds represented by the same letters |
|---|---|---|
| Vowels | mEAt, pEte, pEOple, pennY | vEIn, forEIgn,dEIcer, gEIsha |
| Consonants | SHip, tenSIon, preCIous, naTIon | althouGH, GHastly, couGH, hiccouGH |

***Vowels*** are the sounds produced specifically by the resonance of the vocal tract. Vowels are usually produced with a relatively open vocal tract and the periodic sound source provided by the pulsing of the vocal cords. Vowel sounds are classified according to whether the front or back of the tongue is high or low, whether they are long or short, and whether the lips are rounded or unrounded. The next table shows the vowels classified in this way. In English, all rounded vowels are produced in or near the back of the mouth (*UW, UH, OW, AO, OR, AW*).[69]

---

[68]It turns out there are variations in pronunciation of a phoneme depending on what is around it (take the "o" in "robe" or "robber" for instance). This ***co-articulation*** reflects the way the vocal tract goes through the transitions for the various sounds and results in several sounds for each phoneme, called ***allophones***. There are about 128 allophones in English. For example *DH1* is the sound at the start of *this* or then while *DH2* is the sound at the end of *bathe*. Likewise *KK1* is used in *scream*, *KK2* is used in *speak*, and *KK3* is used in *crane*. The synthesizer chip discussed next is based on allophones. The distinction between allophones may be difficult to recognize without practice, but when developing code for the synthesizer chip, it may help if you try the different variations and listen for the one that "sounds better" than the others.

[69]Try saying the vowels yourself and noticing how your tongue is located to verify this. When it comes to synthesizing and recognizing vowels it is not important to match the phase of the harmonics of the formants. Apparently, the ear is not sensitive to phase. Rather, the critical features of vowels are the values of the formants. Some synthesizers use a digital oscillator and a digital filter whose coefficients simply determine the frequencies of the formants.

**Vowel Phonemes of English**

| | FRONT | CENTRAL | BACK |
|---|---|---|---|
| High | YR (clear)<br>IY (see)<br>IH* (sit) | UW# (to)<br>UH*# (book) | |
| Mid | EY (beige)<br>EH* (end)<br>XR (clear) | ER (fir)<br>AX* (succeed) | OW# (beau)<br>OY# (boy) |
| Low | AE* (hat) | AW# (out)<br>AY (sky)<br>AA* (hot)<br>AR (alarm) | AO*# (aught)<br>OR# (store) |

* = short vowels # = rounded vowels

The next table shows all the ***consonant*** phonemes of English. Creating an occlusion or constriction in the vocal tract that produces an aperiodic sound produces consonants. They differ from vowels in the degree of obstruction of the vocal tract. If the vocal cords are vibrating, then it is a *voiced* consonant. Otherwise the consonant is *unvoiced*.[70] In addition to the role of the vocal cords, sound can be influenced by the lips (*labial*), the lips and teeth (*labial-dental*), the tongue between the teeth (*inter-dental*), the tongue against the roof of the mouth just behind the teeth (*alveolar*), the tongue near the roof of the mouth (*palatal*), the tongue near the back part of the roof of the mouth (*velar*), and by air rushing by the nonvibrating vocal cords (*glottal*).[71]

[70]As an example of the difference, try saying (out loud!) the *FF* in *fat* and the *VV* in *volume* while holding a finger against your larynx. The vibration of the vocal cords should be obvious with the latter case.

[71]When it comes to the synthesis of consonants, phonetic synthesis relies on a noise source in combination with the oscillator and filter of the vowels. For recognition, also, it is possible to separate out consonants by the amount of high-frequency energy relative to the low frequency energy of vowels.

**Consonant Phonemes of English**

| | | LABIAL | LABIO-DENTAL | INTER-DENTAL | ALVEOLAR | PALATAL | VELAR GLOTTAL |
|---|---|---|---|---|---|---|---|
| stops | Voiceless | PP (pow) | | | TT (to) | | KK (sky) |
| | Voiced | BB (busy) | | | DD (do) | | GG (got) |
| fricatives: | Voiceless | WH (white) | FF (food) | TH (thin) | SS (vest) | SH (shin) | |
| | Voiced | | VV (vest) | DH (they) | ZZ (zoo) | ZH* (azure) | |
| affricates | Voiceless | | | | | CH (chin) | |
| | Voiced | | | | | JH (dodge) | |
| nasals | Voiced | MM (milk) | | | NN (thin) | | NG* (tang) |
| resonants | Voiced | WW (wool) | | | RR (rural) LL (lake) | YY (yes) | |

*These do not occur in word-initial position in English.
LABIAL: upper and lower lips touch or approximate
LABIO-DENTAL: upper teeth and lower lip touch
INTER-DENTAL: tongue between teeth
GLOTTAL: glottis (opening between vocal cords)
PALATAL: body of tongue approximates palate (roof of mouth)
VELAR: body of tongue touches velum (posterior portion of roof of mouth)
ALVEOLAR: tongue tip touches or approximates alveolar ridge (just behind upper teeth)

The ***plosives*** or stops complete blockage of the vocal tract for a brief time with the sudden release of pressure following.[72] Interesting examples of plosives on both ends of words are *pip, bib, tit, did, kick,* and *gig*. The differences in the way the sounds are made is most obvious when comparing *did* and *gig*; the tongue comes nowhere near the roof of the mouth for *gig*.

***Fricatives*** and *africates* involve friction of air along the vocal tract. This involves an aperiodic or noise waveform. The narrow passage for the *s*

[72]For recognition the abrupt transition in the waveform could be an identifying feature.

sound produces energy in a higher frequency band than the *sh* sound which constricts the vocal tract further back.[73]

The ***nasal*** sounds are similar to vowels except the resonance involves the nasal cavity rather than the oral cavity.[74] It is necessary to have air flow out the nose for proper production of nasals as anyone speaking with a cold can assert.

The liquids or ***glides*** are called resonants. They are also sometimes called ***laterals***, and include the *r, w, l,* and *y* sounds. The key feature is the change of the vocal cavity *during* the sound. An example is the position of the lips when you say the *w* in *watch*. Likewise, a rolling *r* involves the vibration of the tongue.

The ***aspirate*** (only the letter *h)* is in a class by itself in English. It involves air passing directly through the vocal tract with no hindrance except the narrow passage by the vocal cord area (the glottis). For that reason it is also called a *glottal* sound. There are also aspirated sounds in connection with other sounds as with the *p* in *pen*.

## Phonetic Synthesizer

Several different electronic systems have been available for phonetic speech synthesis. While the actual technique used to produce the sounds has varied, the principles for using the devices are quite similar. Phonetic or allophone speech synthesis gives the ability to synthesize an unlimited vocabulary at a very low bit rate. At least in theory, issuing the appropriate sequence of phonetic sounds and pauses can create *any* English word or phrase. Assuming that speech contains ten to twelve sounds per second, phonetic synthesis requires storing less than 100 bits per second of speech. This is dramatically less than other techniques, but the quality of the speech produced is significantly lower.[75]

---

[73]Again the recognition of these sounds would rely on the amount and distribution of high-frequency energy.

[74]Since the resonances are similar to the vowels, recognition has to rely on the characteristics of the transition between sounds as well as the characteristic during the sound.

[75]One of the features of speech lacking in the SPO256-AL2 synthesizer discussed next is the ability to add *intonation*. Human speech communicates a great deal of additional information in the *pitch* and *loudness* of the speech. Consider the small changes that make a statement into a question or an exclamation. Likewise, intonation can add sarcasm, express doubt of the truthfulness of the statement, or even negate the statement itself. Some of the Votrax synthesizers (not discussed) *do* have the ability to add intonation as additional bits of the phonetic coding.

The General Instrument SPO256-AL2 has coding for fifty-nine discrete speech sounds and five pauses stored in internal ROM. The SPO256 is actually a general-purpose *linear predictive code* (LPC) synthesizer.[76] The -AL2 version is nearly self-contained.[77] A simple amplifier chip and a few discrete low-pass filtering components are all that is required and it can be adequately driven with ports. In the parallel mode, if you ground the *SE* (strobe enable) pin, you put the chip in a mode where *any* activity on the data inputs will be latched in about 1 sec after any address line goes high. In that mode, the codes can be accepted without driving any line at all. Usually, however, a new allophone codes are taken in when the *ALD* (address load) line is pulsed low (driven low and then immediately high) after the allophone address (code) is put out. The *next* code can be sent almost immediately (the input buffer empties—indicated by *LRQ* low—as soon as the given allophone begins to be produced) or when the speech processor finishes (*SBY* is low until the current allophone production ends). The former method is slightly faster, but the delay in loading should not be noticeable.

For a memory-mapped interface, the *ALD* is to be derived from the WR/ signal and an address decoder with the data coming directly from the data bus. To avoid writing codes to the chip before the buffer is ready, it is necessary to add a tri-state buffer to read back the *LRQ* or *SBY* signals (here decoded at the same address so you read the address until LRQ is low or until SBY is high). SE must be tied high so bus activity does not cause the chip to automatically take in false data because of other activity on the data bus.

[76]For the -AL2 version, all the allophones were taken from spoken words and converted to LPC. In a sense, it is a mix of natural and artificial synthesis.

[77]For the -AL2 version, all the speech coding is stored in *internal* ROM, but for other versions an external ROM can be used which loads serially.

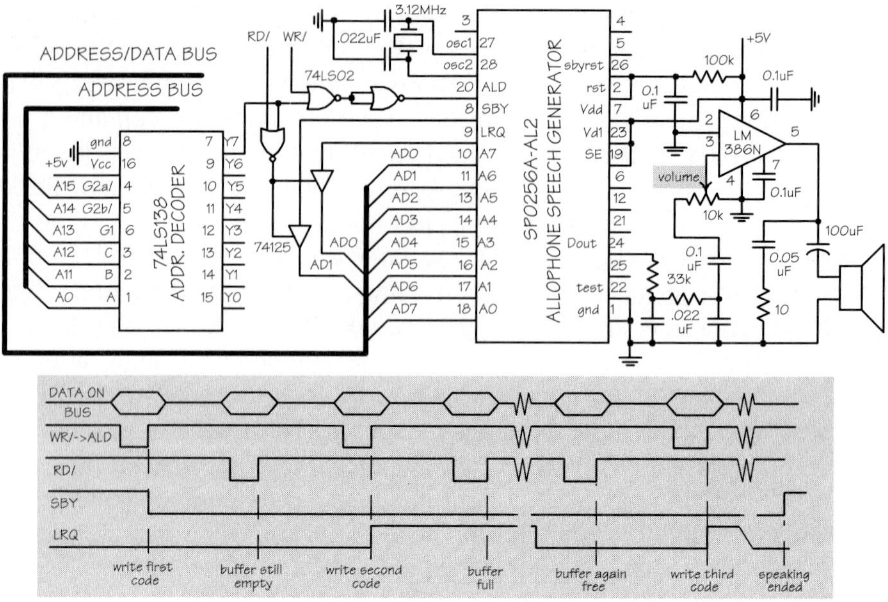


## Memory-mapped Allophone Chip

### SPO256-AL2 Allophone code List

| allophone | sample word | duration (mS) | hex code | allophone | sample word | duration (msec) | hex code |
|---|---|---|---|---|---|---|---|
| AA | hot | 100 | 18 | NG | anchor | 220 | 2C |
| AE | hat | 120 | 1A | NN1 | thin | 140 | 0B |
| AO | aught | 100 | 17 | NN2 | no | 190 | 38 |
| AR | alarm | 290 | 3B | OR | store | 330 | 3A |
| AW | out | 370 | 20 | OW | beau | 240 | 35 |
| AX | succeed | 70 | 0F | OY | boy | 420 | 05 |
| AY | sky | 260 | 06 | PA1 | *pause* | 10 | 00 |
| BB1 | business | 80 | 1C | PA2 | *pause* | 30 | 01 |
| BB2 | business | 50 | 3F | PA3 | *pause* | 50 | 02 |
| CH | church | 190 | 32 | PA4 | *pause* | 100 | 03 |
| DD1 | could | 70 | 15 | PA5 | *pause* | 200 | 04 |
| DD2 | do | 160 | 21 | PP | pow | 210 | 09 |
| DH1 | they | 290 | 12 | RR1 | rural | 170 | 0E |
| DH2 | they | 240 | 36 | RR2 | brain | 120 | 27 |

| | | | | | | | |
|---|---|---|---|---|---|---|---|
| EH | end | 70 | 07 | SH | shin | 160 | 25 |
| ER1 | fir | 160 | 33 | SS | vest | 90 | 37 |
| ER2 | fir | 300 | 34 | TH | thin | 180 | 1D |
| EY | beige | 280 | 14 | TT1 | part | 100 | 11 |
| FF | food | 150 | 28 | TT2 | to | 140 | 0D |
| GG1 | got | 80 | 24 | UH | book | 100 | 1E |
| GG2 | guest | 40 | 3D | UW1 | to | 100 | 16 |
| GG3 | wig | 140 | 22 | UW2 | food | 260 | 1F |
| HH1 | he | 130 | 1B | VV | vest | 190 | 23 |
| HH2 | hoe | 180 | 39 | WH | whig | 200 | 30 |
| IH | sit | 70 | 0C | WW | wool | 180 | 2E |
| IY | see | 250 | 13 | XR | repair | 360 | 2F |
| JH | dodge | 140 | 0A | YR | clear | 350 | 3C |
| KK1 | can't | 160 | 2A | YY1 | yes | 130 | 31 |
| KK2 | sky | 190 | 29 | YY2 | yes | 180 | 19 |
| KK3 | comb | 120 | 08 | EL | saddle | 190 | 3E |
| LL | lake | 110 | 2D | ZH | azure | 190 | 26 |
| MM | milk | 180 | 10 | ZZ | zoo | 210 | 2B |

**Guidelines For Using Allophones[78]**

| | |
|---|---|
| ***SILENCE***<br>PA1 (10 msec)—before BB, DD, GG and JH<br>PA2 (30 msec)—before BB, DD, GG, and JH | PA3 (50 msec)—before PP, TT, KK, and CH, and between words<br>PA4 (100 msec)—between clauses and sentences<br>PA5 (200 msec)—between clauses and sentences |
| ***SHORT VOWELS***<br>(These allophones can be doubled.)<br>IH—sitting stranded<br>EH—extent, gentlemen<br>AE—extract, acting | ***LONG VOWELS***<br>IY—treat, people, penny<br>EY—great, statement, tray<br>AY—kite, sky, mighty<br>OY—noise, toy, voice<br>UW1—after clusters with YY; computer |

[78]These guidelines are suggestions, not laws, and you may want to experiment with other combinations. The allophone set contains two or three versions of some phonemes. Different allophones of a particular phoneme may be needed for initial or final positions in a syllable or word. Speech sounds that have features in common behave in similar ways. For example, the voiceless stop consonants *PP, TT,* and *KK* should be preceded by 50 to 80 msec of silence, and the voiced stop consonants *BB, DD,* and *GG* by 10 to 30 msec of silence. *DD2* sounds good in initial position and *DD1* sounds good in final position, as a daughter and collide. One of the differences between the initial and final version of a consonant is that the initial version may be longer than the final version. To create an initial *SS*, you can use two *SSs* instead of the single *SS* at the end of a word or syllable, as in sister. This can also be done with *TH* and *FF*, and the inherently short vowels (to be discussed below), but with no other

| | |
|---|---|
| UH—cookie, full<br>AO—talking, song<br>AX—lapel, instruct<br>AA—pottery, cotton | UW2—in monosyllabic words: two, food<br>OW—zone, close, snow<br>AW—sound, mouse, down<br>EL—little, angle, gentlemen |
| ***R—COLORED VOWELS*** | ***RESONANTS*** |
| ER1—letter, furniture, interrupt<br>ER2—monosyllables: bird, fern, burn<br>OR—fortune, adorn, store<br>AR—farm, alarm, garment<br>YR—hear, earring, irresponsible<br>XR—hair, declare, stare | WW—we, warrant, linguist scream<br>RR1—initial position: read, write, x-ray<br>RR2—initial clusters; brown, crane, grease<br>YY1—cluster: cute, beauty, yarn, yo-yo<br>YY2—initial position: yes, yarn, yo-yo |
| ***NASAL*** | ***AFFRICATES*** |
| MM—milk, alarm, ample<br>NN—before front and central vowels: YR, IY,<br>IH, EY, EH, XR, AE, ER, AX,<br>AW, AY, UW;—final clusters: earn<br>NN2—before back vowels: UH, OW, OY, CR, AR, AA<br>NG—string, anger | CH—church, feature<br>JH—judge, injure |
| ***VOICED FRICATIVES*** | ***VOICELESS FRICATIVES*** |
| VV—vest, prove, even<br>DH1—word-initial position: this,<br>then, they | FF—These may be doubled<br>TH—for initial position and<br>SH—shirt, leash, nation |

consonants. You will want to experiment with some consonant clusters (strings of consonants such as *str, cl*) to discover which version works best in the cluster. For example, *KK1* sounds good before *LL* as in *clown*, and *KK2* sounds good before *WW* as in *square*. One allophone of a particular phoneme may sound better around back vowels and another around front vowels. *KK3* sounds good before *UH* and *KK1* sounds good before *IY*, as in *cookie*. Some sounds (*PP, BB, TT, DD, KK, GG, CH,* and *JH*) require a brief duration of silence before them. For most of these, the silence has already been added, but you may want to add more. Several pauses are included in the allophone set varying from 10 to 200 msec. To create the final sounds in the words *letter* and *little,* use the allophones *ER* and *EL*. Remember that you must always think about how a word *sounds*, not how it is *spelled*. For example, the *NG* sound is represented by the letter *N* in *uncle*. Some sounds may not even be represented in words by any letters, as the *YY* in *computer*.

As mentioned, there are some vowels which can be doubled to make longer versions for stressed syllables. These are the inherently short vowels *IH, EH, EA, EX, AA,* and *UH*. For example, in the word *extent,* use one *EH* in the first syllable, which is unstressed and two *EH's* in the second syllable which is stressed. Of the inherently long vowels there is one, *UW*, which has a long and short version. The short one, *UW1*, sounds good after YY in *computer*. The long version, *UW2*, sounds good in monosyllabic words like *two*. Included in the vowel set is a group called R-colored vowels. These are vowel +R combinations. For example, the *AR* in *alarm* and the *OR* in *score*. One of the R-colored vowels, *ER*, has a long and short version. The short version is good for several-syllable words with final *ER* sounds like *letter*, and the long version is good for monosyllabic words like *fir*.

You may want to add a pause of 30 to 50 msec between words, when creating sentences, and a pause of 100 to 200 msec between clauses. Every utterance must be followed by a pause in order to make the chip stop speaking the last allophone.

| | |
|---|---|
| DH2—word-final and between vowels: bathe, bathing<br>ZZ—zoo, phase<br>ZH—beige, pleasure | SS—used singly in final position<br>HH1—before front vowels, RY, IY, IH, EY, EH, XR, AE<br>HH2—before back vowels: UW, UH, OW, OY, AO, OR, AR<br>WH—white, whim, twenty |
| ***VOICED STOPS*** | ***VOICELESS STOPS*** |
| BB1—final position: rib—between vowels: fibber—in clusters: bleed, brown<br>BB2—initial position before a vowel: beast<br>DD1—final position: played, end<br>DD2—initial position: down; clusters: drain<br>GG1—before high front vowels: YR, IY, IH, EY, EH, XR<br>GG2—before high back vowels: UW, UH, OW, OY, AX;—clusters: green, glue<br>GG3—before low vowels: AE, AW, AY, AR, AA, AO, OR, ER—meidal clusters: anger;—final position: peg. | PP—pleasure, ample, trip<br>TT1—final clusters before SS: tests, its<br>TT2—all other positions: test, street<br>KK1—before front vowels: YR, IY, IH, EY, EH, XR, AY, AE, ER, AX—initial clusters: cute, clown,<br>KK2—final position: speak—final clusters: task<br>KK3—before back vowels: UW, UH, OW, OY, OR, AR, AQ—initial clusters: crane, quick, clown, scream |

This first program endlessly sends the allophones for "cookie" to the SPO256-AL2.

**Allophone "Cookie" Example**

```
CSEG AT 00H ;COLD START JUMP
 LJMP STRT

COOKIE SEGMENT CODE
RSEG COOKIE
;ASSUMES SBY ON P3.0,ALD ON P3.1,CODE ON P1
SPEAK: JB P3.0, SPEAK ;WAIT 'TILL READY
 MOV P1,A ;CODE PASSED IN ACC
 CLR P3.1
 SETB P3.1
 RET

STRT:;MAIN PROGRAM
 MOV DPTR,#CK
M1:MOV R5,#0
M2:MOV A,R5
 MOVC A,@A+DPTR
 CJNE A,#0FFH,M3 ;TEST FOR END OF STRING
 JMP M1 ;START OVER—DONE
M3:CALL SPEAK
```

```
 INC R5 ;GO TO NEXT ALLOPHONE
 JMP M2

CK: DB 08H,1EH,2AH,13H,04H,04H,0FFH
       ;KK3,UH, KK1,IY, PA5,PA5, EOF
END
```

This second program can take ASCII coded allophones (two- or three-letter spelling rather than single-byte codes), convert them to the codes, and speak them. This allows the speech string to be readable and, combined with keyboard input, could be useful for experimenting with allophone sequences.

**Allophone Driver Program**

```
 #include<reg51.h>
 #define uchar unsigned char
 #define sby 0xfff8
 #define ald 0xfffa

79uchar code codes[]=
 {"PA1PA2PA3PA4PA5OY AY EH KK3PP JH NN1IH TT2RR1AX ",
  "MM TT1DH1IY EY DD1UW1AO AA YY2AE HH1BB1TH UH UW2",
  "AW DD2GG3VV GG1SH ZH RR2FF KK2KK1ZZ NG LL WW XR ",
  "WH YY1CH ER1ER2OW DH2SS NN2HH2OR AR YR GG2EL BB2"};

 uchar cookie[]={0x08,0x1e,0x2a,0x13,0x04,0xff};
 uchar jmp[]= {"gg1owpa5jhaommpppa5ihnn1pa5thaopa5ll
              eykk2"};

80void speakcode(uchar phcode){
 while(sby);
 P1=phcode;
 ald=0;ald=1;
 }
```

[79]The string that makes up this table are entered in double quotes which normally appends a 0 (/0) at the end of the string. It turns out that C specifies that successive strings are *concatenated* so even though the string was broken up, the extra 0 is not added until the end.

[80]This driver function takes the phoneme code, waits for *SBY* to go low, outputs the code, drives *ALD* low and then back high.

```
[81]void speakstring(uchar *string){
 uchar i,j; uchar *ptr;
 for(ptr=string,j=0;*ptr!=0&&j<100;ptr++,j++)
  *ptr&=0xdf; /*to upper, 100 char limit if 0
              missing*/
 for(j=0;*string!=0&&*(string+1)!=0&&j<100;j++){
[82] for(i=0;i<192;i+=3){ /*hunt through table for a
              match*/
  if (codes[i]==*string){ /*first character
              matches*/
   if (codes[i+1]==*(string+1)){ /*second char
              matches*/
[83]  if(*(ptr=string+2)<'1'|| *ptr>'5'||codes[i+2]
              ==*ptr){
     speakcode(i/3); /*array in machine order*/
     i=200; /*found it!—kick out of hunt*/
     for(string+=2;!((*string>='A')&&
              (*string<='Z'))&&
```

[81]This function identifies allophone codes (two- or three-letter) found where string points, converts them to upper case alpha characters, and speaks the allophones. The codes can be packed (two letters when that is sufficient) or spaced out with any non-alpha characters such as spaces or commas. The *speakstring* function is passed a ***pointer*** to a string. For this example, *string* is a variable that holds an *address* of an unsigned character while **string* is the character found *at* that address. Depending on the parameter passing conventions of the compiler, *string* itself is probably on the stack: *void speakstring(uchar *string){*. Notice that the actual string, *"Go jump in the lake,"* is defined as an array of ASCII characters: *uchar jmp[]={"gg1owpa5jhaommpppa5ihnn1pa5thaopa5lleykk2"};* When the function is called, the array is not passed—rather the *address* of the *start* of the array is passed when the simple *jmp* is put in as the parameter to pass: *speakstring(jmp);* It is possible to put the string directly in the parenthesis such as *speakstring ("gg1owpa5");* but it gets awkward to read at times.

[82]When the string pointer is passed, the algorithm works its way through the *codes[]* table for a match with the first character in the string. When a match is found, it checks for a match with the second character and, depending on the specific allophone, the third character. The *codes[]* table is arranged so the numeric position in the table (divided by three of course) *is* the numeric code to be sent to the allophone chip. The allophone designations are two characters with a numeric designator if there are several variants such as *PA1* through *PA5* for various lengths of pause. If the match in the table finds that the third character *in the table* is not numeric, a third character for the entry is not sought or checked. Otherwise, the match must agree to all three characters.

[83]It is possible to make an assignment within an *if* test as shown here. Since the string is to have two added over and over, the value is put temporarily in *ptr*. The assignment happens

```
                    *string!=0;string++);/*next
                    allophone*/
          }
        }
      }
      if(i==192)string++; /*no match—shift and try
                     again*/
      }
    }
  }
}

void main(void){
 uchar i;
 for(;;){
  for(i=0;cookie[i]!=0xff;i++)speakcode(cookie[i]);
                 /*says "cookie"*/
  speakstring(jmp); /*says "go jump in the lake"*/
 }
}
```

**Allophone Day of the week**

| | |
|---|---|
| Sunday | SS SS AX AX NN1 PA2 DD2 EY |
| Monday | MM AX AX NN1 PA2 DD2 EY |
| Tuesday | TT2 UW2 ZZ PA2 DD2 EY |
| Wednesday | WW EH EH NN1 ZZ PA2 DD2 EY |
| Thursday | TH ER2 ZZ PA2 DD2 EY |
| Friday | FF RR2 AY PA2 DD2 EY |
| Saturday | SS SS AE PA3 TT2 PA2 DD2 EY |

before the rest of the testing is done since the precedence of the || is left to right. You may wish to ponder this line for some time to be sure you understand it. It is saying that if the third character *coming in* is not between one and five inclusive or is an exact match with the third character in the table, go ahead and say it. This actually means that the user can define three-character allophones with *two* characters even though the list has only numbered variants with two letters *and* a number; the algorithm will simply pick out the first variant in the list. These features are part of what is called ***robust programming,*** where the deficiencies of the input do not totally disable the system. Some reasonable attempt is made to provide a response to the calling program. The same thing is seen in taking numeric entries from a user. If a three-digit number is expected and the user wishes to input a *1*, must a *001* be entered or can a *1(cr)* be acceptable?

## Allophone Speak Letters

| | | | | | |
|---|---|---|---|---|---|
| A | EY | J | JH EH EY | S | EH EH SS SS |
| B | BB2 IY | K | KK1 EH EY | T | TT2 IY |
| C | SS SS IY | L | EH EH EL | U | YY1 UW2 |
| D | DD2 IY | M | EH EH EM | W | EE2 AX PA2 BB2 |
| E | IY | N | EH EH NN1 | Y | Y1 UW2 |
| F | EH EH FF FF | O | OW | X | EH EH PA3 KK2 SS SS |
| G | JH IY | P | PP IY | Y | WW AY |
| H | EY PA2 PA3 CH | Q | KK1 YY1 UW2 | Z | ZZ IY |
| I | AA AY | R | AR | | |

## Allophone Speak Numbers

| | | | |
|---|---|---|---|
| zero | XX YR OW | sixteen | SS SS IH PA3 KK2 SS PA2 PA3 TT2 IY NN1 |
| one | WW NN1 | seventeen | SS SS EH VV TH NN1 PA2 PA3 TT2 IY NN1 |
| two | TT2 UW2 | eighteen | EY PA2 PA3 TT2 IY NN1 |
| three | TH RR1 IY | nineteen | NN1 AY NN1 PA2 PA3 TT2 IY NN1 |
| four, for, | FF FF OR | twenty | TT2 WH EH EH NN1 PA2 PA3 |
| five | FF FF AY VV | thirty | TH ER2 PA2 PA3 TT2 IY |
| six | SS SS IH IH PA3 KK2 SS | forty | FF OR PA3 TT2 IY |
| seven | SS SS EH EH VV IH NN1 | fifty | FF FF IH FF FF PA2 PA3 TT2 IY |
| eight | EY PA3 TT2 | sixty | SS SS IH PA3 KK2 SS PA2 PA3 TT2 IY |
| nine | NN1 AA AY NN1 | seventy | SS SS EH VV IH NN1 PA2 PA3 TT2 IY |
| ten | TT2 EH EH NN1 | eighty | EY PA3 TT2 IY |
| eleven | IH LL EH EH VV IH NN1 | ninety | NN1 AY NN1 PA3 TT2 IY |
| twelve | TT2 WH EH EH LL VV | hundred | HH2 AX AX NN1 PA2 DD2 RR2 IH IH PA1 DD1 |
| thirteen | TH ER1 PA2 PA3 TT2 IY NN1 | thousand | TH AA AW ZZ TH PA1 PA1 NN1 DD1 |
| fourteen | FF OR PA2 PA3 TT2 IY NN1 | million | MM IH IH LL YY1 AX NN1 |
| fifteen | FF IH FF PA2 PA3 TT2 IY NN1 | | |

## Allophone Months

| | | | |
|---|---|---|---|
| January | JH AE AE NN1 YY2 XR IY | July | JH UW1 LL AY |
| February | FF EH EH PA2 BR RR2 UW2 XR IY | August | AO AO PA2 GG2 AX SS PA3 TT1 |
| March | MM AR PA3 CH | September | SS SS EH PA2 PP PA3 TT2 EH EH PA1 BB2 ER1 |
| April | EY PA3 PP RR2 IH IH LL | October | AA PA2 KK2 PA3 TT2 OW PA1 BB2 ER1 |
| May | MM EY | November | NN2 OW VV EH EH MM PA1 BB2 ER1 |
| June | JH UW2 NN1 | December | DD2 IY SS SS EH EH MM PA1 BB2 ER1 |

## Short Allophone Dictionary

| | |
|---|---|
| alarm | AX LL AR MM |
| bathe | BB2 EH DH2 |
| bather | BB2 EY DH2 ER1 |
| bathing | BB2 EY DH2 IH NG |
| beer | BB2 YR |
| bread | BB1 RR2 EH EH PA1 DD1 |
| by | BB2 AA AY |
| calendar | KK1 AE AE LL EH NN1 PA2 DD2 ER1 |
| clock | KK1 LL AA AA PA3 KK2 |
| clown | KK1 LL AW NN1 |
| check | CH EH EH PA3 KK2 |
| checked | CH EH EH PA3 KK2 PA2 TT2 |
| checking | CH EH EH PA3 KK1 IH NG |
| checks | CH EH EH PA3 KK1 SS |
| cognitive | KK3 AA AA GG3 NN1 IH PA3 TT2 IH VV |
| collide | KK3 AX KK AY DD1 |
| computer | KK2 AX MM PP1 YY1 UW1 TT2 ER |
| cookie | KK3 UH KK1 IY |
| coop | KK3 UW2 PA3 PP |
| correct | KK1 ER2 EH EH PA2 KK2 PA2 TT1 |
| corrected | KK1 ER2 EH EH PA2 KK2 PA2 TT2 IH PA2 DD1 |
| corrects | KK1 ER2 EH EH PA2 KK2 PA2 TT1 SS |
| crown | KK1 RR2 AW NN1 |
| date | DD2 EY PA3 TT2 |
| daughter | DD2 AO TT2 ER1 |
| day | DD2 EH EY |
| divided | DD2 IH VV AY PA2 DD2 IH PA2 DD1 |
| emotional | IY MM OW SH AX NN1 AX EL |
| engage | EH EH PA1 NN1 GG1 EY PA2 JH |
| engagement | EH EH PA1 NN1 GG1 EY PA2 JH MM EH EH PA3 NN2 GG1 EY PA2 PA3 TT2 |
| engages | EH EH PA1 NN1 GG1 EY PA2 JH IH ZZ |
| engaging | EH EH PA1 NN1 GG1 EY PA2 JH IH NG |
| enrage | EH NN1 RR1 EY PA2 JH |
| enraged | EH NN1 RR1 EY PA2 JH PA2 DD1 |
| enrages | EH NN1 RR1 EY PA2 JH IH ZZ |

| | |
|---|---|
| enraging | EH NN1 RR1 EY PA2 JH IH NG |
| escape | EH SS SS PA3 KK1 PA2 PA3 PP |
| escaped | EH SS SS PA3 KK1 PA2 PA3 PP PA2 TT2 |
| escapes | EH SS SS PA3 KK1 PA2 PA3 PP SS |
| equal | IY PA2 PA3 KK3 WH AX EL |
| equals | IH PA2 PA3 KK3 WH AX EL ZZ |
| error | EH XR OR |
| extent | EH KK1 SS TT2 EH EH NN1 TT2 |
| fir | FF ER2 |
| freeze | FF FF RR1 IY ZZ |
| freezers | FF FF RR1 IY ZZ ER1 ZZ |
| freezing | FF FF RR1 IY ZZ IH NG |
| frozen | FF FF RR1 OW ZZ IH NN1 |
| gauge | GG1 EY PA2 JH |
| gauges | GG1 EY PA2 JH IH ZZ |
| gauging | GG1 EY PA2 JH IH NG |
| hello | HH EH LL AX OW |
| hour | AW ER1 |
| infinitive | IH NN1 FF FF IH IH NN1 IH PA2 PA3 TT2 IH VV |
| intrigue | IH NN1 PA3 TT2 RR2 IY PA1 GG3 |
| intrigued | IH NN1 PA3 TT2 RR2 IY PA1 GG3 PA2 DD1 |
| intrigues | IH NN1 PA3 TT2 RR2 IY PA1 GG3 ZZ |
| intriguing | IH NN1 PA3 TT2 RR2 IY PA1 GG3 IH NG |
| investigate | IH IH NN1 VV EH EH SS PA2 PA3 TT2 IH PA1 GG1 EY PA2 TT2 |
| investigated | IH IH NN1 VV EH EH SS AP2 PA3 TT2 IH PA1 GG1 EY PA2 TT2 IH PA2 DD1 |
| investigator | IH IH NN1 VV EH EH SS PA2 PA3 TT2 IH PA1 GG1 EY PA2 TT2 ER1 |

| | |
|---|---|
| investigators | IH IH NN1 VV EH EH SS PA2 PA3 TT2 IH PA1 GG1 EY PA2 TT2 ER ZZ |
| investigates | IH IH NN1 VV EH EH SS PA2 PA3 TT2 IH PA1 GG1 EY PA2 TT1 SS |
| investigating | EH EH NN1 VV EH EH SS PA2 PA3 TT2 IH PA1 GG1 EY PA2 TT2 IH NG |
| key | KK1 IY |
| legislate | LL EH EH PA2 JH JH SS SS LL EY PA2 PA3 TT2 |
| legislated | LL EH EH PA2 JH JH SS SS LL EY PA2 PA3 TT2 IH DD1 |
| legislates | LL EH EH PA2 JH JH SS SS LL EY PA2 PA3 TT1 SS |
| legislating | LL EH EH PA2 JH JH SS SS LL EY PA2 PA3 TT2 IH NG |
| legislature | LL EH EH PA2 JH JH SS SS LL EY PA2 PA3 CH ER1 |
| letter | LL EH EH PA3 TT2 ER1 |
| litter | LL IH IH PA3 TT2 ER1 |
| little | LL IH IH PA3 TT2 EL |
| memory | MM EH EH MM ER2 IY |
| memories | MM EH EH MM ER2 IY ZZ |
| minute | MM IH NN1 IH PA3 TT2 |
| nipped | NN2 IH IH PA2 PA3 PP IH NG |
| nips | NN1 IH IH PA2 PA3 PP SS |
| no | NN2 AX OW |
| physical | FF FF IH ZZ IH PA3 KK1 AX EL |
| pin | PP IH IH NN1 |
| pinned | PP IH IH NN1 PA2 DD1 |
| pinning | PP IH IH NN1 IH NG1 |
| pins | PP IH IH NN1 ZZ |
| pledge | PP LL EH FH PA3 JH |
| pledged | PP LL EH EH PA3 JH PA2 DD1 |
| pledges | PP LL EH EH PA3 JH IH ZZ |
| pledging | PP LL EH EH PA3 JH IH NG |
| plus | PP LL AX AX SS SS |

| | |
|---|---|
| ray | RR1 EH EY |
| rays | RR1 EH EY ZZ |
| ready | RR1 EH EH PA1 DD2 IY |
| red | RR1 EH FH PA1 DD1 |
| robot | RR1 OW PA2 BB2 AA PA3 TT2 |
| robots | RR1 OW PA2 BB2 AA PA3 TT1 SS |
| score | SS SS PA3 KK3 OR |
| seat | SS SS WW EH EH PA3 TT2 |
| second | SS SS EH PA3 KK1 IH NN1 PA2 DD1 |
| sensitive | SS SS EH EH NN1 SS SS IH PA2 PA3 TT2 IH VV |
| sensitivity | SS SS EH EH NN1 SS SS IH PA2 PA3 TT2 TT2 IH VV IH PA2 PA3 TT2 IY |
| sincere | SS SS IH IH NN1 SS SS YR |
| sincerely | SS SS IH IH NN1 SS SS YR LL IY |
| sincerity | SS SS IH IH NN1 SS SS EH EH RR1 IH PA2 PA3 TT2 IY |
| sister | SS SS IH IH SS PA3 TT2 ER1 |
| speak | SS SS PA3 IY PAe KK2 |
| spell | SS SS PA3 PP EH EH EL |
| spelled | SS SS PA3 PP EH EH EL PA3 DD1 |
| speller | SS SS PA3 PP EH EH EL ER2 |
| spellers | SS SS PA3 PP EH EH EL ER2 ZZ |
| spelling | SS SS PA3 PP EH EH EL IH NG |
| spells | SS SS PA3 PP EH EH EL ZZ |
| start | SS SS PA3 TT2 AR PA3 TT2 |
| started | SS SS PA3 TT2 AR PA3 TT2 IH PA1 DD2 |
| starter | SS SS PA3 TT2 AR PA3 TT2 ER1 |
| starting | SS SS PP3 TT2 AR PA3 TT2 IH NG |
| starts | SS SS PP3 TT2 AR PA3 TT1 SS |
| stop | SS SS PA3 TT1 AA AA PA3 PP |
| stopped | SS SS PA3 TT1 AA AA PA3 PP PA3 TT2 |
| stopper | SS SS PA3 TT1 AA AA PA3 PP ER1 |
| stopping | SS SS PA3 TT1 AA AA PA3 PP IH NG |
| stops | SS SS PA3 TT1 AA AA PA3 PP SS |
| subject (noun) | SS SS AX AX PA2 BB1 PA2 JH EH PA3 KK2 PA3 TT2 |
| subject (verb) | SS SS AX PA2 BB1 PA2 JH EH EH PA3 KK2 PA3 TT2 |
| sweated | SS SS WW EH EH PA3 TT2 IH PA3 DD1 |
| sweater | SS SS WW EH EH PA3 TT2 ER1 |
| sweaters | SS SS WW EH EH PA3 TT2 ER1 ZZ |
| sweating | SS SS WW EH EH PA3 TT2 IH NG |
| sweats | SS SS WW EH EH PA3 TT2 SS |
| switch | SS SS WH IH IH PA3 CH |
| switched | SS SS WH IH IH PA3 CH PA3 TT2 |
| switches | SS SS WH IH IH PA3 CH IH ZZ2 |
| switching | SS SS WH IH IH PA3 CH IH NG2 |
| system | SS SS IH IH SS SS PA3 TT2 EH MM |
| systems | SS SS IH IH SS SS PA3 T2 EH MM ZZ |
| talk | TT2 AO AO PA2 KK2 |
| talked | TT2 AO AO PA3 KK2 PA3 TT2 |
| talker | TT2 AO AO PA3 KK1 ER1 |
| talkers | TT2 AO AO PA3 KK1 ER1 |
| talking | TT2 AO AO PA3 KK1 IH NG |
| talks | TT2 AO AO PA2 KK2 SS |

| | |
|---|---|
| thread | TH RR1 EH EH PA2 DD1 |
| threaded | TH RR1 EH EH PA2 DD2 IH PA2 DD1 |
| threader | TH RR1 EH EH PA2 DD2 ER1 |
| threaders | TH RR1 EH EH PA2 DD2 ER1 ZZ |
| threading | TH RR1 EH EH PA2 DD2 IH NG |
| threads | TH RR1 EH EH PA2 DD2 ZZ |
| then | DH1 EH EH NN1 |

| | |
|---|---|
| time | TT2 AA AY MM |
| times | TT2 AA AY MM ZZ |
| uncle | AX NG PA3 KK3 EL |
| whale | WW EY EL |
| whaler | WW EY LL ER1 |
| whalers | WW EY LL ER1 ZZ |
| whales | WW EY EL ZZ |
| whaling | WW EY LL TH NG |
| year | YY2 YR |
| yes | YYS EH EH SS SS |

## Text-to-speech Chips

By combining the CTS256A-AL2 code-to-speech chip with the SP0256A-AL2 speech-processor chip (just discussed), you have a system which uses letter-to-sound rules to say any word that can be spelled. This "unlimited" vocabulary sets it apart from fixed-vocabulary devices like the Digitalker, but there is a certain reduction in clarity and "naturalness" that results.

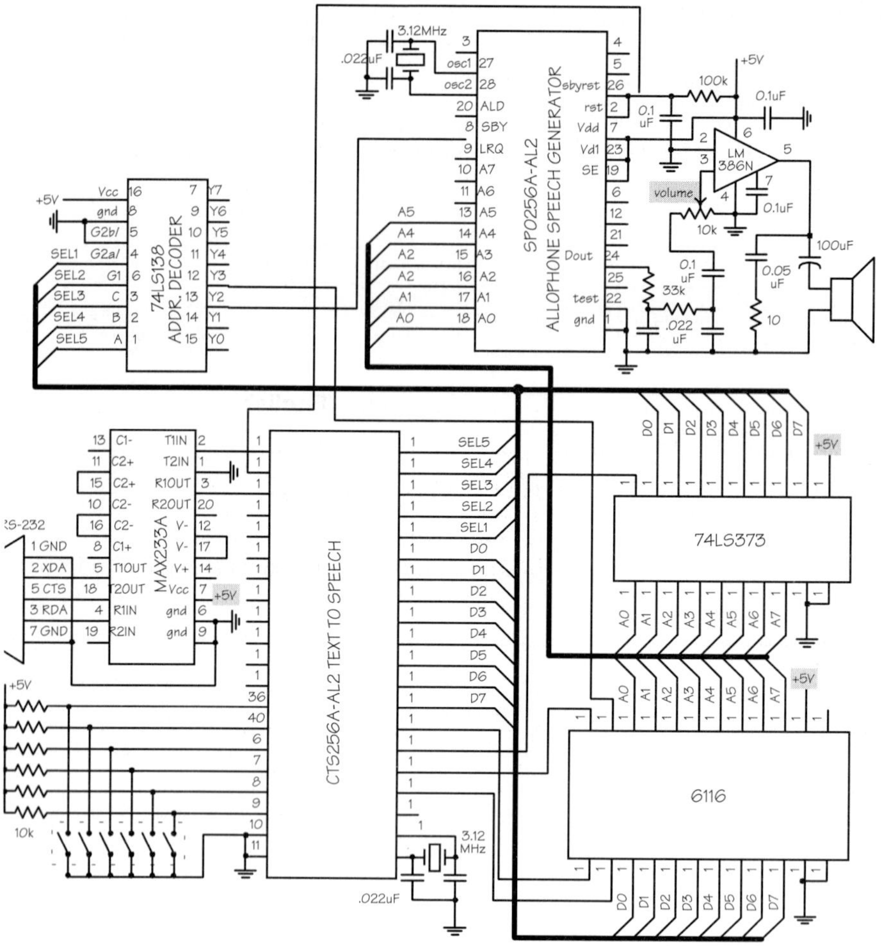


**Text-to-speech Circuit**

You may interface either serially (RS-232/ASCII coded characters) or in parallel (strobe and acknowledge/also ASCII coded characters) to the CTS256 chip. There are options to cause it to say each word as the word is completed or to wait until a carriage return and then repeat an entire phrase or sentence. The baud rate or parallel mode are selectable as well. An external RAM buffer (included) may be selected. The buffer allows much longer sentences to be assembled before speaking. The schematic is incomplete since I seem to have lost the data sheet.

As mentioned, the output chip must be fed allophone codes, yet the user supplies the codes for English letters that spell words. The translation (since English is not phonetic in its spelling) is done by the CTS256A-AL2 chip—actually a pre-programmed 8-bit microcomputer. The table below shows a simplified version of the rules for the translation process. By working through a word letter-by-letter, the allophones/phonemes can be derived. For instance, following the text-to-phoneme rules attached, CAT would see that "C" is not followed by H, not followed by E, I, or Y, hence it becomes "K." A is not final, not followed by R, nor L so it becomes "AE1." Finally, "T" is not THE, TO, THAT, TH, TI, so it becomes "T." The phonemes are thus: K, AE1, T; K as in trick, AE as in after and T as in tap. There are considerably more rules if you want a more realistic result.

**Characters to Allophone Conversion Rules (English Language)**

| | |
|---|---|
| context symbols: | ! = non-alphabetic character<br>: = zero or more consonants<br>$ = one or more consonants<br># = one or more vowels<br>+ = e, i or y<br>; = terminate rule—go to next rule or character |
| A | if (A)! then AA<br>if (AR) then AA, R<br>if #:(AL)! then AA. L<br>if (AL)$ then AO, L<br>if (A) then EY |
| B | if (B) then BB1 |
| C | if (CH) then CH<br>if (C)+ then SS<br>if (C) then KK3 |
| D | if (D) then DD1 |
| E | if #:(E)! then ;<br>if !:(E)! then IY<br>if #:(E)D! then ;<br>if (ER) then ER1<br>if #:(E)S! then ;<br>if: (EE) then IY, IY<br>if (EA) then IY<br>if (E) then EH |
| F | if (F) then FF. |
| G | if (G) then GG1 |
| H | if (H)# then HH1<br>if (H) then ; |
| I | if !(IN) then IH, NN2<br>if (I)$+:# then AY<br>if (I)$+ then IH<br>if (I) then IH |
| J | if (J) then JH |
| K | if (K) then KK1 |
| L | if (L)L then ;<br>if {L) then LL |
| M | if (M) then MM |
| N | if (NG) then NG<br>if (N) then NN1 |
| O | if (OF)! then AO, VV<br>if (OR) then AW, RR2<br>if I(ON) then AO, NN2 |
| P | if (P) then PP |
| Q | if (QU) then KK3, WH<br>if (Q) then KK2, IY, UW2 |
| R | if (R) then RR1 |
| S | if (SH) then SH<br>if #(SED)! then ZZ, DD1<br>if S(S) then ;<br>if (S) then SS |
| T | if !(THE)! then TH, AX<br>if (TO)! then TT2, UW2<br>if (THAT)! then TH, AE, TT1<br>if (TH) then TH<br>if (TI)O then SH<br>if (T) then TT1 |
| U | if (U)$$! then AX<br>if (U) then YY1, UH |
| V | if (V} then VV |
| W | if !(WAS)! then WW, AO, ZZ<br>if (W) then WW |
| X | if (X) then KK3, SS |
| Y | if #:$(Y)! then EY<br>if (Y) then IY |
| Z | if (Z) then ZZ |
| pause | if ( ) then PA4 |

With the CTS265, each time the system is fed characters, there must be delimiters to mark the boundaries between words. These can be . , ; : ?, space, or carriage return. Switch 5 selects whether speech comes after each word or only after carriage return. Two control characters also have significance: ESC (1B) causes current buffer to clear and silences any speech in process; BACKSPACE (08) erases previously sent characters one at a time beginning with the most recent.

The baud rate can be selected with switches 1, 2, 3 as shown in the table below.

**Baud Rate Select for Text-to-speech Chip**

| BAUD | parallel | 50 | 110 | 300 | 1200 | 2400 | 4800 | 9600 |
|---|---|---|---|---|---|---|---|---|
| S1 | 0 | 0 | 0 | 0 | 1 | 1 | 1 | 1 |
| S2 | 0 | 0 | 1 | 1 | 0 | 0 | 1 | 1 |
| S3 | 0 | 1 | 0 | 1 | 0 | 1 | 0 | 1 |

The schematic shows hard wiring for 7 bits plus 2 stop-bits and no parity (which is the same as 1 stop-bit and mark parity which is commonly used). Note that the jumper block sets up *DTE* mode (pins 2, 3 of DB-25 connector) with jumpers in center mode with jumpers on ends.

Note the switch settings for parallel input. The parallel strobe must go high while valid parallel data is on the data inputs. A positive acknowledge pulse (or "busy") is produced but is potentially too short to recognize in software. If port addressing, don't require a high acknowledge before proceeding—if it is found to be high, wait, but otherwise go ahead as though the acknowledge has already come and gone.

I show no program examples but you would use it by sending ASCII codes either serially or in parallel much like driving a display.

## Natural Speech Reproduction

As mentioned earlier, several answering-machine chips are readily available to store and play back speech. I believe the chips include a non-volatile (flash?) memory so the stored speech can stay even when the power is removed. Such an arrangement could be quite versatile in storing realistic sentences, words, or even other sounds, but I am not personally familiar with the details.[84]

[84]One student project involved training parrots to talk, and I believe it included stored tones as well as phrases.

## STORAGE

In the same way that power supplies are not exactly input or output, so storage fits neither category. It is technically internal hardware, involving neither input nor output, but since you can think of first writing *to* memory, I put it here.

### LCD as Memory

It is a seldom-considered fact that most alphanumeric LCD modules have more display storage than is shown on the display. Specifically, the standard driver chips handle 4 × 20 characters even when they only have an LCD module for 2 × 20 or even 1 × 16 characters. Often users will play games with writing to the non-showing memory and then issuing a shift command to scroll characters into view. But if you are in a situation where you need temporary storage beyond the 64 bytes of the 80C750 or 751 and are using an LCD module, this display RAM can be used for other purposes. It is possible to both write to and read from this space even though many users work in a write-only mode. Thus, by writing to the non-showing space, you can have additional memory of 40 or 50 bytes. Admittedly this would not have very fast access, but it could make the cost difference for a minimal system and might even constitute the power-down storage memory since the LCD modules draw extremely small amounts of current.

### Serial EEPROM

In addition to the normal memory devices such as those described in *Volume I,* there are interesting storage devices well suited to tiny systems and limited I/O pins. The basic device has four signal lines—data bits in, data bits out, a clock and an enable. This should work directly in the shift-register mode of the standard serial UART on the base 8051.

# 6

# Algorithms

An algorithm is the way to do something. For example, let's play the game, "*I'm thinking of a number between 1 and 100.*" After a bit of thought you guess *"50?" "Higher." "75?"* And so on. You have intuitively established an *algorithm*—divide the intervals in half, because that narrows in on the number with the fewest guesses.

The sections that follow present ways of doing things in software that, hopefully, are more efficient than your first approaches. Actually, there is a whole world of algorithms out there for things like sorting data and records, solving complex mathematical expressions, and signal processing. I will not go into the more complex ones— small microcontrollers are seldom appropriate for such applications—but I will touch on fuzzy logic and describe digital signal processing (DSP).[1] I focus on the simple problems such as getting numbers into binary or out in displayable decimal digits. Along the way, I intend to convince you that a lookup table is usually better than an equation. The examples I give are targeted to small, efficient applications—for the other algorithms you need a different book.[2]

## NUMBER CONVERSION

When you think of a number (a "variable"), think of a bucket full of marbles—whatever way you group the marbles, there are still the same number of marbles in the bucket. Whatever way a number is represented, the impor-

[1]Strangely enough, despite the existence of very efficient DSP processors, there is no inherent reason why the same thing cannot be done (more slowly) with a conventional microcontroller.

[2]Excellent articles on algorithms regularly appear in *Programmer's Toolbox*, Jack W. Crenshaw's column in *Embedded Systems Programming* (Miller Freeman, San Francisco). To the extent that I dabble in math-oriented algorithms, many of the ideas come from that source.

tant thing is the "how many" that the number holds. My students frequently ask how to get back and forth between number systems. While all sorts of systems are possible, there are only four that have significant use—***binary*** (2), ***octal*** (8), ***hexadecimal*** (16), and ***decimal*** (10).

Binary, octal, and hexadecimal notations describe the *same* arrangement of flip-flop states (bits) in the computer. There really is no "conversion" involved. A $3E_{16}$ (hexadecimal) is the same as $00111110_2$ (binary) or $076_8$ (octal)—the only difference is in how you choose to group the bits.

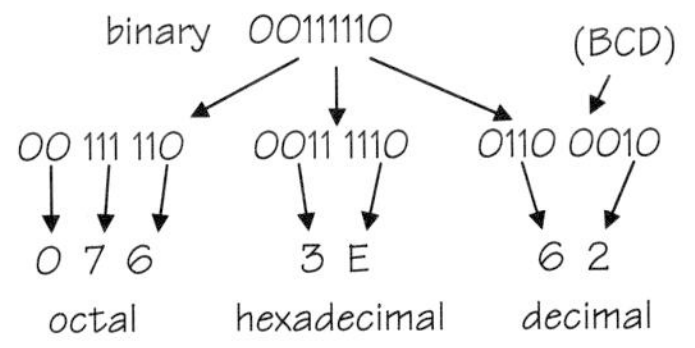


**Number Representation**

In a digital computer, all four notations describe logic levels in a string of flip-flops (a register)—the value of a variable. The first three notations directly indicate that the two most significant bits (msb) are at logic 0 (low level), the next five are at logic 1 (high level), and the least significant bit is a 0 (low level).

Decimals are another story. First, picture a binary count chain going from right to left.[3] Imagine pulses coming into the chain. The first negative-going edge at the input toggles the right-most flip-flop giving $0000\ 0001_2$ ($01_{16}$). The next negative edge toggles the that flip-flop back to 0, but that flip-flop's transition toggles the second flip-flop, giving $0000\ 0010_2$ ($02_{16}$). Keep going up to $0000\ 1111_2$ ($0f_{16}$) and the next negative edge clears the entire first IC, but, because of the interconnect wiring, also toggles the first flip-flop of the second device in the chain, giving $0001\ 0000_2$ ($10_{16}$). The states of the flip-flops represent the number of negative edges that have happened at the input. The highs and lows are the ***binary*** representation of the number.

Now replace your ICs with 4-bit decade counters.[4] Now, as the count goes up, from $0000\ 1001_2$ ($09_{10}$), the next count in will cause the right-most

---

[3]If you are familiar with TTL ICs, imagine, for this example, a string of 7493s arranged from right to left (backwards from the order of good electrical schematics!) with the output of one (4 flip-flop) device connected to the input of the device to its left and so on up the chain to the left.

[4]If you are circuit-oriented, take 7490s and arrange the single (÷2) flip-flop to the right. Each IC still has four flip-flops, but internal logic causes the upper three flip-flops of the chain to reset to 000 0 after 100 1 ($9_{10}$). You now have a ***decade*** count chain.

counter to reset to $0000_2$ and the transition of the fourth bit toggles the next counter in the chain, giving 0001 $0000_2$ ($10_{10}$). You now have a chain that counts in decimal and the counters show the count in ***binary-coded decimal* (*BCD*)**.[5]

*Computers* work best with binary numbers but *people* are familiar with decimal numbers. *Efficient* computer software is written to convert back and forth to decimal *only* when a human user is involved. When a number comes in from a user, your best solution is to switch *to* a binary representation as quickly as possible and switch *back* to decimal representation only at the last possible moment before the number goes to the printer or display that the user will see.[6] How to do those conversions is presented in the next few sections.

Although the 8051 instruction set *does* support one or two BCD instructions, *don't use them*. With C this is no problem since the language has no decimal variables at all. If you use assembly to carefully carry out a BCD addition or subtraction, you will be in trouble when you need to multiply or divide. BCD multiplication and division are not supported in the 8051 instruction set.

**USE BINARY:** When working with decimal numbers from a user or supplying decimal display outputs, *always* convert to binary and *stay* in binary as long as possible because binary math operations are readily supported by C compilers and are the inherent representation within any modern digital computer

[5]If you wired 7447s and seven-segment displays to the outputs of the 7490s, you could see the count in a human-readable decimal form (see the section on displays for circuit specifics).

[6]In the early days of digital computing, some systems computed strictly in decimal (BCD) to satisfy the financial world. I think it was due to mistrust of computers after experience with adding machines—perhaps a fear that computing in binary might lead to rounding in the "wrong" direction from previous practice. If you change to BCD, you have added some inefficiency. In the TTL count chain described above, two devices can now only hold up to 99 rather than 255, and there is additional circuitry inside each of the devices. If you have studied half-adders and full-adders in a digital electronics course, you can appreciate that doing math with BCD is much more complicated than with binary. Incidentally, the two-numbers-to-a-byte arrangement is called ***packed BCD***. BCD notation can represent fewer numbers, and it can't efficiently represent non-numeric characters, punctuation, or codes for computer instructions.

## Binary to Decimal

Conversion from binary to decimal is quite straightforward for integers.[7] *Integer* division is different from what you have on a pocket calculator because the division throws away (truncates) the remainder.[8] For example, in integer math $25_{10}$ / $2_{10}$ is exactly $12_{10}$. The half is lost.

The C "mod" (%) operator takes the remainder after a division. For example $16_{10}$ % $10_{10}$ is $6_{10}$. In this case, the operator removes tens from a number until there are no more left, leaving six ones. In the same way $8674_{10}$ % $10_{10}$ is $4_{10}$. Notice that I'm using decimal notation even though the number is actually in stored in binary and can also be expressed in binary or hexadecimal. People still think in decimal. If you prefer, 0x10 % 0xa = 0x6, or even, $10000_2$ mod $1010_2$ = $110_2$.

First, let's convert a ***single unsigned byte (char)*** into its decimal digits. How large can the number be? A byte is no bigger than $255_{10}$, so three digits will suffice. Suppose the number you need to convert is in a *char* variable *X* and the result is to go into three *char* variables *hundreds, tens,* and *ones*. Assume X holds a value of $123_{10}$. First compute *hundreds* = *X / 100.* Remember that integer (or char) math truncates (cuts off) any remainder, so $123_{10}$ / $100_{10}$ = $1_{10}$ (*NOT* 1.23!). Next discard the hundreds from *X* with *X* = *X* % 100. This leaves the remainder by removing all the hundreds, here leaving 23 in *X*. So the next step is to get the tens with *tens* = *X / 10.* In this case, it gives $2_{10}$ because $23_{10}$/$10_{10}$ truncates to $2_{10}$. Finally *ones* = *X % 10* gives $3_{10}$ because that is the remainder obtained by removing all the possible tens from the number. In program form, this sequence is shown next.

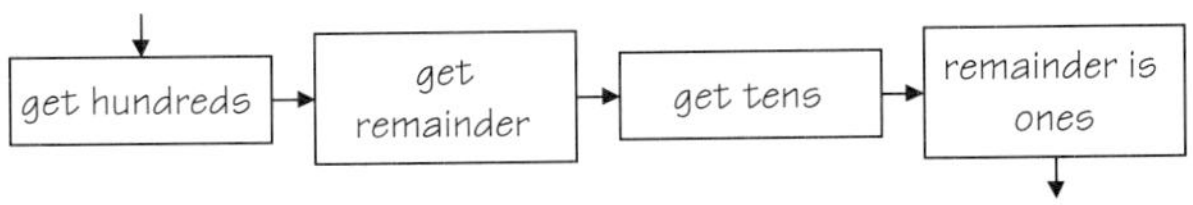


**Binary Byte (char) to Decimal Conversion**

---

[7] I am not going to go into assembly conversions. Binary math is not difficult but, if you are creating your own high-resolution variables you can write your own routines to divide by shifting-and-adding or by successive subtraction. I went through an example on page 203 of *Volume I,* if you enjoy such things.

[8] While even ordinary calculators handle fractions, all scientific calculators either work in or can revert to floating-point math. If they stay in floating point for their (binary) representation of numbers, those calculators then convert to fixed point for the user's display when appropriate.

```
            Byte to Decimal Conversion
uchar x,hundreds,tens,ones;
hundreds=x/100;
x=x%100;
tens =x/10;
ones=x%10;
```

If you have a ***16-bit unsigned integer*** to convert, you can get up to $65{,}535_{10}$ or five digits. The process is an extension of that for a byte and is shown next.

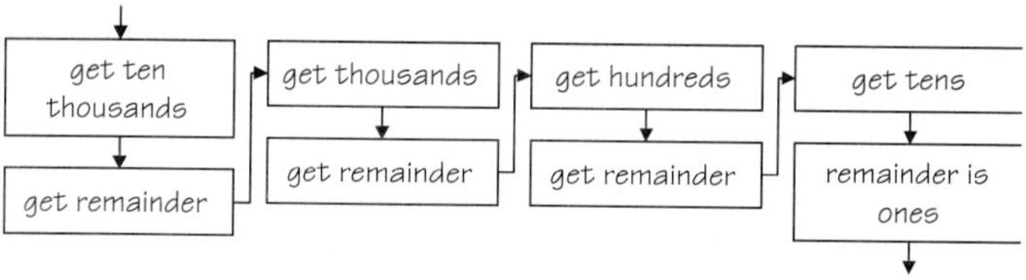


**Straight-line Integer to Decimal Conversion**

```
            Integer To Decimal Conversion
uint x;
uchar ten_thousands,thousands,hundreds,tens,ones;
ten_thousands=x/10000;
x=x%10000;
thousands=x/1000;
x=x%1000;
hundreds=x/100;
x=x%100;
tens =x/10;
ones=x%10;
```

The conversion process so far could become tedious and, if you are looking for code size efficiency, you will want to go to a ***looping 16-bit unsigned integer*** conversion routine like the next one. This routine happens to put the results in an array. Note that the array index, *n*, is incremented *after* the index value is used, so the first digit goes in *digit[0]*. Also the *do while* loop does the changing of *m after* the first value is used. *m* goes down by tens until it reaches 1 and then the next divide brings it to zero which exits the loop. Convince yourself that the loop has the same result as the previous straight-line program above.

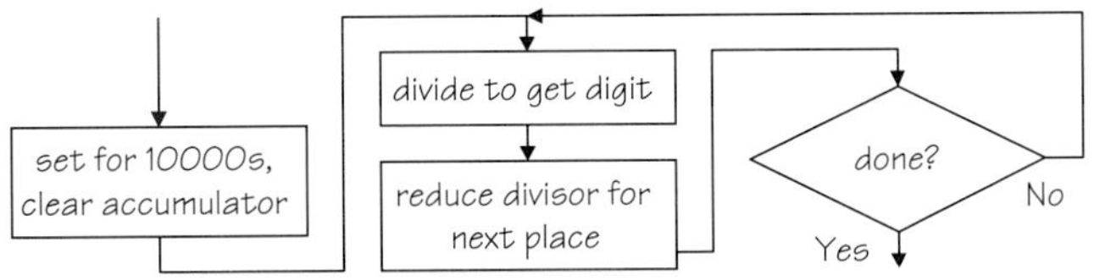


**Looping Unsigned-integer to Decimal Conversion**

**Looping Integer to Decimal Conversion**

```
uint x;
uchar digit[5];
uint m=10000;
uchar n=0;
do{
  digit[n++]=x/m;
  x%=m;       /* same as x=x%m   */
}while ((m/=10)>0); /* same as m=m/10 */
```

## Binary to ASCII

The most common need for decimal representation of numbers is when they are sent out to a display, usually in ASCII[9] coding if it is an LCD module. Adding 0x30 converts a decimal digit to its ASCII equivalent (0x30 is the ASCII code value for zero and can also be written as '0' in C—the *single* quotes cause the compiler to use the ASCII code for the digit or letter).

> **BCD to ASCII:** It is simplest to convert *decimal* digits to ASCII code by adding 0x30 ("0") to the digit's binary value. If you want to convert binary 4-bit groups to *hex* you can add 0x37 to the weights above 9 to get A through F (0xA + 0x37=0x41="A"). A lookup table can do the transformation to hex without a decision.

[9]ASCII (American Standard Code for Information Interchange—the II is not a roman numeral 2!) is a representation of numbers, upper- and lower-case letters, and some punctuation developed for Teletype machines many years ago. While there have been other codes, notably that of IBM, ASCII has become *the* standard. The codes involve only 7 bits, but characters are usually transmitted with either a leading 0 or a parity bit.

As the designer, you will have to decide what to do about displaying leading zeroes. Are the numbers going in fixed locations so you have to insert blanks for leading zeroes? Do you instead display the number left-justified beginning with the first non-zero digit? The next example assumes you are going out to an LCD module and want leading blanks. If not, just delete the line with the ‘ ‘ in it.

The following display function uses the conversion approach of the previous example but assumes there is a function named *lcddat* that sends a single character to an LCD module's memory (along with *lcdcmd* that sends a character to the module's command registers). The example specifically allows you to have a decimal place ahead of the last two digits if, for example, you want to show dollars and cents or if your integer represents the number of hundredths in the number.

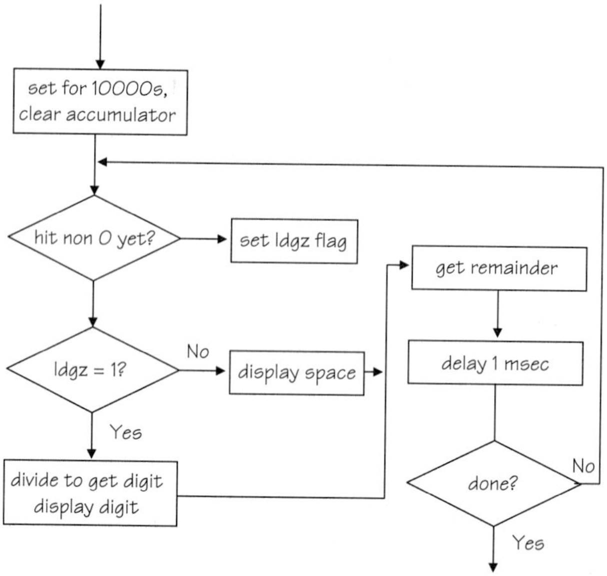


**Multi-digit Conversion and Display**

---

**Function to Display Integer on LCD**

```
void numdsp (uint num, bit dpt){
  uint m=10000; /*divisor to work down by digit*/
  bit lz=0;/*leading zero flag:1= show zeroes*/
  do{
    if(((num/m>0)||(m==1))||((m==100)&&(dpt==1)))
            lz=1;
```

```
      if(lz==1)lcddat(num/m+'0');
      else lcddat(' ')
      num%=m; /* same as num = num % m; */
      if((dpt==1)&&(m==100)lcddat('.');
   }while((m/=10)>0);
}
```

You can make more flexible display routines that, for example, put the decimal point wherever you wish, choose whether to blank leading zeroes, display them, or right justify (as the example above). You can make a function where you pass in a parameter to indicate where to position the number on the display. The more general-purpose you make the display routine, the less obvious the routine is to the casual reader, but the more useful it is when you have a variety of numbers to display.

> **GENERAL-PURPOSE FUNCTIONS:** You are better off in the long run to make the most general-purpose function you can. Then all your applications can use the same one.

**LCD Display Functions for Text and Numbers**

```
#include<reg51.h>
#include<absacc.h>
#define forever for(;;)
#define uchar unsigned char
#define uint unsigned
#define lcddat XBYTE[0xc000]   [10]
#define lcdcmd XBYTE[0x8000]
#define pcmd XBYTE[0x2003]
#define PORTA XBYTE[0x2000]
bit newkeyflg,newnumbflg,goflg;
struct {uint abstime;uchar pattern;} next [20];
uchar i,j,entry,x;
uint nowtime,scan,oldsc,newsc;

void display(uchar startloc,uchar *s){
lcdcmd = 0x80 | startloc; msec(1);
  while(*s){
```

[10] The assumption *here* is that the LCD is memory-mapped at two addresses where one of the high address bits is tied to the *cmd/da* input. If you are port-addressing, you will need to write *functions* for *lcddat* and *lcdcmd*.

```
    lcddat = *s++; msec(2);
   }
  }

  void numdsp(uchar startloc,uint number,bit dpt){
   lcdcmd = 0x80 | startloc; msec(1);
   uint m = 10000;
   bit ldgzero = 0;
   do{
    if ((((number/m)>0) || (m == 1)) || ((m == 100) &&
                (dpt == 1))) ldgzero = 1;
    if (ldgzero == 1) lcddat = number/m+'0';
    else lcddat = ' ';
    msec(1);
    number %= m;
    if ((dpt == 1) && (m == 100)) {
     lcddat ='.';
     msec(1);
    }
   }while((m /= 10)>0);
}
```

There is a later section (*User Entries*) that discusses taking in user entries in decimal and converting them to binary (or hexadecimal!).

## SCALING AND VARIABLE SIZE

Many people confuse pocket calculators with microcontrollers. Most of these calculators at least use fractions, if not floating-point representation. While they give high precision, they need not be very fast. After all, how quickly can a user enter numbers by pushing keys? While microcontrollers *can* be programmed to process floating point numbers, for the 8051 at least, its result is a lot of additional software from libraries to carry out the operations. This consumes more code space, along with more computer time. A large percentage of embedded applications don't *really* need fractions or floating point numbers.[11] It is much more efficient to stay in one- or two-byte representation (*char* or *int*) if you can maintain sufficient accuracy.

[11]Again, this is based on experiences with students who have just survived extensive math-oriented courses. The saying is, "If the only tool you have is a hammer, you tend to treat everything as a nail."

**AVOID SIGNED OR FLOAT VARIABLES:** It is almost always more efficient to keep variables *as small as possible* to avoid needing extra library functions and to have the fastest calculations.

And *that* is the key. What is *sufficient* accuracy? Take, for example, a thermostat-setting for a home. It is highly unlikely that any user will want a resolution of greater than 1-degree Fahrenheit (perhaps 1/2 degree Celsius?). If the range of comfort in a home is from 50 to 90 degrees Fahrenheit, that breaks down into a resolution of only forty units. That easily fits in *1 byte* per reading. Assuming your software knows that the base is 50, a reading could fit in 6 bits.[12] But an inefficient programmer would keep the readings in integers, using *2 bytes* per reading. A *very* inefficient programmer would use floating-point variables because that makes fractions easier. Now each reading takes *4 bytes*! If efficiency is your goal, consider how big a number you *really* need.

Often your projects take in values from external sensors. A temperature sensor, a humidity sensor, a distance measurement that arrives as a time-delay value—the list goes on and on. How can you get from the "raw" number to the number to send to a display? The calculation for linear data is not complex, but if you set it up carefully you can avoid fractions and floating-point variables.

## Gain and Offset Scaling—Linear Conversion

Suppose you have a temperature sensor that comes into an 8-bit A-D converter. Assume you get in 4.5 V when the temperature is 95°F and 2.1 V when it is 43°F. Further suppose you want to have the temperature stored to the nearest degree in a byte (*char*) variable. How can you process the A-D reading to get the number you want? The mathematics involves getting the *slope* and *offset* (or the *zero* and *span* for you controls folks).

[12]If there were *very* limited variable storage, these readings could even be packed to get *four readings in 3 bytes.*

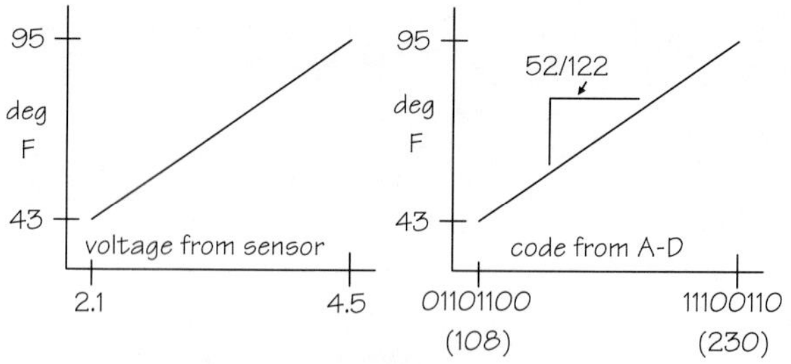


**Linear Scaling Example**

First convert the voltages to the numeric values expected from the A-D converter. Assuming full scale is 5.0 V,[13] the expected binary readings from an 8-bit converter are:

$$4.5/5.0*256=230_{10}=11100110_2 \text{ and } 2.1/5.0*256=108_{10}=01101100_2$$

Next determine the slope:

$$( 95 - 43 ) / (230 - 108 ) = 52 / 122$$

So you have a spread of 52°F giving an output reading change of 122°F. For a new reading, the amount above 108°F is the amount above 43°F. In other words:

```
temperature = 43 + (52/122)(temperature_rdg-108)
```

Now this is all well and good, but how do you compute this in your microcontroller without using fractions (requiring floating point)? The key is in the *order* of the operations. As long as your biggest value will not overflow, always do the multiplication *before* the division. The scaling operation above (in C) becomes:

```
uchar x, t;
t=((uint x-108)*52/122)+43;
```

[13]Some A-D converters might give either 5.12 V or 2.56 V for full scale, giving a basic step in voltage of 0.01 or 0.02. If so, you would have to use that value in place of 5.0.

Since a value up to 256 could legitimately come from the A-D, the intermediate value when the multiplication by 52 has happened could get larger than a byte (actually $7996_{10}$ in this case) but would be well within an *unsigned int* value.[14] This calculation would give a temperature of 73° F for an input binary code of $180_{10}$. A more mathematically precise answer, if you really needed more digits, is 73.69. If the answer should be expressed to the *nearest* degree, let the integer truncation *round* by adding a $61_{10}$ before dividing by $122_{10}$—in this case the answer then comes out 74° F.

```
uchar x, t;
t=(((uint x-108)*52)+61)/122)+43;
```

Use this same process for *any* linear scaling. Always be sure to test a few sample values to make sure the answers are reasonable—particularly at the extremes where overflow might happen. This is where simulation software such as DS51[15] can be most convenient.

> **MULTIPLY BEFORE DIVIDING:** To preserve accuracy in multiplying by a fraction with integer variables, always multiply before dividing to maximize the resolution.

### Scale the Math to Just Fit

If you come upon a scaling situation where there isn't enough variable space to fit the multiplication *before* the division without overflow, it may be appropriate to pick smaller values for the numerator and denominator. While the fit happened to be no problem in the example above, you could scale down the $52_{10}$ and the $122_{10}$ in proportion—$26_{10}$ and $61_{10}$ would lose nothing and $13_{10}$ and $30_{10}$ would give sufficient accuracy in that example. I know those of you from the pocket-calculator generation hate to allow any calculation to add even the slightest bit of "error," but you should remember that all the digits in the world will not make the answer more accurate than sensor accuracy can justify. You can even *discard* accuracy if it satisfies user needs! If your humidity sensor is only accurate to 10%, you're fooling

---

[14]The *cast* making *x* an unsigned integer may be necessary if *x* is brought in as a *char* so that the math will be done with 16-bit *int* values. Some compilers would automatically go to *int*, but a good efficient one should do only what you request.

[15]Described in *Volume I* (and included on the CD in both volumes.)

yourself (and everyone else) if you express the scaled reading to 0.1%. Even if it *is* accurate to 0.1%, the user may only be interested in the reading to the nearest 1% or 5%—why clutter up the display with irrelevant details? The best way to check accuracy of a "chopped down" scaling operation is to run some sample data through the equation and then compare the results using a calculator. Efficient design and programming should avoid computation if it isn't needed to meet the error budget.[16]

## LOOKUP TABLES

If the resolution of your input is coarse, you will always be faster in execution and usually more efficient of code space using a look-up table. You might choose a table even where the input/output relationship is linear, but the lookup table is especially useful where the relationship is non-linear.

When would you have a non-linear relationship? Perhaps you have an input of light transmitted through a coating and want to convert that to a density measurement. I think that is a logarithmic relationship. Or you might decide to use an inexpensive thermistor for a temperature sensor by putting it in a resistor divider circuit (definitely not linear). I think the thermistor is logarithmic, but included in a leg of a voltage divider it is better to test the idea experimentally before going to equations. The point is, with many sensors you have a non-linear (but hopefully predictable and repeatable) relationship to include in your data processing.

**PREFER LOOKUP-TABLES:** If the calculation is at all complex, tables are faster and often entail less code space.

Another non-linear relationship arises for trigonometric functions—sine, cosine, tangent, arcsine, and so on. These show up in relationships involving angles such as power factor for a power meter, coordinate conversions for a rotating platform or robot, and so on. Why not just use the trig functions that come with your C library (obviously not with assemblers!)?

---

[16]You could argue that the processor loading is so light in your application that there is more than enough free time to handle floating-point calculations. If you also have lots of EPROM space unavoidably free as well, I can't argue against you except to say that things might change in the next project that includes your already-developed software, making the efficiency important *then*.

For starters, if you were using the 750/1 processor, the missing instructions might make it impossible to use the libraries. But whatever relationship you are using, you might want to totally avoid use of floating point to increase computation speed or reduce code size. If the resolution or accuracy requirements are not too high, a simple table of a few values might be sufficient.

Later, I'll discuss adding linear interpolation between points and refer to several articles that cover the more elaborate algorithms. If you think about it, even the compiler's library functions have to use *some* algorithm and, with a limited range of values to process, you might be able to do even better.

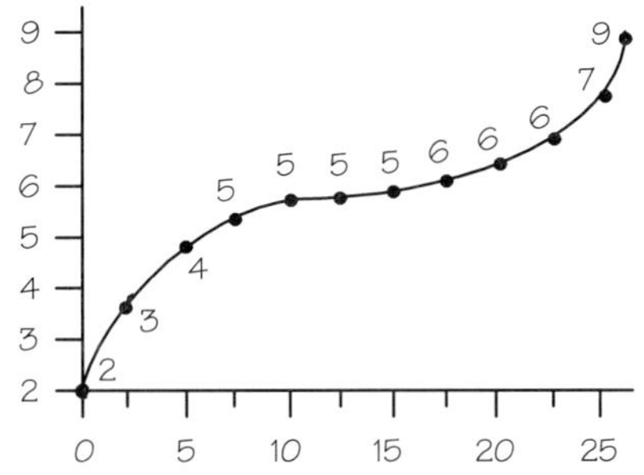


**Non-linear Conversion with Regular Spaced Input**

Faced with a non-linear relationship, if you come from an engineering background your first instinct is to try to obtain from the manufacturer the characteristic equations for the devices involved. Failing in that, you might get samples of the devices and run experiments to get empirical data on them (hoping all the time that the devices will behave the same in future lots). In any case, when the process finishes, assume you have data or an equation describing the desired output as a function of the input that is *not* a straight line.

Many engineers would start looking for a best-fit polynomial equation. The issues surrounding these equations and the acceptable degree of fit are beyond the scope of this book and are to be avoided in a micro wherever possible! Remember such math works well in calculators, but is inefficient in a microcontroller. The size of the floating-point library could be quite large, and the execution could be quite slow.

### Tables with Regular *Input* Steps

If the output data for a lookup table can be expressed as a function of ***regular, linear input* steps**, you can use a single column table (a one-dimension array) listing the resulting values for each step of the input. The index through the table is the increment of the incoming value. Computing this is very fast because you do not have to search through the table—the index takes you directly to the answer. The index can be regularly spaced without covering every step as shown on the previous page where the data is only for every 25th value.[17] A function to do the conversion is shown below. The number 12 is added so the integer division will round to the *nearest* point rather than to the *lower* point. This is a very fast conversion algorithm since it compiles to one 8-bit division and a fetch from an array—16 μsec at 12 MHz—and only 12 bytes of program code space (or 25 bytes if you count the table itself).

**Non-linear, Non-interpolated Conversion**

```
#define uchar unsigned char
uchar code d[12]=
              {20,38,48,53,57,58,59,61,63,69,78,90};
uchar convert(uchar x){
  return d[(x+12)/25];
}

; FUNCTION _convert (BEGIN)
;---- Variable 'x' assigned to Register 'R7' ----
0000 EF                 MOV     A,R7
0001 240C               ADD     A,#0CH
0003 75F019             MOV     B,#019H
0006 84                 DIV     AB
0007 900000   R         MOV     DPTR,#d
000A 93                 MOVC    A,@A+DPTR
000B FF                 MOV     R7,A
000C 22                 RET
; FUNCTION _convert (END)
   CODE SIZE            =     13
   CONSTANT SIZE        =     12
```

### Example: Golf Driving Range

An actual example came up recently with a student who was developing a senior project in the Electrical Engineering Technology (EET) program here at Purdue. This student had a set of sensor beams—one at the tee

[17]The curve is the same as in a later section (page 332) where linear interpolation is used between points.

and additional sensor beams 5 feet out—to detect a golf ball at the start of its travel and a short distance along its path (it is actually trapped in a net just after the 5-foot point). By timing the travel of the ball and determining the angle, the project should compute the anticipated distance for the drive. The result should be expressed to the nearest 25 yards up to 300 yards. The student's problem, when I was approached, related to a detail of the floating-point math that was "necessary." It seems the problem related to the conversion of numbers in and out of floating point.

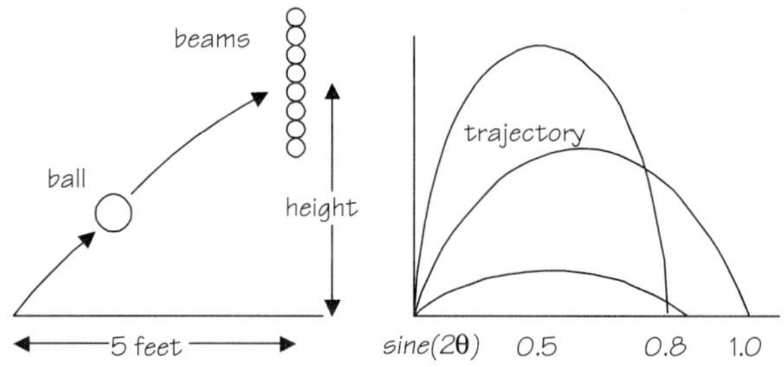


**Golf-Game Trajectories**

The student was using an equation for the distance:

$$\text{DIST} = V^2 \,\text{sine}(2\theta)/g$$

where *DIST* is the horizontal distance of travel, *V* is the velocity at the start, θ is the angle of initial travel up from horizontal *(45° × 2 = 90°* and *sine(90°) = 1* which is the maximum distance), and *g* is the acceleration of gravity.[18]

Since absolute accuracy was not the goal and high precision was irrelevant, a lookup table cried out to be used. Assume the table only gives answers for 25-yard increments—that gives thirteen entries to cover the 300-yard range. The best approach is to pre-calculate the table entries to the most useful form. I persuaded the student to use a 250 μsec interrupt driving a software counter of time (easily done with the 8-bit auto-reload mode for

[18] As is often the case with mathematical formulae, I am quite sure this expresses the distance assuming the same elevation for the end point as the start and ignores any wind resistance—a drive in a vacuum if you wish. This doesn't take into account the friction of the wind, the shape or size of the ball, or any wind velocity. At this point, that is too much to worry about, and besides, since the ball never leaves the office, no one can prove you wrong! The only *real* purpose of the project is to give opportunity to improve one's drive, and *relative* improvements will be readily shown.

the standard 8051 timers). The time to travel the 5-foot distance then turns out to be about 110 of these ¼-msec intervals for a velocity that gives a travel of 300 yards. Notice that I don't have to square anything—just solve the equation for *V* at each of the *DIST* values of interest and convert those to times to travel five feet in time units of ¼-msec. The programmer does all the calculation, in advance, with one of those pocket calculators that thrives on floating-point math. The software in the microcontroller just looks up the right answer.

Trusting that the 110 is right for 300 yards (I'm relying on the student's math), all the other answers relate by:

$$T \propto \frac{1}{\sqrt{DIST.}}$$

The program fragment that follows shows how the table can be used.

**Lookup Table for Optimum Distance**

```
uint dt[]= {0xffff,381,269,220,191,170,156,
                144,135,127,121,115,110};
uint dist;
uint T;
uchar I;
19for(I=0;dt[I]>T;I++);
dist=25*I;
```

Next I need to take the *sine(2θ)* into account. Ideally at 45° the ball goes the furthest, and higher or lower angles reduce the distance of travel. Since high accuracy isn't needed, again this is a prime candidate for a table. In the case of this particular student, the sensing of angle is determined by a series of horizontal beams at the 5-ft point. The steeper the angle of travel, the higher the ball will be at that point. By spacing the beams and detectors right, the ball has to interrupt one at least one beam—the particular one interrupted can correlate to an angle. I don't know the specifics of the student's arrangement, but lets assume there are sixteen beams, half above and half below the 45 point (centered about 5 ft up, at 5 ft from the tee). Again, math and geometry are incidental to the methodology here—all that is needed is the ability to establish the angle. I'll assume that the beams are

[19]This *for* loop walks through the table until it gets to the time that is smaller than the target value. Since the table entries are for each 25-yard increment, the returned value is multiplied by 25 to give the distance. Note that the *for* loop has nothing in the loop itself—just the semicolon.

wide enough to give a detectable signal at a 4-in spacing (perhaps detecting a small dip in brightness because the ball doesn't totally obstruct the beam).

**Distance Multiplier (Ht @ 5 ft)**

| ht | rad | dist | ang | ht | rad | dist | ang |
|---|---|---|---|---|---|---|---|
| 30 | 0.464 | 0.8000 | 26.6 | 62 | 0.802 | 0.9995 | 45.9 |
| 34 | 0.516 | 0.8579 | 29.5 | 66 | 0.833 | 0.9955 | 47.7 |
| 38 | 0.565 | 0.9040 | 32.3 | 70 | 0.862 | 0.9882 | 49.4 |
| 42 | 0.611 | 0.9396 | 35.0 | 74 | 0.889 | 0.9784 | 51.0 |
| 46 | 0.654 | 0.9657 | 37.5 | 78 | 0.915 | 0.9665 | 52.4 |
| 50 | 0.695 | 0.9836 | 39.8 | 82 | 0.939 | 0.9531 | 53.8 |
| 54 | 0.733 | 0.9945 | 42.0 | 86 | 0.962 | 0.9385 | 55.1 |
| 58 | 0.768 | 0.9994 | 44.0 | 90 | 0.983 | 0.9231 | 56.3 |

Basically I need the *sine of twice the arctangent* for 30/60 up through 90/60 as the values in the lookup table. Then depending on the particular beam that is interrupted, the angle is known and thus the multiplier. To avoid fractions, the table values are multiplied by 100 corrected with a divide by 100 at the end of the calculations—integer math should be perfectly adequate for the desired resolution in this project.

The program fragment then would be:

```
            Angle Correction for Distance
uchar am[]={80,86,90,94,97,98,99,99,99,99,99,
                98,97,95,94,92};
uint beam_pattern;
uchar I;
for(I=0;beam_pattern&1==0&&I<16;I++);
dist=(dist*am[I])/100;
```

This is revisited later when linear interpolation is added.

## Tables With Regular *Output* Steps

If the input is non-linear, but the outputs are only in regular steps, you can have a *single* column table of the *input* values with the resulting outputs corresponding directly to how far through the array you had to travel. This wouldn't be as fast as a single column of *output* values, but it still reduces the size of the array.

## Two-Column Lookup

Where the data is irregularly spaced on *both* axes or there are significantly different slopes over different portions of the curve, to improve accuracy you might want to use a two-column table. The two-column table can be represented in C as a two-dimensional array, an array of two-element structures, or two one-dimensional arrays of identical size traversed by the same index. The software walks down the first column of the table to the value nearest the current input and returns the corresponding output value. The number of table entries depends on your necessary resolution. For example, if you were converting a voltage from a humidity sensor into a reading accurate to the nearest 10 percent, you would need no more than nine entries. If the input came from an 8-bit A-D converter, at the very worst you would need 256 entries.

Usually by the time you have gone to two columns, you are looking for higher accuracy between points, so interpolation is discussed next.

## Linear Interpolation

Suppose you need better accuracy than is possible with a short look-up table. You can, of course, just make a much longer table with a finer resolution, but somewhere the space exceeds the space required for the mathematical approach you were trying to avoid. While a huge table might be faster than using math algorithms, you may be able to fit the curve to the data points of a table in a "piecewise linear" fashion to get sufficient accuracy with a small table.

If you plot the input/output conversion curve, no matter what the shape, a series of short straight-line segments can approximate it. If the curve is changing slowly, only a few segments might suffice. Where it changes rapidly, you may need shorter segments. Whatever the shape, the end points of the segments can describe the straight-line, and can be put in a lookup table.[20] Anywhere between the end points of a segment can be generated by linear interpolation—nothing more than multiplication and division as was used earlier in the section on scaling. Again, if you work it right, the math can probably be simple integer math that is faster and uses less code space.

---

[20]The end points can be purposely put off the actual curve so the points near the middle of the segment come closer to the curve.

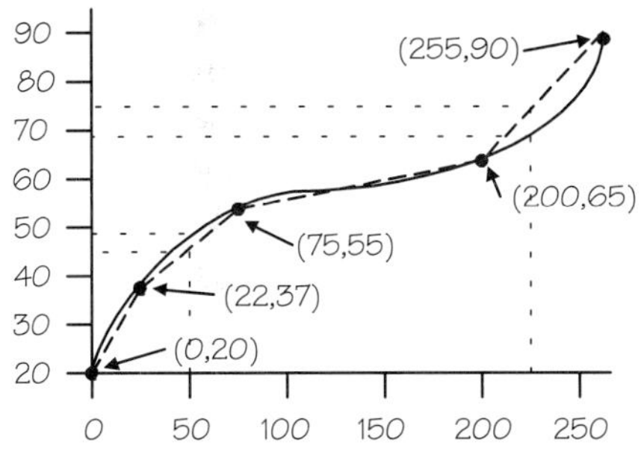


**Piecewise Linear Fitting**

For a specific example of linear interpolation, assume the same curve as on page 325. The assumed piecewise linear fit with the endpoint coordinates is shown in the graph above. The two program fragments that come on the next page show both a two-dimensional array (that may be more difficult to follow) and an array of two-element structures.[21] Notice that, as discussed earlier, the math is arranged to multiply up the value before dividing so integer math is sufficient. Again, the first variable is cast to an integer so the entire math sequence will use *int* (16-bit) rather than *char* (8-bit) values.

If the input reading is $50_{10}$, a visual estimate of the corresponding output value is about $48_{10}$. Based on the linear interpolation between the two line segment end points, (22,37) and (75,55), the value should be about 46.5 (truncating down to $46_{10}$, because of the way the integer math is done).

```
37 + ((50-22)/(75-22))(55-37)= 37+(28/53)18 = 46.5
```

This error, assuming the curve and my visual estimate are exact, is about 4 percent of the reading or about 2 percent of full scale. Is that a problem? That depends on the justifiable accuracy and resolution of the transducer and converter as well as the desired accuracy of the output. Most physical measurements associated with small, embedded systems don't need more accuracy. Many applications don't need scientific precision—just an approximate answer is sufficient. For the curve we just considered, over near input values of 225 the error will be even worse for the line segments shown. The fit could be improved by choosing endpoints not exactly on the

[21]You could also use two identically-sized one-dimension arrays that might speed up the actual calculations—test the results each way if speed is critical.

curve so the values near the middle of the segment would come closer to the value of the curve. Likewise, more segments could be used to give a closer fit.

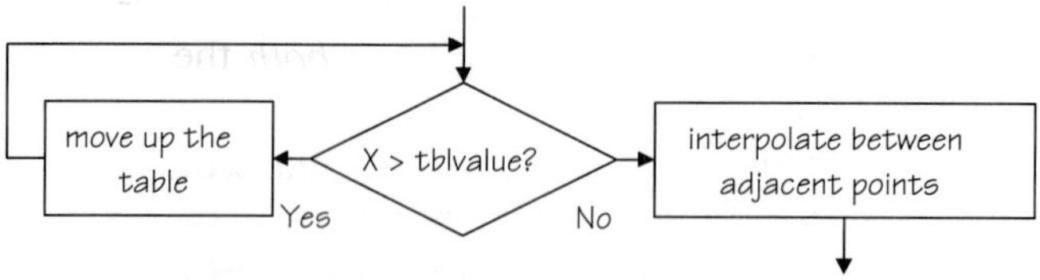


**Flowchart for Linear Interpolation in Two-dimensional Array**

**Linear Interpolation—Two-dimension Array**

```
uchar convert(uchar x){
22uchar d[5][2]=
                {{0,20},{22,37},{75,55},{200,65},{255,
                90}};
  uchar i=0;
  while(x>d[i++][0]); /* find segment */
  return(d[i][1]-d[i-1][1])*(x-d[i-1][0])/
                (d[i][0]-d[i-1][0])+d[i-1][1];
}
```

**Linear Interpolation—Array of Structures**

```
struct pair{uchar x; uchar y;}
23struct pair d[5]=
                {{0,20},{22,37},{75,55},{200,65},{255,
                90}};
uchar conv(uchar num){
  uchar j=0; uchar i;
  while(x>d[j++].x);
  i-j-1; /*find seg*/
  return(uint(d[j].y)-d[i].y)*(num-d[i].x)/ (d[j].x-
                d[i].x)+d[i].y;
}
```

One additional step that could reduce some of the run-time computation is to fill the table with the *difference* values rather than the absolute val-

[22]This two-dimensional array holds the data points. Note that the data is inserted by careful use of two sets of curly braces {{}}.

[23]Here a specific instance of the structure template *pair* is defined with a name of *d* for the specific instance. Remember that a structure reference involves a period.

ues. The first value in the table could be the absolute value and the rest would become (22, 17), (53, 18), (125, 10), (55, 25). During the search for the right segment, the values could be built up of the differences and when the correct entry is found, the differences are already available.

A final speed-up step would be to have *both* the absolute values and the differences pre-computed in the table. In the given example the additional 8 or 10 bytes of code space used in the table would probably be offset by the reduced code involved with the math.

### Example: Golf Revisited

Now lets revisit the golf project's (on page 336) distance and angle tables and add some interpolation. The new procedure is to get the bracketing distances, interpolate to get a finer distance estimate, multiply by the angle factor, and then quantize the result to express a realistic resolution.[24] Below you see the interpolation added to the first program fragment and then combined with the angle multiplier. It makes a good illustration of how to make C incomprehensible to the casual reader—comments would be strongly recommended!

**Table-Based Golf Distance Computation**

```
uint dt[]= {0xffff,381,269,220,191,170,156,144,135,
            127,121,115,110};
uchar angmult[]={80,86,90,94,97,98,99,99,99,99,
            99,98,97,95,94,92};
uint beam_pattern; uint dist; uint T; uchar I;

for(I=0;dt[I]>T;I++);
25dist=25*(I-1)+(25*(dt[I-1]-T))/(dt[I-1]-dt[I]);
for(I=0;beam_pattern&1==0,I<16;I++);
dist=(dist*angmult[I])/100;
```

## MATH ALGORITHMS

I had planned to discuss some math algorithms with ways to implement limited versions for specific purposes, but I think that the library functions that come with the compiler are sufficient for most applications. If you need

[24]It would be very misleading to round to the nearest 25-yard value *first*, multiply by the angle factor, and *then* express it to the nearest 25-yard value!

[25]The computation of the interpolation factor involves backing up from the selected table entry by one—the segment the actual time falls on—and doing a linear interpolation.

exponents and trig functions, the *math.h* header below describes the standard C functions available. Part of development efficiency is to not develop code yourself if equivalent or better code is already available.

```
        MATH.H: Prototypes for Mathematic Functions
extern char  cabs  (char  val);
extern int   abs   (int   val);
extern long  labs  (long  val);
extern float fabs  (float val);
extern float sqrt  (float val);
extern float exp   (float val);
extern float log   (float val);
extern float log10 (float val);
extern float sin   (float val);
extern float cos   (float val);
extern float tan   (float val);
extern float asin  (float val);
extern float acos  (float val);
extern float atan  (float val);
extern float sinh  (float val);
extern float cosh  (float val);
extern float tanh  (float val);
extern float atan2 (float y, float x);

struct FPBUF  {
  unsigned char save[16];
};

extern void fpsave (struct FPBUF *);
extern void fprestore (struct FPBUF *);

extern float ceil  (float val);
extern float floor (float val);
extern float modf  (float val, float *n);
extern float pow   (float x, float y);
```

There are situations where the standard functions are not suitable. If you are working with a very limited memory space and must stay in char or int notation, you may have to develop your own math functions. Likewise, if speed of execution is crucial and you know, in advance, that you have only a limited range of numbers to handle, you can design-in shortcuts. That, of course, requires that you write your own algorithms in place of the more general ones.[26]

[26]While the *general* library functions are probably highly optimized, they must handle *all* cases. Usually they end up using floating-point variables to ensure that overflow and under-

If you come to the end of all this and *still* don't have enough accuracy, you're stuck with floating-point variables. The better compilers support floating point and will take care of the details if you use the right library. On the previous page is the header file that Keil provides to describe all the available functions. Note that they almost all involve floating-point variables. I believe conversion to or from other variable types is invoked automatically by assigning a *float* variable to an *int* or vice versa. If you use any floating point variables, be sure the *c51fps.lib, c51fpl.lib,* or *c51fpc.lib* file is included in the linking process.

If you only use certain functions, most linkers will leave out unneeded code, saving some space. If you *must* have floating point functions for one part of the job, all the efficiency rules change and it costs no more code space (other than RAM for storing 4-byte variables) to use them over and over. Your scrimping and squeezing to work with *unsigned chars* and *unsigned ints* is largely wasted, although you still take the speed penalty of computing in floating point. The main argument against using floating point is that probably 99 percent of the embedded applications that use it do so in ignorance of the alternatives. Most applications don't need such accuracy because most inputs aren't very accurate anyway, and the end user probably doesn't have any use for so much detail.

**RE-USE OF FUNCTIONS:** If you *must* have a given floating-point math function, you add no more code to use it multiple times. However, each different function used adds code—the whole library is not included unless you use all the functions.

## SMOOTHING AND AVERAGING

It may be useful to maintain a sliding average of readings to add noise rejection. In the simplest form you just save numbers in a first-in-first-out (FIFO) buffer with a looping pointer.[27] If you want to get fancier and are processing

---

flow are not a problem. In doing so, they immediately expand the size of the code markedly and reduce code efficiency.

[27] I might add that just these sort of operations form a key to digital signal processing (DSP). There are all sorts of algorithms and hardware designs to improve the efficiency of such calculations and allow them to come in real time. If you find that you are averaging very many numbers in a hurry, you may have a good candidate for DSP.

data from a signal that should not change rapidly, you can reject readings that are too different from their neighbors.[28]

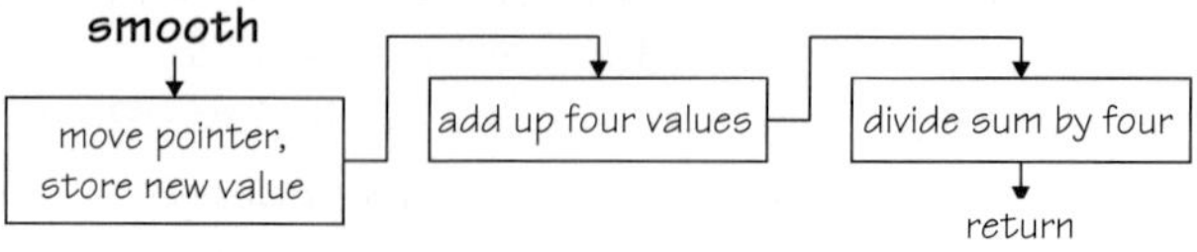


**Averaging Function**

---

**Routine to Average Last Four Readings**

```
#define uchar unsigned char
#define uint unsigned int
uchar sdat;

uchar smooth(uchar newest){
 static uchar data rdgs[4];
 uint data accum,i; static uchar data ptr;
 rdgs[ptr=++ptr&0x03]=newest; /*save reading*/
[29]for(accum=i=0;i<4;accum+=rdgs[i++]);
 return accum>>2;
}

void main(void){
 for(;;){
  sdat=smooth(0x80);
 }
}
```

---

## RANDOM NUMBERS

You may want random numbers for a game ("*I'm thinking of a number between 0 and 100*") or to even up the chances of equal-priority things to happen (collision detection in Ethernet requires each node to pause for a random time so they do not both try again at the same time). Most random

---

[28]A lot of this sort of processing is involved in image processing as well. You can eliminate spots (presumably noise) if they are radically different in value than the neighbors. Likewise, there are algorithms for locating edges or boundaries of images. Again, this is a field outside the scope of this book.

[29]Of course, you could do the summing outside the third condition of the for and could avoid the += operator so the line would be more obvious to a casual reader *for(accum=i=0;i<4;i++)accum=accum+rdgs[i];*

numbers are bounded—you don't want 10,000,000 for the 0 to 100 game, and you don't want an Ethernet node to wait two weeks to try again.

Most of the mathematical random numbers come from algorithms that return a *pseudo*-random number in response to a seed number. They are not truly random in the sense that the same seed provides the same result. Rather, the sequence of returned numbers appears random as you get successive numbers in the sequence or supply different seed numbers. The problem, of course, is that each time the computer boots up the initialization sequence will supply the same seed. Even if you use the power-up contents of RAM, it may be possible that the "garbage" will come up the same much of the time.

If you want a more *truly* random number, if you have the luxury of a clock-calendar chip in the system that remains operational when the power is off, you might derive a seed number from the day and date to produce a different startup each time.

Otherwise, using human input is very effective. Suppose you are coding the "guess the number" game. At the start, after you display a "push the button to start" message, you initiate a rapidly cycling counter that stops when any user input is sensed. Whatever the current count, that is what is processed to become the number to guess. By cycling through 100 numbers at, say, 1 MHz, the user is quite unable to control where in the count they will push.[30] The entire range of numbers passes by in 100 μsec. The precise instant the switch contacts probably cannot be controlled to better than about 50 msec—1/20 sec. As long as the time spent on each number is equal, the probability of each number is equal.

Rather than resetting the count when $100_{10}$ is reached (making that number more likely than 0 through $99_{10}$), either count to a full 16 bits ($65535_{10}$) and take the mod $100_{10}$ result[31] or take the mod $100_{10}$ result *each* time, as shown next.

---

[30]I have confronted this issue in the case of electronic slot machines. The display is actually "rotating" past at about fifteen symbols per second in nine windows. The illusion intended is to suggest that the player can watch the blur of symbols and control the outcome, but the harsh reality is that no human can process information that quickly, nor can they synchronize the timing of a switch activation to that precision. The blur, on analysis with a video camera, is a true representation, but it might as well be a pre-determined random number!

[31]Each number above 35 has 655 chances to come up and each number below 35 has 656 chances—the lower numbers have about a 0.1% higher probability of occurring. If you find that unacceptable (As I have heard said, "You really can't see it from the road."), you can short-cycle the counter and add some delay between counts so the reset to 0 when 100 is reached takes the same as a dummy delay between other counts. Finally, you could count to 101 and then reset with the proviso that if 101 comes up you ask for another user input.

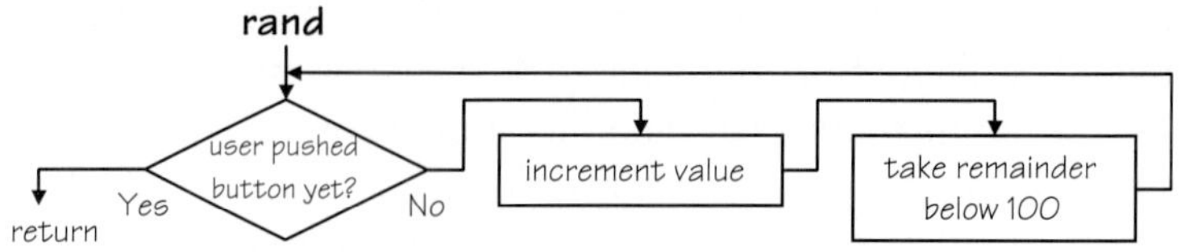


**Random Number from User Input**

**Random Number (0-99) Function**

```
#include<reg51.h>
#define uchar unsigned char
/*assumes user input on lsb of P1*/
uchar rand(void){
  uchar val;
  while(P1 & 1)val=++val%100;
  return val;
}

void main(void){
  uchar x;
  for(;;){
    x=rand();
  }
}
```

## TIME MEASUREMENTS

Measuring time with any 8051-family microcontroller can be very little burden on the processor. Since you have a crystal oscillator already running, it is just a matter of relating the time interval to the number of oscillator cycles. Unfortunately, the first approach taught to most students is to make a software loop and poll the input until the time expires. If you have read Chapters 2 and 3 (or Chapters 11 and 12 of *Volume I*), you recognize the drawbacks of this approach. Nothing else can happen while the software loop times the pulse or time between pulses. Unless there is *absolutely* nothing else needing doing, you are better off measuring time with a real-time clock running off the timer. With a regular interrupt that manages such things, it is a simple matter to count interrupts and maintain a continuous indication of time duration (or time of day) in some global variable. For example, if the real-time interrupt comes every 10 msec, the reaching 100 could signal the incrementing of the *sec* variable. Sixty of those could cause the incrementing of the *min* variable, and so on.[32]

### Software Delays

Looping constructs can produce delays because instruction execution consumes a known amount of time. Examples of such time delay are shown next. While the time is based on the total number of clock cycles, it will also depend on the crystal frequency. Chapter 3 of *Volume I* lists the number of clock cycles required for each machine instruction. With a 12-MHz crystal, twelve-oscillator-period instructions take 1 μsec, and twenty-four-oscillator-period instructions take 2 μsec. All that remain are the *mul* and *div*, which each take 4 μsec. To generalize about individual instructions beyond this is difficult but all 3-byte instructions take 2 μsec, all branching instructions take 2 μsec, all 1- or 2-byte logical and arithmetic instructions take 1 μsec, and the register-to-register moves take 2 μsec. In summary, your best bet is to look it up![33] The example that follows is a delay routine that takes about 1 msec for each unit passed to it. The example using time delay passes a value of 50 to get a delay of about $50 \times 100 = 5{,}000$ μsec = 5 msec.

---

[32]Most of the material here was condensed from *Volume I* (pages 298–302). If you want to keep a calendar you have to decide how accurate the clock needs to be. Is it going to run for days or months? Is it just going to time some non-critical functions? Would it be better to use a photocell and decide when it is daylight by that input? Is the system turned on and off frequently—if so, the use of a *separate* clock may be necessary to avoid resetting the time each power-up (like the old PCs without a battery-driven clock). The accuracy of the time will depend on the crystal accuracy, but that is usually quite good—about 100 PPM (parts per million—1 in $10^4$— about one second per week or one minute per year. That is probably good enough for anything but the bureau of standards!

The bigger problem in generating a software-based calendar is the odd number needed for the divide when the 11.059 crystal is used to get the correct clock rate for 9,600-bd serial transmission. That gives 921,583 counts per second instead of 1,000,000. If you want some even divisor of that to trigger scanning of keyboards, updating of displays, nothing comes out even. The easiest solution is to trigger the 10-msec count every 9,216 clock pulses and then subtract one from the 10-msec counter every fifth second (or something else like that to fine-tune the count much like our calendar adds days on leap years and century years).

[33]Or, alternately, write the program and look at the cycle count as you run it on the simulator—a much better way if you have a lot of instructions or are a lazy, sloppy person like me.

**Delay by Software Looping**

```
[34]void msec(unsigned int x){
  unsigned char j;
  while (x– > 0){
[35] for (j=0;j<125;j++){;}
 }
}
```

**DELAY BY SOFTWARE LOOPING**

```
[36]PUBLIC MSEC
 MSECM SEGMENT CODE
 RSEG MSECM
 MSEC: JZ X ;QUIT IF A = 0
[37] MOV R0,#250 ;250 * 4 = 1000
[38]Z: NOP
  NOP
  DJNZ R0,Z ;4 USEC PER LOOP
  DJNZ ACC,MSEC
 X: RET
END
```

## Measuring Intervals

**Infrequent Events and Low Frequencies.** When events come at a rate of 10 Hz or less, poll at each real-time interrupt using ordinary port lines. By checking at the real-time clock interval (in the example below it is at every millisecond) the software can determine if the incoming line(s) has changed. When the input line first is found to have gone *low*, clear a flag (call it *was_high*) and set a variable (call it *ms_count*) to zero. Increment

[34]This routine can take integer values to produce long delays. It is not precise as written although it is approximately correct based on an analysis of the assembly showing the most inner loop with *j* taking 8 μsec. Different compilers may come out with quite different times and the value of 125 could be adjusted to compensate empirically. Loops are often an inefficient way to get delays. For example, interrupts will stretch out time for anything using software delay loops since the loop will stop incrementing during the interrupt routine.

[35]The null statement is not necessary within the braces.

[36]This module is made public so it could be linked with a program needing a time delay.

[37]The delay could be fine-tuned to compensate for the overhead at the start and the testing the accumulator for the end. Technically the delay is still not precise since there are 4 microseconds lost for the *LCALL* and *RET*.

[38]The *NOP*s bring the inner loop up to 4 μsec so the index (*R0*) can fit in a byte for a 1-msec delay. Note that this routine destroys the value in *R0*.

*ms_count* at each real-time clock tick. At the tick where the line is found to have gone *high*, set *was_high*. At the tick when both the line is found low *and was_high* is set, transfer *ms_count*'s value to the *most_recent_period* variable and reset *ms_count* to zero. Any time the main program needs the most recent period reading, it can immediately access the global variable, *most_recent_period*. Given a 1-msec interrupt, the count will be the number of milliseconds between negative edges (the period) of the incoming signal.

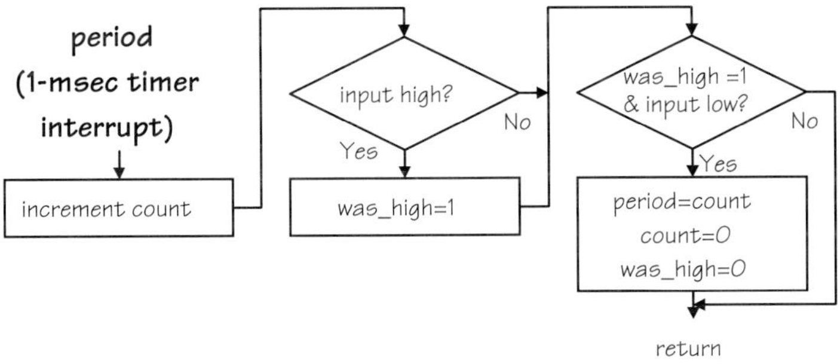


**Low-frequency Measurement**

---

**Measuring Intervals—Low Frequencies**

```
#include<reg51.h>
#define uchar unsigned char
#define uint unsigned
uint most_recent_period;
sbit PIN=0x91;   /*here P1.1*/
/* Assume the incoming signal is on P1.1*/

void msec(void) interrupt 3 using 1{
 static uint ms_count;
 static bit was_high;
 TH1=-1000>>8; /* reset the timer for 1000*/
 TL1=-1000&0x00ff;
 ++ms_count;
 if (PIN>0) was_high=1;
 if (was_high && !PIN){
  most_recent_period=ms_count;
  was_high=0;
  ms_count=0;
 }
}
```

```
void main(void){
 TMOD=0x01; /*initialize real-time interrupt*/
 TH0=~(1000/256);
 TL0=-(1000%256);
 TR0=1;
 ET0=1
 for(;;){
 }
}
```

If you want a number for frequency, the period can be inverted, but avoid introducing floating-point math just so you can have readings in a preconceived way. Even if you must have frequency, careful scaling may allow the math to be done in fixed-point form.

From 10 Hz down, this approach will give at least a 1 percent resolution. If you have very low frequencies, the period measure could overflow a 16-bit variable in 65 sec. To avoid overflow, you could test before the increment each interrupt to see if an overflow is imminent and carry to a second variable (or use a *long* in C). Alternately, you could have a counter that you reset every 1,000 interrupts, at which time you increment a *seconds* variable. You can carry this process to *hours, days, weeks, months, years, centuries,* and so on.

**Fast Events and High Frequencies.** You can count fast events, up to about 500 kHz (2 µsec apart) in the micro's on-chip hardware counter. If there is a 1-msec real-time interrupt, read the count of external events (negative-going edges of the signal) at the start of every interrupt to get the count of events happening in a millisecond. For an incoming frequency of 100 kHz, at each interrupt the count will be 100.

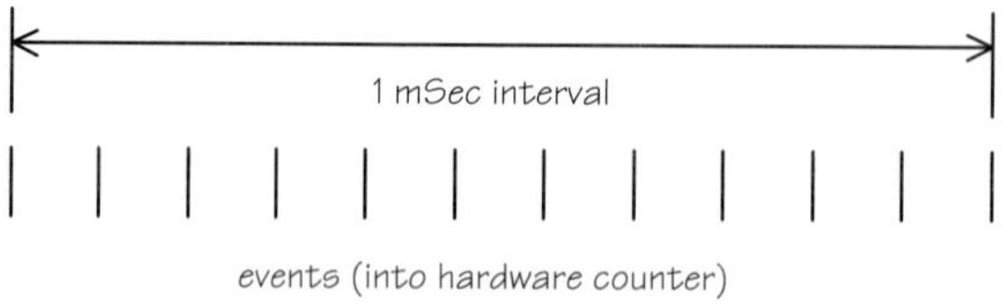


**Counting Events During an Interval**

For most ongoing counting and time-duration operations, do not reset the timers. When the counter rolls over, let it keep on counting. With *unsigned integer* math, subtracting a value obtained *before* the overflow from one *after* the overflow still gives the correct difference! If you read the value in the counter at the *start* of a time interval and then *subtract it* from the value at the *end*, you get the duration of the interval.[39] When the counter rolls over, there is no math problem as long as you treat the readings as 16-bit *unsigned* integers.[40]

If you are measuring a 100-kHz signal, using a 1-msec interval you get a 1 percent resolution. If you need a higher resolution, let the counter accumulate for a longer time. If you subtracted the readings every 100th interrupt (100 msec), the 100 kHz could build up to a count of 10,000 or a resolution of .01 percent. That may be unnecessary accuracy, but it *would* give a 1 percent resolution for a 1-kHz frequency. Notice that this only gives ten readings a second. Somewhere in this frequency range, you could shift to measuring the *duration* of *one* pulse rather than counting pulses for a fixed duration. I discuss this next.

**Measuring Fast Events**

```
uint count;
void tracking (void) using 1{
  uint static newcount, oldcount;
  newcount=TH0<<8|TL0;
  count=newcount-oldcount;
  oldcount=newcount;
}
```

**Infrequent Events and Low Frequencies.** When events come at a rate of 10 Hz or less, it is usually better to write software to watch them at ordinary port lines. Check at the real-time clock interval (here every millisecond). In the interrupt, the software checks whether the incoming line has changed. When the line goes low, say it sets a variable to zero and clears a flag. Each real-time clock interrupt after that, the software increments the

[39]Suppose you use a timer to measure the duration of a signal from a V-F (voltage to frequency) converter. If the counter value when a logic 1 arrives is 3754, and the count when the next logic 1 arrives is 4586, then the period of the V-F converter is 832 clock units or 832 μsec (1.202 kHz), with a 12-MHz crystal (if that count had come with an 11.059-MHz crystal, it would equal about 903 μsec or 1.107 kHz).

[40]Some of the 8051 relatives have a 16-bit ***capture register*** to avoid the possibility of catching the 2 bytes of the count at different states. Otherwise reading the low-byte of 00000001 $11111111_2$ first could obtain $11111111_2$ and, after another count and carry, reading the high-byte of 00000010 $00000000_2$, could result in a wrong result of 00000010 $11111111_2$. For slow count rates, this would happen very seldom, but if it is a major issue and the capture register is unavailable, your software could compare several readings to discard inconsistent ones.

variable. At the first interrupt when the line is found to have gone high, it sets the flag. At the interrupt when the line is found low *and* the flag is set, the software transfers the variable's value to a "most recent period" variable and starts over. Anytime the main program wishes to know the current reading it can pick up the stored value at once. Given a 1-msec interrupt, the reading will be the number of milliseconds between negative edges (the period) of the signal.

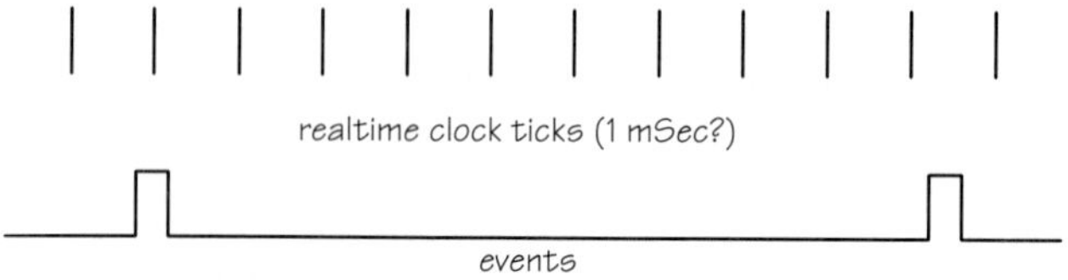


**Counting Intervals Between Events**

**In-between Frequencies.** If you have been alert, you realize that, for a 1-percent resolution, there is a band between 1 kHz and 10 Hz that is missing. The rate is too slow to measure in a reasonable time by counting and too fast to measure by counting the 1-msec interrupts. What are the choices?

First, you could speed up the interrupt to perhaps 100 μsec. In this case, you accumulate the period in 100-μsec units instead of 1-msec units. This might mean the overhead eats up most of the processor time—especially if there are many other things to do each interrupt.

Second, you can count events over a longer time and accumulate the frequency over a longer time. A 100-Hz signal can be resolved to 1 percent in one second. If you do not need frequent readings, this may be quite acceptable. It does not divert processor time from other activities. You can program the real-time clock to count 1,000 or even 10,000 interrupts before reading the counter.

Third, the event could trigger an interrupt of its own. At the instant the interrupt comes, the internal timer can be read (remember it is counting at 1 MHz so it resolves to the nearest μsec). Since you already have a timer generating the 1 msec interrupt for the clock, just capture the current value in that hardware timer and combine it with the count being accumulated in the real-time clock's variable. Essentially you get a high-resolution time stamp for each event and can subtract to find the period. The math may not be trivial, but it is *do*able. In this case you have freed up the second counter, but tied up one of the external interrupts.

**Broad Range Frequencies.** If you must handle a broad range of frequencies, I recommend the time stamp just described. Combining the instantaneous counter reading with the real-time clock variables,[41] the interrupt can capture the exact instant to a resolution of 1 μsec on out to eternity or to when the system is turned off, whichever comes first. With such a configuration, the hardware hookup does not change. You can even use external hardware to switch in various sources to the single interrupt for successive measurements of various frequencies.

**Interval Measurement—Time Stamp Method**

```
#include<reg51.h>
#define uint unsigned
extern uint msecs;
uint countusecs, countmsecs;

void event(void)interrupt 0 using 1{
 static uint usecsold, msecsold;
 uint usecsnew;
 usecsnew=TH1<<8|TL1;
 countusecs=usecsnew-usecsold;
 countmsecs=msecs-msecsold;
 usecsold=usecsnew;
 msecsold=msecs;
}
```

## Real-time Clock Chip

This is a choice to avoid if possible. It entails an extra chip and a battery to maintain the counting, as well as adding the issue of hardware-interface to the microcontroller. I find that many students include one because they are afraid of the software involved with maintaining day, date, and hour in the controller. The only really valid reason for a separate chip is the need to maintain correct time even when power goes off for extended times. (Even then it may be possible to maintain the processor in operation with batteries if the power drain is not high). The clock chips are very good at tracking time and recording days, months, and years with even leap year covered. They do take very little battery power.

I want to take you through two of these chips to give you some idea of the interfacing and use of such chips. First, I describe the Dallas DS 1302.[42] It is an 8-pin device with a 3-wire serial interface common to many new de-

[41] I assume you have variables that accumulate seconds, minutes, hours, and so on.

[42] An 8-page data sheet is available over the web—it is from that that the following example is developed.

vices. It adds a charger feature (over the DS 1202) so the normal power supply can recharge the battery that keeps the clock alive when the power is off. If you are concerned with keeping track of time on a battery, this device can do so drawing no more than 300 nA at 2 V (more when addressed or running off 5V).

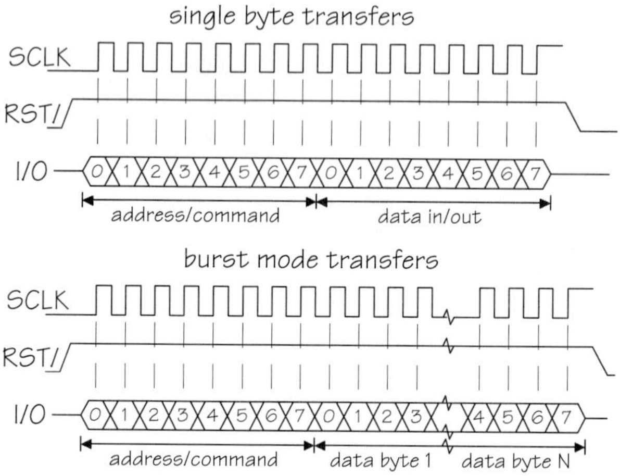


**Serial Interface Waveforms for DS1302**

The serial interface, shown in the waveform diagrams, involves bringing the *RST*-line high and sending a first (command) byte a bit at a time in synchronism with the clock bits also supplied from the microcontroller. Then, depending on the mode indicated by the first bit of the command, the second byte is either received by the clock (write mode) or driven on to the data line by the clock chip (read mode). In either case, the clock signal must continue to come from the microcontroller. The addresses and functions are shown in the next table.

The easiest way to send or receive data is through the standard 8051 UART in the shift-register mode (mode 0).[43] The *TxD* line sends out the clock, and the *RxD* either sends out or receives bits. The RST line has to be driven from another port line. A schematic follows, showing just the I/O lines in use from the microcontroller. The rechargeable battery or super capacitor can be charged from the normal power supply and will take over running the chip when the power is off. Obtain the full data sheet for more details of the software configuration of the trickle charger built into the chip.

---

[43]The shift register mode is described in a little more detail in *Volume I,* page 310. It may be possible to use the $I^2C$ bus interface where supplied, but I do not have any experience with that yet.

| register address | | | | | | | | | register contents | | | | | | | | |
|---|---|---|---|---|---|---|---|---|---|---|---|---|---|---|---|---|---|
| bit | 7 | 6 | 5 | 4 | 3 | 2 | 1 | 0 | bit | 7 | 6 | 5 | 4 | 3 | 2 | 1 | 0 |
| ***clock functions*** | | | | | | | | | | | | | | | | | |
| sec | 1 | 0 | 0 | 0 | 0 | 0 | 0 | r/w | 00-59 | ch* | 10 sec | | | sec | | | |
| min | 1 | 0 | 0 | 0 | 0 | 0 | 1 | r/w | 00-59 | 0 | 10 min | | | min | | | |
| hr | 1 | 0 | 0 | 0 | 0 | 1 | 0 | r/w | 01-12<br>00-23 | 12/<br>24 | 0 | A/P,<br>10 hr | | hr | | | |
| date | 1 | 0 | 0 | 0 | 0 | 1 | 1 | r/w | 01-<br>28/29,<br>30, 31 | 0 | 0 | 10<br>date | | date | | | |
| month | 1 | 0 | 0 | 0 | 1 | 0 | 0 | r/w | 01-12 | 0 | 0 | 0 | 10<br>m | month | | | |
| day | 1 | 0 | 0 | 0 | 1 | 0 | 1 | r/w | 01-07 | 0 | 0 | 0 | 0 | day | | | |
| year | 1 | 0 | 0 | 0 | 1 | 1 | 0 | r/w | 00-99 | 10 year | | | | year | | | |
| control | 1 | 0 | 0 | 0 | 1 | 1 | 1 | r/w | | wp | 0 | 0 | 0 | 0 | 0 | 0 | 0 |
| trickle charger | 1 | 0 | 0 | 1 | 0 | 0 | 0 | r/w | | tcs | | | | ds | | rs | |
| clock burst | 1 | 0 | 1 | 1 | 1 | 1 | 1 | r/w | | | | | | | | | |
| ***RAM functions*** | | | | | | | | | | | | | | | | | |
| RAM 0 | 1 | 1 | 0 | 0 | 0 | 0 | 0 | r/w | | RAM data 0 | | | | | | | |
| ... | | | | | | | | | | ... | | | | | | | |
| RAM 30 | 1 | 1 | 1 | 1 | 1 | 1 | 0 | r/w | | RAM data 30 | | | | | | | |
| RAM burst | 1 | 1 | 1 | 1 | 1 | 1 | 1 | r/w | | | | | | | | | |

r/w = 0—write to clock chip; 1—read from chip

*ch = clock halt

wp —1-write protect; 0—can alter clk or RAM

tcs—only 1010 to alter; ds—1 or 2 diodes;

rs—2-K, 4-K, or 8-K series resistor

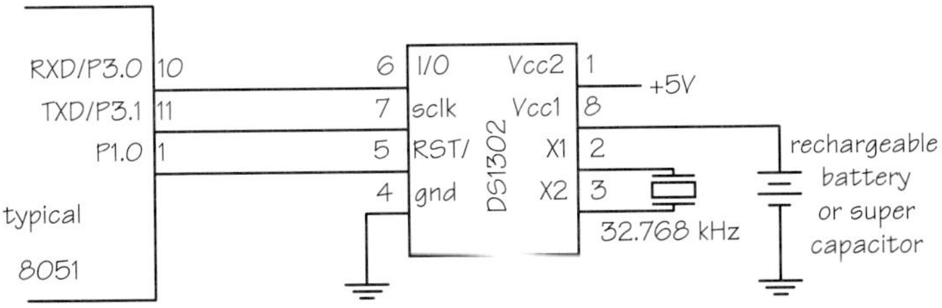


**Serial Real-time Clock Connections**

Having waded through all that hardware-specific stuff, let me put together a sample program to set the chip. I'll assume a specific day and date are loaded into variables from some user interface function (probably with a keypad and LCD). On normal power up you would never do this since the chip would have been ticking off time from its battery back-up. A second only-prototyped function displays the current time.

---

**Program to Read From/Write to Clock Chip**

```
#include <reg51.h>
#define uchar unsigned char
#define uint unsigned
sfr RST = 0x90; /*P1.0*/
[44]uchar sec,tsec,min,tmin,hr,thr,date,tdate,mo, tmo,
              yr,tyr,dwk;
void calset(void); /*user interface to set variables*/
void tdisplay(void); /* fn to show date, time on LCD*/
uchar entry(void); /* yes=1 or 0=no */
void display(uchar *string); /*display function*/

void clkwr(uchar addr,dat){ /*write to clk chip*/
  RST=1; TI=0; /*clear last bit sent*/
  SCON=addr ;while(TI==0);/*wait till last bit sent*/
  TI=0; SCON=dat; while(TI==0);
  RST=0;
}

uchar clkrd(uchar addr){ /*read from clk chip*/
  uchar dat;
  RST=1; TI=0; /*clear last bit sent*/
  SCON=addr ;while(TI==0);/*wait till last bit sent*/
  RI=0; dat=SCON; while(RI==0);/*till last bit rcvd*/
  RST=0; return dat;
}

void clkset(void){
[45]clkwr(0x80,tsec<<4|sec);
clkwr(0x82,tmin<<4|min);
```

---

[44]Note that all the data in the chip is kept in packed BCD—the lower 4 bits holding the base-10 ones value and the upper few bits holding the tens digit value.

[45]I suspect this could be done more efficiently in the burst mode, but I am confused by the data sheet showing 8 bytes to send while I only can identify 7—perhaps they include the control at the end.

```
    clkwr(0x84,thr<<4|hr);/*using 24 hr mode*/
    clkwr(0x86,tdate<<4|date);
    clkwr(0x88,tmo<<4|mo);
    clkwr(0x8a,dwk);
    clkwr(0x8c,tyr<<4|yr);
  }

  void clkfetch(void){
    char temp;
    sec=(temp=clkrd(0x81)&0x0f,tsec=temp>>4);
    min=(temp=clkrd(0x83)&0x0f,tmin=temp>>4);
    hr=(temp=clkrd(0x85)&0x0f,thr=temp>>4);
    date=(temp=clkrd(0x87)&0x0f,tdate=temp>>4);
    mo=(temp=clkrd(0x89)&0x0f,tmo=temp>>4);
    dwk=clkrd(0x8b);
    yr=(temp=clkrd(0x8d)&0x0f,tyr=temp>>4);
  }

  void main(void){
    SCON=0x10; /*set up mode 0--shift register*/
    clkwr(0x8e,0); /*ensure WP is low after power up*/
    for(;;){
      clkfetch(); /*get latest clock info*/
      display("want to change clock?");
      if(entry()==1)clkset();
    }
  }
```

---

I suggest you explore the Oki MSM6542 real-time clock chip if you need a periodic interrupt at up to a 10-min interval or an alarm function.[46] You could, for a very-low power system, set this real-time clock chip to reactivate the microcontroller only occasionally or at a set time. The clock chip can run on 2 to 6 V drawing as little as 5 μA (must be 5 V when being addressed by the microcontroller—then it draws 30 μA). The open-drain interrupt/alarm output(s) could be combined with additional circuitry to wake a microcontroller out of a low-power mode or even turn on the power totally in a battery-powered system where minimizing power consumption is crucial and the microcontroller need not be active most of the time. The biggest drawback is the fact that the chip has a parallel interface so you either have to address it as a mem-

[46]The fifty-four-page data sheet can be obtained over the web (start at *oki.com*) and gives a large amount of application information.

ory peripheral on the expansion bus (no trouble if you've added RAM or EPROM) or give up at least a port for port addressing.

### Example: Two-Wheel Direction Control

For an example of algorithms involving interval measurements, consider a vehicle that moves and turns by two separate traction motors. These might drive two independently driven wheels (with one or two castor wheels to keep it from tipping), or there might be two separate treaded tracks like on a tank. In either case, varying the relative speed of the two motors does the steering. This example assumes shaft rotation sensors that output a high signal during a fraction of a full revolution. They could be magnets taped to the wheels and magnetic sensors or reed switches, reflective strips taped on the rims of the wheels with optical emitter-detector pairs, or mechanical switches picking up bumps from screw heads. A higher resolution system might have several marks around the wheels or even shaft encoders to provide speed and distance information more frequently. (See the shaft encoder section of the input chapter for further discussion.)

What is needed is a control algorithm to give a correction to the motor drives based on the time difference for the two wheels' turns. Later software development could relate to turning to the left or right, but let's start with just keeping the vehicle moving straight ahead. Assume that the vehicle is going straight if both wheels complete turns in the same amount of time,[47] as indicated by signals from two rotation sensors that each put out a pulse once for each turn of their wheel. Assume the feedback signals exist, and the goal is to process them.

The problem is a common closed-loop control one—keep two wheels turning at the same speed by adjusting the drive voltage (or pulse width for on/off drive) to two independent motors. Represent the motor drive amounts by two variables. The speed of the two wheels is to be determined from a single pulse per revolution. Assume the pulses from the sensors are high for 5 percent of the turn.

A few quick estimates are necessary. I'll choose a small electric vehicle with two wheels (as opposed to a tank) so the wheel probably turns no more than 5 turns per second giving minimum periods of the incoming signals of 200 msec and minimum pulse widths of 10 msec. That can easily be detected with the 1-msec real-time clock (Chapter 3; page 78) while still

---

[47]Don't even think about different diameter wheels! If additional alignment reference is needed, consider the section of Chapter 4 on magnetic inputs. It would be nice to have some external navigation system—perhaps a gyroscope—but that would get very expensive.

giving reasonable resolution. Presumably the tracking algorithm should work for considerably slower speeds, at least down to one wheel turn in two seconds. That gives a period of 2,000 msec and a pulse width of 100 msec. To handle that *range* of values, I'll use readings in units of milliseconds stored as an unsigned integer (16-bit). That allows 65,535 msec or about 65 sec. If the wheel is slower than that, consider it stopped!

Watching the pulse edges can be done by the real-time clock tick (interrupt), which can make the most recently completed period values available to the main program at once. Add a flag for each new reading and perhaps a flag to alert the program to a possible stopped condition.

The best algorithm for the corrections is open to debate. A PID algorithm might be a good solution, but it involves significant math and may be unnecessarily complex. For a simpler first try, consider the following algorithm. Assume that there are two motor drive *numbers* somewhere else in the program; this routine merely adjusts the relative drive number for each motor to keep the actual speed the same for both wheels. Ideally the same drive number should produce the same speed to each, but it is possible to have small changes that cause the motors to differ in speed. Assume the motor drive numbers are each a byte where 255 is full speed and 0 is stopped. The numbers might actually result in a pulse width or an analog voltage at DC motors or stepping rates for stepper motors. The adjustment will need to speed up (increase the drive) to the slower motor (having the larger period measure) or slow down the faster motor. For a really robust algorithm, the adjustments near full on should slow the faster wheel to avoid data overflow. Near stopped, the algorithm should speed up the slower motor to avoid under-flow folding to full on. In between, it would make sense to adjust the two drive numbers up and down by about the same amount to maintain an average drive value close to that specified by the other software. For simplicity, the routine shown here will follow the middle approach and just add tests to avoid fold-over.

How should drive relate to speed difference in milliseconds? Should the drive adjustment vary depending on the absolute time difference or should it depend on the relative difference? These questions relate to the linearity and gain of the control system and can be complicated for thorough solutions. Without specific information about gain, transfer functions, or response speed, I assumed that the adjustment will be based on percentages. If a given wheel is faster than the other by 10 percent, the current drive numbers for that wheel is decreased *from its most recent value* by 5 percent and the other wheel's drive is increased *from its most recent value* by 5 percent. That way the adjustments should progressively bring the wheels to matching speed—a PI sort of control where the adjustment is proportional to the error and the accumulating adjustments reflect an integral effect. The big question will be the appropriateness of adjusting the drive percentage the same as the speed error. That would have to be empirically determined.

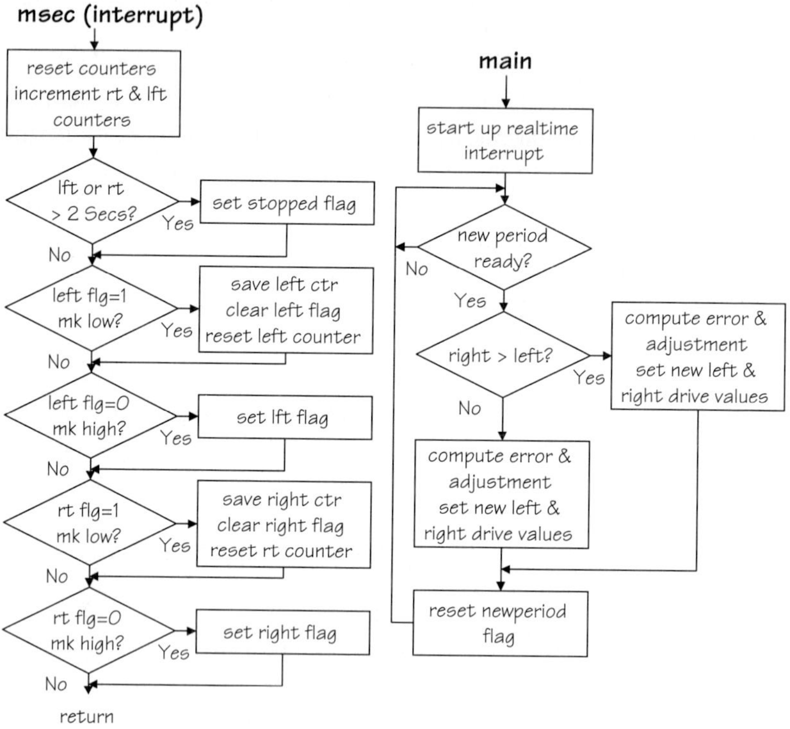


**Two-wheel Direction Control**

---

```
              Two-wheel Direction Control
#include<reg51.h>
#define uchar unsigned char
#define uint unsigned int
sbit lmark=0x90;/*P1.0*/
sbit rmark=0x91;/*P1.1*/
uint msecs, lpd, rpd;
uchar drive, ldrive, rdrive, err, adj;
bit stopped, new;

void msec(void) interrupt 1 using 1{
  static uint lpdtmp, rpdtmp;
```

```
    bit llow,rlow;
    TH1=-1000>>8;
    TL1=-1000&0x00ff;
    ++lpdtmp;
    ++rpdtmp;
    if (lpdtmp>2000||rpdtmp>2000)stopped=1;
    else stopped=0;
    if(llow&&lmark){
      lpd=lpdtmp;
      new=1;
      llow=0;
      lpdtmp=0;
    }
    else if(!llow&&!lmark) llow=1;
    if(rlow&&rmark){
      rpd=rpdtmp;
      new=1;
      rlow=0;
      rpdtmp=0;
    }
    else if(!rlow&&!rmark) rlow=1;
}

void main(void){
    TMOD=0x11; /*free run 16-bit timers*/
    TH1=-1000>>8;
    TL1=-1000&0x00ff;
    TR1=1; /*start timer1—realtime clock*/
    IE=0x88; /*ET1 enabled*/
    for(;;){
      if(new){ /* wheel completed turn*/
        if (rpd>lpd){ /*go more to left*/
          err=(100*(rpd-lpd))/lpd;
          adj=(drive*(err))/100;
          if (ldrive>adj)ldrive=ldrive-adj;
          else ldrive=0;
          if((uint)rdrive+adj<256)rdrive=rdrive+adj;
          else rdrive=255;
        }
        if (rpd<lpd){ /*go more to right*/
          err=(100*(lpd-rpd))/lpd;
          adj=(drive*(err))/100;
          if (rdrive>adj)rdrive=rdrive-adj;
          else rdrive=0;
```

```
            if((uint)ldrive+adj<256)ldrive=ldrive+adj;
            else ldrive=256;
        }
        new=0;
    }
  }
}
```

# USER ENTRIES

## Decimal to Binary (from Buttons)

If going out to a display is one of the most common embedded applications, coming in from a keypad has to be a close second. Here is where the first primitive entry techniques can benefit most from some more user-friendly algorithms.

**Fixed-Digit Count.** The most common way for decimal numbers to come in is from a user entering them from a keypad or keyboard. The weakest conversion method is to take the successive input digits and multiply them by, say, $1000_{10}$, $100_{10}$, $10_{10}$, and 1. This requires the user to enter leading zeroes such as *0 0 2 5* for 25. In the first example assume the digits come from a *keyin* function that waits until the user pushes a button. The process of assembling the incoming digits into a binary number is shown next.

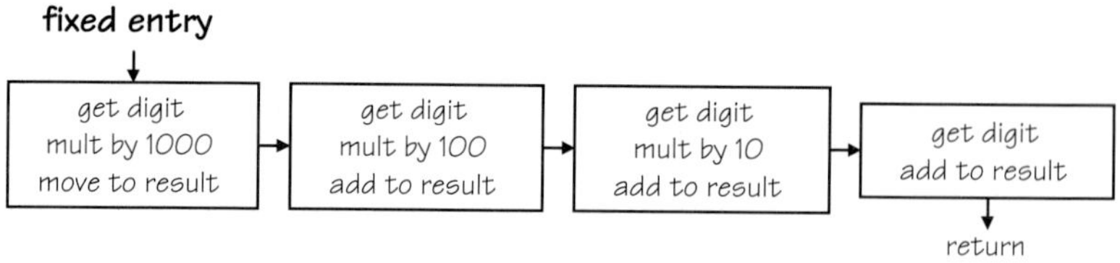


**Fixed Decimal Entry-to Binary Routine**

```
        Simplest Decimal-to-Binary Conversion
uint result;
result=keyin()*1000;
result=keyin()*100+result;
result=keyin*10+result;
result=keyin()+result;
```

Here your user must enter in a fixed number of digit. If the user forgets the leading zeroes (quite common) the program allows no going back to revise or edit the entry before going on.

**Variable Digit Count with *ENTER*.** It is friendlier to let a number build up as long as the user keeps pressing numeric keys. Making a non-numeric entry—perhaps pushing one of the A-F keys on a 16-key keypad or an *enter* key on a keyboard—finishes the number. For each new key push, the software should assume that the newest key is the *ones* place and all the previous entries move up by a factor of 10. If your user pushes a 3, it has a value of 3 unless they then push a 2. In that case, the 3 takes on a weight of 30. If your user then pushes a 4, the previous 32 becomes 320, and so on. The only caution is that the repeated pushing of keys can overflow the variable. If overflow matters and will cause serious problems, then you will have to add some sort of test and choose some action to take for that error condition. The program is shown next:

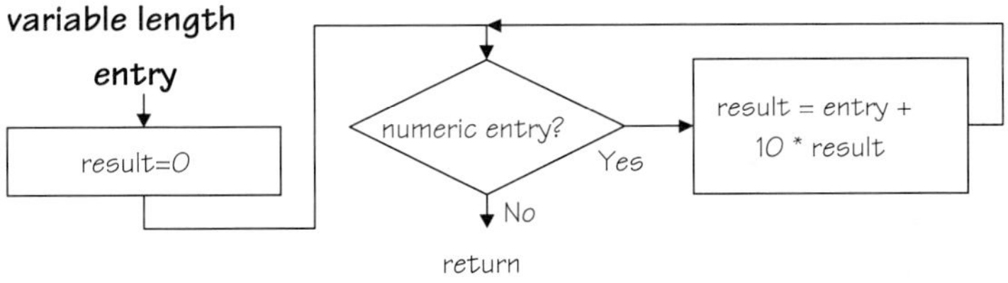


**Variable Length Numeric Entry Flowchart**

**Decimal to Binary Entry**

```
uint result=0;
uchar keyin(void);
uchar tmp;
while((temp=keyin())<=9)result=10*result+tmp;
```

More elaborate key-entry functions could allow correcting mistakes or at least allow the user to either accept the completed numeric entry or else reject the number and start over entering the entire number. You might have to save each key entry until the entry is done to allow correction.

## Parsing

Parsing is breaking something up into the component parts, as some of you may have had to do in grammar classes where you broke a sentence up into the subject and verb. Now suppose you are developing a user-friendly system where the user can enter the dimensions of a part from an alphanumeric keyboard.[48] To keep it simple, assume the part is a bolt that is being inspected for flaws as well as measured for diameter and length.

The rules for user entry should be very forgiving, so assume that the dimension for the bolt's length could be either preceded *or* followed by *l, L, length, Length,* or *len.* Likewise the diameter could have *D, d, diameter,* or *diam.* Flaws are designated by a number code—perhaps *1* for *mis-threaded*, *2* for *broken*, *3* for *no threads*, *4* for *other flaws* (I have no clue as to other flaws that could happen in manufacturing a bolt). In addition, assume finding a flaw eliminates the need to make any dimension measurements, so a flaw requires only the entry of a single-digit number (but *can* be with an *f, F,* or *flaw*). Just to be safe, have the system confirm the entry in a standard format and request the user to hit the *Carriage Return (Enter)* if correct or any other key if incorrect.

There is a parsing problem that sounds complex enough to be real.

***Parsing Algorithm in Words***

1. Clear and then build up keyboard entries in a buffer as they are typed. When the entries end (*CR* or *LF*), go to step 2.
2. Work through the buffer until a non-space non-punctuation character is encountered. Go to step 3.
3. Categorize the character encountered. Go to step 4.
4. If it is a number, continue processing into a binary number until the numeric digits run out. Then go to step 3.
5. If it is a letter, check for case-insensitive f, d, or l. Then go to step 6.
6. If it is an approved letter, flag the type of entry and eat up (skip over) the rest of the alpha entry. Go to step 3.
7. If the character is a return, store or return the data if any is completed and go to step 9.
8. If none of those, reject and give an error message to re-enter. Go back to step 1.
9. Return to main program.

The program that follows executes the algorithm, but the order of the steps was changed so it didn't exit out before the last entry was processed. The routine will take a string of entries on one line or process just one entry per line.

---

[48]A really low-cost, compact system might use only a numeric keypad with some of the A-F keys re-labeled with D, L, F, and enter

It will take the letter before or after the number but allows no non-numeric entries in the middle of the number (you might want to allow a decimal point).

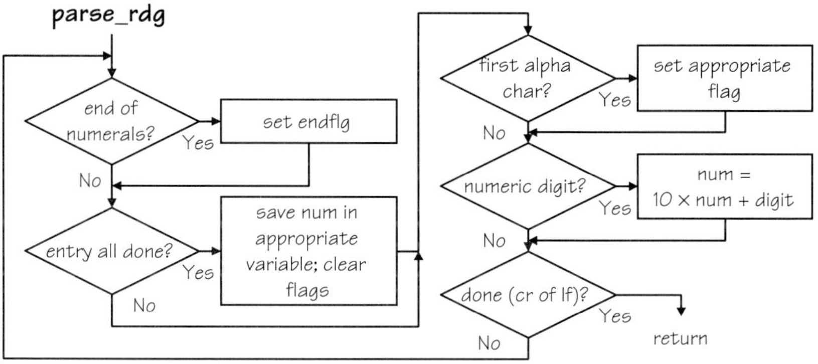


**Bolt Data Parsing Flowchart**

---

**Bolt Data Parsing Function**

```
#define uchar unsigned char
#define uint unsigned int
uchar diam,leng,flaw;
bit flwflg,diaflg,lenflg,numflg,ltrflg,endflg;
uchar buf[30];

void parse_rdg (uchar buffer[]){
  uchar num=0;
  uchar bptr,ch;
  flwflg=diaflg=lenflg=numflg=ltrflg=0;
  for(bptr=0;bptr<30;bptr++){
    ch=buffer[bptr];
    if((numflg==1)&&((ch<'0')||(ch>'9')))endflg=1;
    if(ltrflg&&numflg&&endflg){/*one entry done*/
      if(lenflg)leng=num;
      else if(flwflg)flaw=num;
      else if(diaflg)diam=num;
      num=ltrflg=numflg=endflg=0;
    }
    if(ltrflg==0){ /*letter processing*/
      if((ch=='d')||(ch=='D'))diaflg=ltrflg=1;
      if((ch=='f')||(ch=='F'))flwflg=ltrflg=1;
      if((ch=='l')||(ch=='L'))lenflg=ltrflg=1;
    }
    if((ch>='0')&&(ch<='9')){/*numeric processing*/
      num=10*num+(ch-'0'); numflg=1;
```

```
    }
    if(ch==0x0d||ch==0x0a)return;
  }
}

void get_kbd(void){ /*dummy data entry routine*/
  buf[0]=' ';buf[1]='1';buf[2]='7';buf[3]='D';
  buf[4]='9';buf[5]='3';buf[6]='1';buf[7]=0x0d;
}

void main(void){
  for(;;){
    get_kbd();
    parse_rdg(buf);
  }
}
```

Whenever you develop a parsing algorithm, be sure to check all the stupid, illogical entries you can invent to be sure you reject bad entries in a safe way.

## ADJUSTING READINGS

### Automatic Gain Control (AGC)

Particularly obvious with audio signals, but also important for television and many other applications, gain control is often desired. Consider a car radio. In the presence of ambient noise, you have to turn the volume up to hear quiet passages, but then the loud passages drive you out—this is particularly true with classical music; the popular-music stations adjust volume to keep the level of modulation as high as possible. What you need is some way to automatically increase the volume for the quiet passages and yet quickly bring it back down the moment a loud passage arrives. Obviously, if you stored and delayed the signal you could look ahead in time and see a loud passage arriving in time to correct.[49] This is a natural application for

[49]The more mundane approach is an analog circuit one of rectifying and integrating and peak detecting the signal with small leakage to produce a voltage where in the presence of a large signal it rises quickly but in the presence of a small signal the voltage drops slowly. A sudden large signal pushes the gain down at once while the gain during a quiet passage is only cautiously increased. In designing such a circuit, a good bit of empirical work is needed to decide how fast to drop the gain—should a single spike cause it or can it take several cycles before you push it down? What is objectionable to the listener? Likewise, how long should you wait to bring the gain up and how rapidly can you do it without being noticed?

digital signal processors if they are already taking in the digitized audio for other processing.

Another gain-adjust situation is when you are sampling a signal and need to cover a large dynamic range.[50] Returning to Chapter 4 on Inputs, you can choose a very high-resolution A-D converter, but you might prefer to add a programmable gain stage ahead of the converter. There are amplifiers that can give a gain of 1, 10, or 100 by grounding certain pins. Now you can write an algorithm to change the gain if the readings are either too close to full scale or too close to zero. You can develop a "smart" instrument that way.

## Self-calibration of Sensors

It is a fact of electronics that things drift with temperature and sometimes with aging of components—particularly capacitors. The fundamental reason for the success of digital electronics is the ability to correct for these shifts. With analog electronics, any errors that creep in are permanently part of the signal from then on. Noise has a cumulative effect. Particularly when you want to maintain high accuracy while reducing cost, it is easiest to self-calibrate rather than adjust analog circuits for perfect accuracy. Particularly if you wish to reduce thermal drift, you can either purchase expensive temperature-insensitive devices and components, or you can design with one accurate element and make everything scale to that.

As a simple example, take a temperature sensor such as an LM 335 from National Semiconductor. The device is highly linear—essentially a temperature-sensitive Zener diode. The problem is that the offset and slope can vary from device to device. One option is to add a potentiometer circuit to correct for the offset variation. If the gain also is to be adjusted, then you need a second pot as well as perhaps two operational amplifiers. The calibration requires the sensor to first be brought to a known temperature and the calibration pot set for the offset. Then at the other end of the temperature range, the second pot can be set. Hopefully, the adjustments are not interactive.

An alternative is to have a way to set an offset and a gain factor in software. Taking a reading (uncorrected) at the first calibration point and then taking a second reading at the high point, a straightforward math routine can

---

[50]You should know that there are non-linear A-Ds, usually called *codecs,* which convert on a logarithmic scale. These are common on digitized audio for telephones because the perceived loudness of a sound is logarithmic and needs less resolution per change in voltage for large signals than for small ones.

do the corrections.[51] If the correction is small, it may be necessary to work in floating point, but the hardware costs would be much lower.

## CONTROL ALGORITHMS

### Bang-bang Control

Essentially, this is a control where the output is either full on or full off. Later on in the fuzzy logic section, I go through a control system for a furnace. Most furnaces do not have a variable level of heat output—perhaps a variable-speed fan or a variation in delay, but seldom a variation in gas level. So ultimately a controller for such a system is an all-on/all-off one. There is no proportional output. Then the key is in smoothing the time so the furnace comes on for a long enough time to be effective. This usually entails a hysteresis band so the furnace comes on when the temperature is low and goes off after the temperature has risen several degrees. To try to keep the temperature precisely set would force the furnace to go on and off much more frequently, with a corresponding inefficiency.

### Two-Wheel Direction Control

Refer back to page 350.

### PID Algorithms

This is a common control algorithm where the correction is a function of the error (proportional), the rate of change of error (derivative), and the long term error (integral). The determination of the correction has historically involved analog circuits with op-amps, although it could also be computed mathematically. As was discussed early with other math functions, it should be possible to simplify the process for specific situations. The entire field of control systems thrives on mathematics, so I stop here!

---

[51]It would probably be desirable to store the correction factors in some sort of non-volatile memory since the calibration process takes time and requires some other highly accurate temperature measurement equipment. Other calibrations can be done automatically.

## DIGITAL SIGNAL PROCESSING

This area is one of the fastest growing ones to be found these days. Just as the microcontroller has been replacing relay and analog control functions for decades, so the DSP chips have been replacing others. Signal processing by digital means is generally so compute-intensive that the real-time processing is done by specifically designed chips with hardware and instructions that efficiently carry out multiple additions and multiplications as well as providing circular buffers. It is not that DSP chips do anything that a microcontroller could not do, given enough time, but rather that they are optimized for the sort of operations common to the mathematical operations of DSP. So what are these operations?

Most DSP involves operations on a string of samples of an analog signal. The operations are the combining of weighted multiples of the most recent digitized sample back through ten, twenty, forty, or even one hundred samples previous. The picture below shows the configuration of a finite-impulse-response (FIR) filter. The current output is the weighted sum of the past inputs. The response is finite in time duration because when zeroes start coming in, after enough time delays the last non-zero sample will have been shifted out and no longer has any effect.

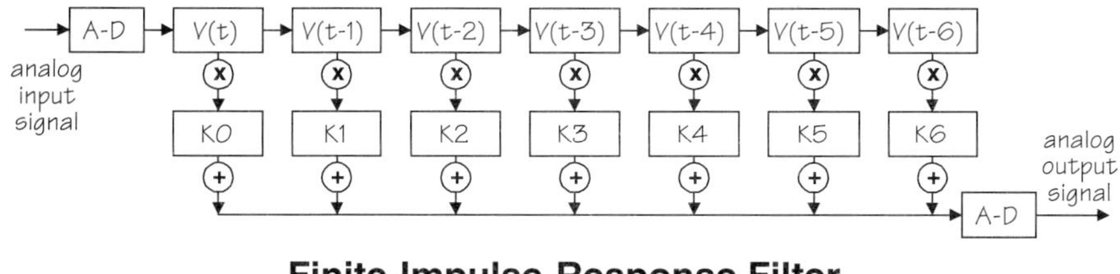


**Finite-Impulse-Response Filter**

The second type of filter on the next page is called an infinite-impulse-response (IIR) filter because, by feeding the output back as one of the inputs, a signal can cycle around indefinitely, even after the new inputs go to zero. In general the same results can be obtained with fewer stages, but there is a higher risk of going unstable, much like a system with positive feedback.

Having sketched the typical hardware for digital filters, I have to say that the mathematics to compute the multiplication constants is beyond the scope of this book.[52] While I explored hand-calculated examples many years ago,

[55]I tried to get my associates who work in this field to give me some simple examples but they seemed to think that was an oxymoron for DSP.

without the mathematics I suspect there is very little that can be done. In addition, for real-time computations the math is usually done in signed-integer form (or floating point on the newer processors) which makes an 8-bit processor even slower.[53] For the only practical example of the process, I refer you back to the *Smoothing and Averaging* section early in this chapter where the last four samples are combined with equal coefficients of 0.25.

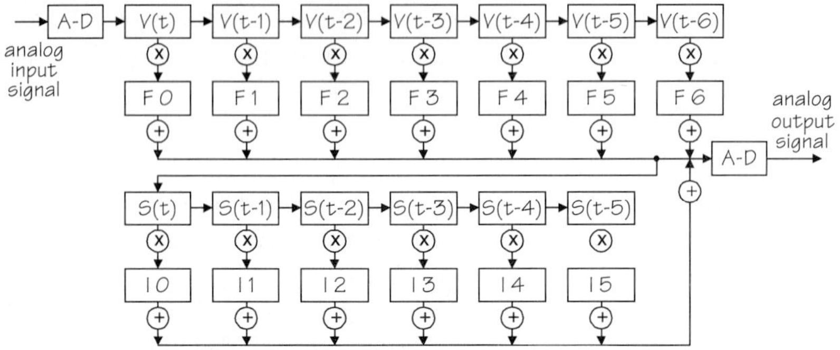


**Infinite Impulse Response Filter**

As I hinted, if you are willing to give up the real-time feature, then you can do almost anything. I can envision using DSP algorithms to compute Fourier transforms of a stored waveform to get the frequency components of the signal. Perhaps you are designing a very low-cost ambient noise analyzer for measuring environmental safety compliance. The acoustic levels could be sampled in bursts—perhaps triggered by a threshold detector on the front end—and the burst of samples could be analyzed slowly for frequency content and energy levels. It couldn't be continuous, but it could be often enough for the intended purpose.

Likewise, an automatic squelch control on an audio mixer could sample the signal and watch for a sudden increase in level that signals excessive feedback. At that point, it could sample a burst and do some quick processing to determine the approximate frequency and reduce the gain in the appropriate equalizer section.[54] You can do a lot with a microcontroller if the

[53]I believe Philips has advertised use of a high speed version of their 16-bit step-up processor for combined microcontroller and DSP applications. I assume the architecture is more suited to DSP while having all the microcontroller features you are already using. I can't address the sort of DSP algorithms involved in their claim, but experience with student DSP projects suggests that DSP chips make very poor general purpose microcontrollers.

[54]This might or might not be fast enough to avoid the ear-splitting shriek. Of course, if an equalizer function was already being done by DSP processing, this function could be built into the existing processing.

real-time requirement can be waived. Don't assume only DSP chips can do DSP.

## FUZZY LOGIC

If you have heard of this term or of *expert systems*, you may have wondered if they were something you should know about in solving problems with embedded controllers. I find it difficult to sort out what is useful to me from what is has the attention of researchers. I am inherently suspicious because most researchers are used to using large computers and probably have very little interest, I suppose, in small, efficient, embedded systems. On further investigation though, I believe fuzzy logic is one of those areas that can probably be embedded nicely for many problems.

What *is* fuzzy logic? It is not poor logic or vague, indefinite logic that varies from time to time. Instead it is logic that takes into account the fact that many things are only *partly* true or false. If I were 6-ft tall, I would be *sort of* tall, but not tall next to a 6½-ft tall basketball player. If I were 5½-ft tall, I would be even less a member of the *tall* set. But there is no sharp line dividing *tall* from *not tall*. You have the same distinctions regarding measures of things like *hot, humid, dry, windy, fast, slow, cold, old, smart, dirty* or *expensive*.

Fuzzy logic doesn't claim to be the *only* solution for problems with varying inputs. It deals with verbally understandable inputs such as *is increasing rapidly, is large,* and *has been large for a long time.* The key for the much more mathematical PID algorithms is to determine the coefficients of the terms to produce a stable system, which can be mathematically described if enough is known about the system.

The reason that fuzzy logic is useful is that it doesn't have to have everything defined in detail. It works more the way that humans think. In fact, that is where the approach comes close to that of expert systems. You can quickly describe the rules in words, the way an expert might describe them.

### Example: Furnace Controller

For example, suppose you were trying to describe the way a furnace should work. For a first trial your rule might be:

*IF the temperature is cold turn ON the furnace.*

But after thinking a bit you might add:

*IF it ran recently turn OFF the furnace.*

*IF it ran very recently always turn OFF the furnace.*

A little later you might decide to add:

*IF the temperature has dropped a little below the desired temperature turn ON the furnace (*so it doesn't make short, inefficient cycles).

Your next rule might have to do with safety:

*Only IF the pilot or igniter operates turn ON the furnace.*

You could go on and on making up rules, but the next formal step, called fuzzification, is to spell out what you meant by the terms you used in expressing your rules—turning *crisp* values into memberships in fuzzy *membership functions.*

### Fuzzification

This is the process of defining meaning of the words. Start with the tentative rule, *IF the house is cold turn ON the furnace.* Take the term *the house is cold.* What is *cold?* If I have set the thermostat to 70°F then I expect that nearly everyone would agree that 60°F is *cold.* Most would perceive 65°F as *sort of cold* and surely 75°F is *not at all cold.* So the *temperature* membership functions might be described graphically as shown below:[55]

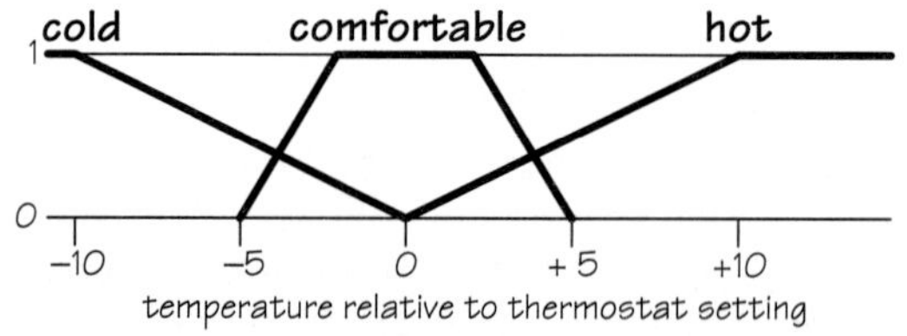


**Membership Functions for "Temperature"**

[55]In the "for what its worth" category, cold is a Z-type, comfortable is a Pi-type and hot is an S-type membership function. A function with only a point on the top is a Lambda-type.

Taking the second rule, *IF it ran recently turn OFF the furnace,* I imagine 5 min would usually qualify as *recent* for a furnace, but 10 min would be long enough a time to wait between cycles. So for a first try, I would define membership functions for *ran* as shown above.

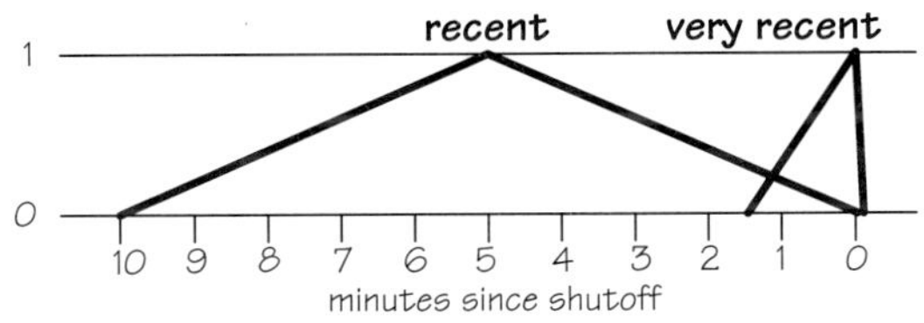


**Membership Functions for "Ran"**

### Inference

Inference is the process of determining which rules fit the given situation. For the furnace the following might be a set of rules to be applied (note that they turn out to be inadequate at the end of this exercise!).

*Turn ON the furnace IF the temperature is cold.*

*Turn OFF the furnace IF it ran recently.*

*Always turn OFF the furnace IF it ran very recently.*

Taken separately these rules would be disaster. If you take the second rule by itself you have the situation where the furnace would run every so often even in the summer! Instead I combined the functions to get this set of rules:

Rule 1: *IF the temperature is cold AND the furnace has run NOT recently turn ON the furnace.*

Rule 2: *IF the furnace has run very recently turn OFF the furnace.*

Rule 3: *IF the temperature is hot turn OFF the furnace.*

Rule 4: *IF the temperature is comfortable turn OFF the furnace*

This is a mess of rules, but lets try to take these four rules and apply them to the furnace.

Suppose the furnace ran 8 min ago and the room temperature is with a thermostat setting of 70° F. To what extent do each of the rules apply (called *aggregation*)?

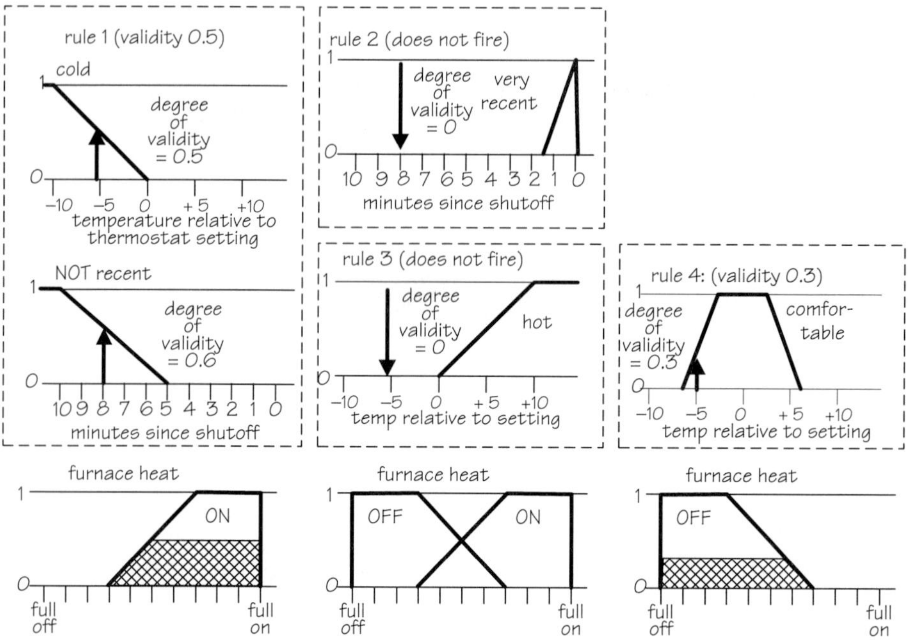


**Furnace Control Rules for Specific Condition**

In the diagrams above, you can see the graphical application of the rules to the specific conditions.[56] Combining the results is called *composition*.[57] In this specific case it is almost trivial, but if several rules fired to give overlapping areas, the total height at any one point along the horizontal axis would be the sum of the values from each rule. The result in this case is the graph shown next. An actual numeric result can be obtained in several ways as discussed in the next section.

## Defuzzification

How do we get from the rule to the binary on/off condition of the furnace? The process is called defuzzification. The furnace cannot be at a range of heat outputs—even if it was continuously variable in output, it would have to be at a specific value. For most furnaces there are no choices other

[56]There are two inference methods—*MIN* and *PRODUCT*. The former is the traditional interpretation of the fuzzy and (I don't know why) picks the smaller of the individual membership functions, while the latter multiplies the individual rules' degrees of truth together.

[57]Interestingly enough, once you have determined the individual rules' degrees of truth, you can either get the composite output from the clipped off function shown or you can multiply the individual shapes by their degree of truth.

than full on and full off so in the end I would assume that the furnace should be full on if the weight of the heat function exceeds 50 percent. The composition below is for the specific instant where (limited to only rules about the temperature and time since last on) the temperature is 65 (with a thermostat setting of 70) and the furnace has been off for 8 min. The center of mass of the sum is a common method, but it is easier to use the centroids to compute.

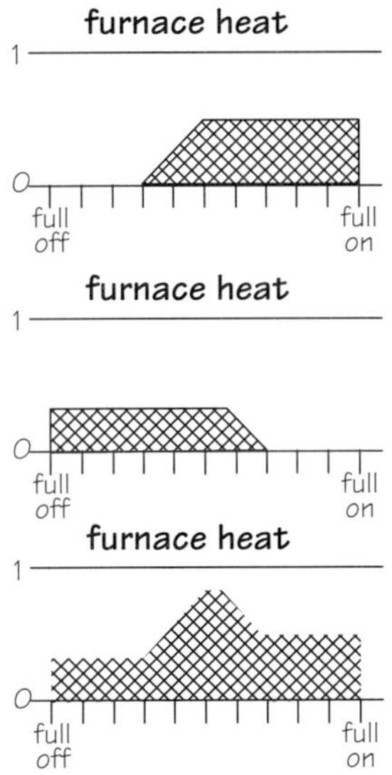


**Building the Composite Result**

Finding the center of mass forced me to dig into an ancient calculus book to discover the simple formula that the center of mass of a triangle is one-third of the way from the base to the apex. Obviously, the center of a rectangle is midway. Much as I hate math, I had to use a bit of algebra to determine that the center of mass (X) along an axis where various shapes bear down ($W_1$, $W_2$...) at different points ($X_1$, $X_2$...) is given by the formula:

$$X = (X_1 * W_1 + X_2 * W_2 + X_3 * W_3 \ldots) / (W_1 + W_2 + W_3 \ldots)$$

For the case above the center of mass turns out to be 0.519 or just a bit above midway. By my rule of 50 percent, that means that the furnace would be full ON at that instant.

If you are going to process this precisely and mathematically in a microcontroller, determining the center of mass is going to require a significant amount of calculation. Each of the four output functions must be described mathematically and the calculation to determine the area and center of mass established. Nothing more than simple multiplication and division is involved, but I rapidly came to the conclusion that there has to be a better way.

Bearing in mind that high precision is probably unnecessary, I shifted to a numeric integration technique. It sounds awesome, but in a microcontroller it is quite easy to implement. I start by describing the two furnace heat functions (*ON* and *OFF*) in ten discrete positions from *full off* to *full on,* resolved only to the nearest integer with eight for *full on* (I hate fractions). The two output functions can then be expressed by arrays of ten *uchar* (8-bit) values as shown in the table below. Then, in the specific instance illustrated before, the rules gave one-third *OFF* and one-half *ON*. With only eight levels, the truth level of one-third becomes 3 and the truth level of one-half becomes 4. Walk through the appropriate function, summing into a new array the *smaller* of the array value and the truth level. For each rule that fires, do the same thing (if more than eight rules could fire, you could overflow the *uchar* array). The array now holds the profile of the sum, as in the third graph on the previous page. Now walk through the sum array generating the sum of the weights and also the sum of the weights multiplied by the array position. Then divide the position-multiplied sum by the straight sum and you have the center of mass. In this case, I get 249/43 = 5.7 which would turn the furnace on.

| position | 1 | 2 | 3 | 4 | 5 | 6 | 7 | 8 | 9 | 10 | full off to full on |
|---|---|---|---|---|---|---|---|---|---|---|---|
| ON | 0 | 0 | 0 | 1 | 3 | 5 | 7 | 8 | 8 | 8 | in permanent array |
| OFF | 8 | 8 | 8 | 7 | 5 | 3 | 1 | 0 | 0 | 0 | in permanent array |
| 1/3 OFF | 3 | 3 | 3 | 3 | 3 | 3 | 1 | 0 | 0 | 0 | smaller of 3 or value |
| 1/2 ON | 0 | 0 | 0 | 1 | 3 | 4 | 4 | 4 | 4 | 4 | smaller of 4 or value |
| sum | 3 | 3 | 3 | 4 | 6 | 7 | 5 | 4 | 4 | 4 | sum = 43 |
| weighted sum | 3 | 6 | 9 | 16 | 30 | 42 | 35 | 32 | 36 | 40 | sum = 249 |

This may seem horribly crude, but you must remember that the logic is a significantly more refined one than a simple binary 1/0. It is not that the approach is endowed with some divine authority—the only real test is what works well enough. If ten outputs or a height resolution of eight is insufficient, you can always increase the resolution, but notice that for eight rules or less, this can always be done with bytes and the use of $8 \times 8$ multiplication which is supported in the hardware. I show a function to compute all four rules based on the two input below.

**Compute Function for Fuzzy Example**

```
#include <reg51.h>
#define uchar unsigned char
#define uint unsigned
#define ON 1
#define OFF 0
uchar data sum[10];
uchar data set;
uchar data i,j;
uchar xdata result[12][21];

void compute (uchar validity,output){
  uchar code ONA[]={0,0,0,1,3,5,7,8,8,8};
  uchar code OFFA[]={8,8,8,7,5,3,1,0,0,0};
  uchar i;
  uchar code *array;
  if(output==ON)array=&ONA;/*which output rule
              applies?*/
  else array=&OFFA;
  for(i=0;i<10;i++){/*add in smaller of output fn and
              validity*/

              sum[i]=(validity<array[i])?validity+su
              m[i]:array[i]+sum[i];
  }
}

uchar fuzzy(uchar temp,tentime){
  char tdif;uchar i,x;uint strtsum,wtsum;
  uchar data validity[4];
  validity[0]=0;/*get validity of rule 1*/
  if((tdif=temp-set)<0){
    if(tdif>-10)validity[0]=(8*-tdif)/10;
    else validity[0]=8; /*at least 10deg low*/
  }
  if((tentime<100)&&(tentime>0)){
    tentime=(tentime>50)?((tentime-50)*8/10):0;
              validity[0]=tentime<validity[0]?tentim
              e:validity[0];
  }
  validity[1]=0;/*get validity of rule 2*/
  if((tentime<15)&&(tentime>0))validity[1]=(15-
              tentime)*8/15;
```

```
    validity[2]=0;/*get validity of rule 3*/
    if(tdif>0){
      if(tdif>10)validity[2]=8;
      else validity[2]=tdif*8/10;
    }
    validity[3]=0;/*get validity of rule 4*/
    if(tdif>-7){
      if(tdif<-4)validity[3]=(-tdif-4)*8/3;
      else if(tdif<4)validity[3]=8;
      else if(tdif<7)validity[3]=(7-tdif)*8/3;
    }
    for(i=0;i<10;i++)sum[i]=0;/*clear sum array*/
    compute(validity[0],ON);/*add in effect of rule 1*/
    compute(validity[1],OFF);/*add in effect of rule 2*/
    compute(validity[2],OFF);/*add in effect of rule 3*/
    compute(validity[3],OFF);/*add in effect of rule 4*/
    for(i=0,strtsum=0,wtsum=0;i<10;i++){
      strtsum+=sum[i];/*add up individual values*/
      wtsum=wtsum+(i+1)*sum[i];/*add up weighted
                  values*/
    }
    if (strtsum>0)x=10*wtsum/strtsum;/*compute center of
                  mass*/
    else x=0xff;/*if no rule applied*/
    return x;
  }

  void main(void){
    set=70;
    for(i=60;i<81;i++){/*get entire control array*/
      for(j=0;j<12;j++){
        result[j][i-60]=fuzzy(i,10*j);
      }
    }
  }
```

Using the above program I got the values out (×10) for the entire control surface and show them in the table that follows. The columns are for the temperature (around a set point of 70) and the rows are for the delays in minutes (the top row is for the furnace presently on). The shaded cells would be the ones where the furnace is on.

| | *60* | *61* | *62* | *63* | *64* | *65* | *66* | *67* | *68* | *69* | *70* | *71* | *72* | *73* | *74* | *75* | *76* | *77* | *78* | *79* | *80* |
|---|---|---|---|---|---|---|---|---|---|---|---|---|---|---|---|---|---|---|---|---|---|
| *0* | 78 | 77 | 76 | 75 | 52 | 61 | 44 | 41 | 37 | 31 | 31 | 31 | 32 | 33 | 33 | 34 | 36 | 34 | 33 | 32 | 31 |
| *1* | x | x | x | x | 34 | 37 | 31 | 31 | 31 | 31 | 31 | 31 | 32 | 33 | 33 | 34 | 36 | 34 | 33 | 32 | 31 |
| *2* | x | x | x | x | 34 | 37 | 31 | 31 | 31 | 31 | 31 | 31 | 32 | 33 | 33 | 34 | 36 | 34 | 33 | 32 | 31 |
| *3* | x | x | x | x | 34 | 37 | 31 | 31 | 31 | 31 | 31 | 31 | 32 | 33 | 33 | 34 | 36 | 34 | 33 | 32 | 31 |
| *4* | x | x | x | x | 34 | 37 | 31 | 31 | 31 | 31 | 31 | 31 | 32 | 33 | 33 | 34 | 36 | 34 | 33 | 32 | 31 |
| *5* | x | x | x | x | 34 | 37 | 31 | 31 | 31 | 31 | 31 | 31 | 32 | 33 | 33 | 34 | 36 | 34 | 33 | 32 | 31 |
| *6* | x | x | x | x | 34 | 37 | 31 | 31 | 31 | 31 | 31 | 31 | 32 | 33 | 33 | 34 | 36 | 34 | 33 | 32 | 31 |
| *7* | 65 | 63 | 62 | 60 | 48 | 53 | 42 | 40 | 37 | 33 | 33 | 33 | 33 | 34 | 34 | 35 | 36 | 35 | 34 | 33 | 33 |
| *8* | 78 | 77 | 76 | 75 | 52 | 61 | 44 | 41 | 37 | 31 | 31 | 31 | 32 | 33 | 33 | 34 | 36 | 34 | 33 | 32 | 31 |
| *9* | 78 | 77 | 76 | 75 | 52 | 61 | 44 | 41 | 37 | 31 | 31 | 31 | 32 | 33 | 33 | 34 | 36 | 34 | 33 | 32 | 31 |
| *10* | 78 | 77 | 76 | 75 | 52 | 61 | 44 | 41 | 37 | 31 | 31 | 31 | 32 | 33 | 33 | 34 | 36 | 34 | 33 | 32 | 31 |
| *11* | 78 | 77 | 76 | 75 | 52 | 61 | 44 | 41 | 37 | 31 | 31 | 31 | 32 | 33 | 33 | 34 | 36 | 34 | 33 | 32 | 31 |

Looking at the table, I would have to say the result doesn't look very smart. The entire rule set could effectively become:

*If the furnace has been off for at least 7 min and the temperature is at least 5° below the set point, turn the furnace on.*

*If the furnace is running and the temperature is at least 4° below the set point, turn the furnace off.*

Aside from being almost trivial, my set of rules would never keep the furnace on long enough to reach the set point! I suspect the OFF rules need to be revised to not take effect when time is 0 and possibly be given less weight than the ON rules. There are too many OFF rules firing so several OFF rules outweigh one ON rule. Also, I need to decide what to do if no rule fires. Here you see some of the problems of fuzzy logic design.

## ARTIFICIAL INTELLIGENCE

### Tracking and Hill-climbing

Imagine you have some device that takes in a signal and can be aimed. I am familiar with a photodetector having two stepper-motors that move it in polar coordinates, but it could be some sort of radar tracking or even an A-D algorithm. The goal is to find the best (biggest, smallest, or closest) value.

One way is to climb the hill until the largest value is found.[58] Compare each reading with the one before and move in the direction of the bigger reading. That is good as long as the increasing values lead to the *true maximum*, but it is possible to find a *local maximum* without finding the absolute maximum, as shown in the next diagram.

**Example of Local Maximum**

The job becomes even more complicated if there are several variables to change. With two dimensions it is possible to have a saddle where the maximum in one direction is not the maximum in the other dimension.[59]

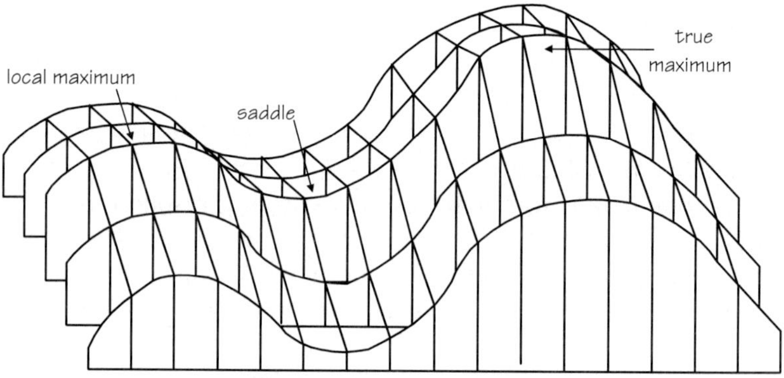


**Example of Saddle**

Having located the brightest spot, you need to follow it as it moves. If the photodetector were a "quadrant detector" having four detectors, it would be straightforward to compare the four signals to know which direction to

[58]Incidentally, these same methods apply to artificial intelligence algorithms where a variety of parameters are adjusted until the most optimum combination is found.

[59]In artificial intelligence algorithms, it can be virtually impossible to explore the entire space. Following all the branches of a tree may take longer than even a computer can accomplish—think of all the possible moves in a chess game, for example. An algorithm to pick the best move at any point has to rely on some assessment of the strength of the next position relative to the current one. Presumably the one that gives the best score is the move to make. You can imagine, however, that a different sequence of moves might lead to a stronger final position, so you can see that a local maximum might not be the best long-term strategy.

move. But, failing in that, you will need to move the photodetector slightly in a pattern to see if a nearby point is brighter than the current point. If so, that becomes the new "center." If it is not brighter then the scan should return and go the other direction. This will not be wholly satisfactory because, when the scan is directly overhead, a large azimuth change is needed to get a square or crossed pattern.

## DYNAMIC MEMORY ALLOCATION

In *Volume I,* there was some discussion (page 144) of linked lists and sparse arrays with a suggestion that this volume would cover memory allocation. Traditionally all embedded applications run programs that assign fixed locations to variables. The only dynamic use of space is in the automatic variables defined within a C function that can be overwritten when the function is not in use. This overlay feature is transparent to the programmer.

Dynamic allocation of space is not common in small systems because there isn't very much memory to start with and most applications have a clear idea of the amount of memory needed.[60] If there isn't enough memory, there is usually some dire consequence—lost data, or a need to stop acquisition until data can be processed or transmitted to a remote host system. The exception is with real-time operating systems as discussed in Chapter 3. Here there are commands to obtain resources from a resource pool—often the resource being memory.

So what would a small, embedded system do with dynamically allocated memory? I recall BITBUS systems using the memory for messages. Incoming messages were put in a block of RAM shared with the serial interface controller. Outgoing messages were transferred to a block of allocated memory before transmission. Unfortunately there was only room for about six messages, so it was quite easy to run out of space.

Similarly, it would be possible to allocate off-chip RAM for a synthetic stack or for a temporary storage for a copy of on-chip stack when tasks are switched. I can imagine a dynamic use of memory for data if there was a multi-part processing algorithm that stored intermediate results that were later discarded. I specifically remember such an algorithm for a project where camera data was received and them progressively boiled down until

---

[60]This of course is just the opposite of PCs and other systems where various jobs run at different times and where any overflow of RAM space can be shifted off to a hard drive or other large storage device. Swapping files and data in and out of RAM is a common activity of such systems.

there was a set of numbers representing the misregistration of a CRT. Big "hunks" of data came in and had to be sifted until only a few numbers were left giving the desired answers.

Given the assumption that various functions (or tasks) temporarily need space to keep a significant amount of data, how could it be implemented? Assume that a function is established to manage 1,000 bytes of off-chip RAM and make 100-byte subblocks available on request. The managing function should "own" the memory and only supply the other users with pointers to the particular blocks. To reduce overhead, I'll assume the protection from overflow into the next block is handled by the user functions. I will name the dispensing function *allocate*, and the restoring on *deallocate*. These two functions need to share access to a list of possible blocks and their status.

For now, I will assume all the subblocks are the same size (100 bytes) so I needn't keep track of the specific start addresses as long as I have the start address of the 1,000-byte overall space. All I need then is a 10-*bit* array of the status of each block. Unfortunately, the Keil compiler doesn't allow that. Instead, I will set up a 10-*byte* array in *off-chip* RAM.[61] That array will store a 1 for a free block and a 0 for an already allocated block.

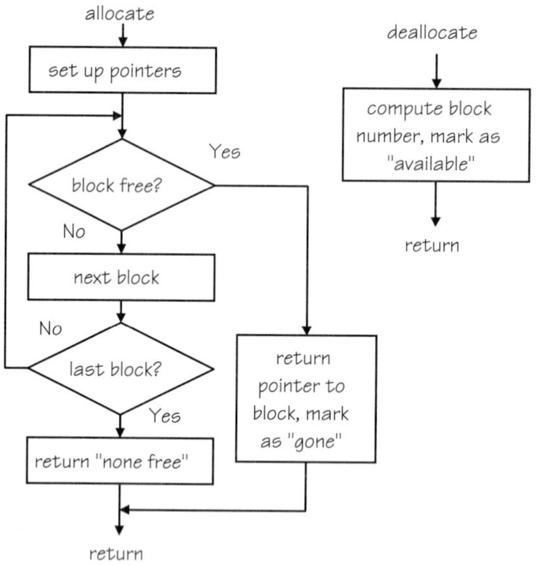


**Flowchart of Memory Allocate and Deallocate Functions**

[61]It is less efficient this way, and slower access, but we have already established that there are at least 1,000 bytes of off-chip RAM that will be used. By using entire bytes, I avoid the cumbersome manipulation to access bits within an 8- or 16-bit variable.

When a task needs some memory, it will call the *allocate* function which returns a pointer to the start of a free block, or, lets say, a zero, indicating no blocks are free. In the latter case, something else will have to be done, but the error handling will have to be written into the task—perhaps it waits and then retries after some time in hopes that some other task has returned a block to the pool. When the task is done with the memory block, it should give it up by passing the address to the *deallocate* function. Below are the flowcharts for the functions and then the code in C.

**Allocate and Deallocate Functions**

```
#define uchar unsigned char
uchar xdata blk[1000];
[62]uchar xdata sblks[]={1,1,1,1,1,1,1,1,1,1};

uchar xdata allocate(void){/*returns ptr to block*/
 uchar I;
 for(I=0;I<10 && sblks[I]==0;I++);
 if(I<10){ /*found a free block*/
  sblks[I]=0;
  return &blk[100*I]; /*no blocks free*/
 }
 else return 0;
}

void deallocate(uchar xdata *ram){
 uchar I;
 for(I=0;ram>&blk[100*I]&&I<10;I++);
 sblks[I]=1; /*sets block available*/
}

void main(void){
 uchar xdata *ramblock;
[63]ramblock=allocate();
[64]*(ramblock+8)=0X77;
[65]deallocate(*ramblock);
}
```

[62]This array holds a 1 for any free block and a 0 for any that are in use.

[63]Here the pointer for *ramblock* is loaded from the function that allocated the block

[64]In this usage example, the 9$^{th}$ element of the block is filled with a 0x77. Remember that arrays start with 0.

[65]This returns the block to the free pile.

# 7

# Efficiency at Work

Having spent all the previous space discussing various hardware and software tricks, I want to follow through with more elaborate examples. I start with a system to control the distribution of heat (or cooling) in a house serviced by a duct and blower system such as is common in centrally-heated houses in the United States of America. I discuss some of the tradeoffs to consider and present the solution essentially as I planned it about ten years ago.

The second example, is a hypothetical light meter system for use with photography. Here the goal is very low power consumption and size while keeping costs low.

Be aware that *neither* example is all finished and working! Both of them illustrate the problem of software growth. In the first case the display functions have grown and grown, from the start when I just wanted to put something up on a screen, to the multi-screen version filling 4 × 20 character displays with a severe sense of crowding. As I planned the system the hardware simplicity was the chief goal—let the software pick up the slack since it is a one-time cost! That is well and good for the hardware developer, but it doesn't help the programmer much. I have struggled most with the display functions (although I am quite proud of the result) and discarded about as much code as I have kept.

## ZONE HEAT SYSTEM

### What Does the Customer Want?

One of the cardinal rules of any design is to make sure that the solution addresses the problem that the customer really has. Often the customer may not know what is *really* the problem and may have preconceived solutions in mind. Let me take the customer role for a time in order to describe what I think I want.

The type of residential furnace in view includes both heating and air conditioning, sharing the same blower and duct system, with the air travelling past the heat exchanger and then past the cooling coil assembly. The plenum then has several branches that then spread out to air outlets in each room. Near the outlets the final ducts are usually 6 inch diameter round ones. The actual air outlets are commonly rectangular and located in the floor near the perimeter of the house. There is a return-air system as well, but that is limited to three or four larger floor inlets closer to the center of the house.

The biggest problem is balancing the ducts so the appropriate share of the heat goes to each room to offset loss through the walls and ceiling and maintain the desired temperature in each room. The customer may want some rooms warmer than other rooms. With a strong wind out of one direction, that side of the house is cooler than the other side, but that changes as the wind changes direction. Also, if a separate wood-burning fireplace is used in one room, the extra heat makes that room too warm at first, but the room later gets colder because of heated air lost up the chimney. Finally, the occupant of a room may want it warmer or cooler than the rest of the house or someone wants an unused room closed off to conserve energy.

What I envision is a system that will adjust the airflow into each room so the rooms are maintained at their desired temperatures. The system turns on the furnace (or air conditioner) when needed, and adjusts the airflow so each room reaches its desired temperatures at about the same time. Any adjusting entails only small changes to dampers at infrequent intervals—I will not hear motors moving all over the house all the time. Each room (or zone) will have its own temperature-setting control and a temperature-sense device—I don't want to have to set things at some complicated central console. The balancing will remain in place if there is a power outage and will dynamically change as conditions alter. The furnace runs for a reasonable amount of time—at least 5 min, say—so the furnace has time to heat up and increase in efficiency.

The system must be easy to install in an existing house with a minimum of wire runs. In addition, the installation should allow the conventional thermostat to take over whenever there is a problem.[1] Ideally, the in-

[1] I suspect the fire underwriters approval for such a system would have to be obtained if it were to actually be used, but I expect usual safety features of the furnace will remain active.

stallation should entail no additional cutting of ducts or access to tight spaces.

Let me put these into a list:

Operation
- Individual airflow control to at least six rooms or zones
- Temperature setting controls for each zone (40° to 90°F)
- Temperature sensing for each zone
- System that makes adjustments only occasionally
- Allows full-OFF settings for zones
- Maintains set temperatures within 5 degrees at all times after a 4-hr adjusting time
- Runs damper motors no more often than every 15 min

Installation
- Minimal or no duct work or cutting
- Very simple wire runs
- Makes use of existing thermostat connections
- Powered by 110-V 60-Hz wall outlet

Environment
- Functions over ambient of 30° to 120°F
- Runs off 110-V 60-Hz maximum of 10 Watts
- Not waterproof, main unit wall mounted
  Parts cost under $100 for six zones

### Roughing Out Zone Heat System

Obviously, I already have ideas about what to do. The block diagram here gives an idea of the configuration I envision.

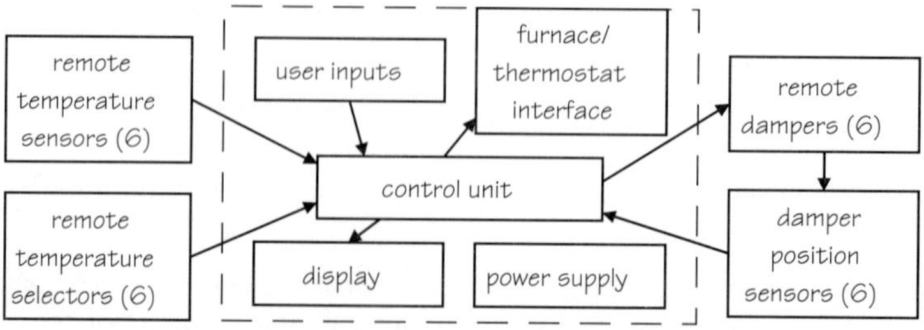


**Block Diagram—Zone Heat System**

Let me discuss each block by itself, exploring some of the options to consider.

**Temperature Sensing.** As was mentioned, you can sense with thermocouples or thermistors as well as solid-state semiconductor devices. A thermocouple is not good for long distances because the signal is so small, unless the thermocouple wire is run the entire way, which is a problem for flexibility and may be a problem for cost.

A thermistor is inexpensive, but non-linear. Unlike the hot water system of Chapter 14 of *Volume I*, here I want a calibrated, moderately accurate temperature indication. I fear the thermistors as received may have variations in resistance for a given temperature as well as presenting a non-linear relationship.

Solid-state temperature sensors are accurate to about 2 degrees as received. Since they are quite linear, I can live with that. For practical purposes devices such as National Semiconductor's LM335 behave as temperature-sensitive Zener diodes. There are devices that give values directly related to the Fahrenheit temperature as well as special circuits for adjusting offset and gain, but, since the data goes by way of a microcontroller, I will go for the minimum circuitry.

There is, I'm told by my students, a temperature sensing device that includes a serial interface and an A-D converter, as well as another one that puts out a digital wave related to temperature (I'm not sure if is frequency or duty-cycle changes). I set these aside for several reasons—first, the cost is higher than the LM 335, and second, they require more than two wires.

**Temperature Selection.** I can imagine temperature selection in each room by digital keypad entry (with a display) or with a potentiometer (either a rotary pot or a slide pot). Of course, temperature could be set at a central location, but that makes the use of the system unattractive to nontechnical people and requires going *to* the central location to make any adjustments.

I see no good reason to select temperature digitally, despite the high-tech "feel" it represents. A keypad would not show the current setting unless a numeric display is included, driving up the cost significantly. I suppose a set of DIP switches could set temperature, but that would require a significant amount of circuitry or extra wires to the central controller. The only other option I see for a digital input would be to have a separate microcontroller for each zone, since the 87C750 can come in under four dollars, I believe. By the time you add switches and a display, the parts costs are up towards fifteen dollars per node. Of course, the individual nodes could then communicate over a serial bus, but each node would need its own power.

From a user standpoint, an analog temperature setting is best. Why would anyone want to set the temperature to the nearest degree? By using a potentiometer, it is possible to choose arbitrary units so the entire range from, say 0 Ω to 10 kΩ represents 40° to 90°F. Not only is it an inexpensive device, it continues to indicate the setting by observing the angle of the knob. A numeric scale can be printed around the knob. Since the resolution is relatively coarse, an added benefit is the fact that the inaccuracies of the sensor calibration will not be visible to the user—a range of 50°F over an angle of about 300°, or about 6° of rotation per °F, is not going to make a discrepancy of 1°F obvious. Anyway, if the room is too cool, whatever number the system assigns to it, the user can just turn it up a little!

**Combining Selection and Sensing.** Some of the other options might be more attractive if I had not come upon a clever way to combine the two functions *over a single wire pair.* I have alluded to this before, but I rely on the fact that the temperature sensor is effectively a Zener, giving a fixed voltage drop for a given temperature. Putting a resistor is series gives a current-dependent voltage component—the selection pot! Now by successively driving two currents through the series circuit, the voltage can be broken down into that due to the Zener—the crossing of the y axis, and the change or slope—that due to the pot resistance, as shown in the next figure.

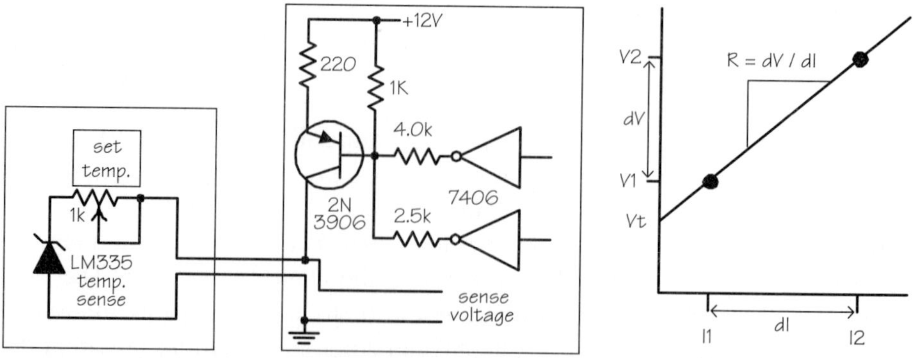


**Combined Select and Sense—Two-wire**

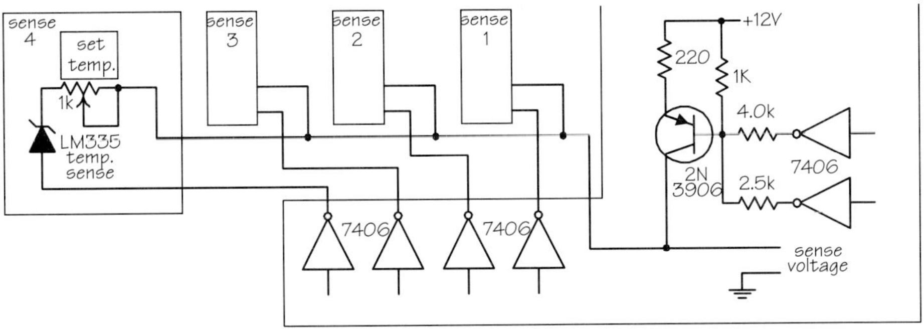


**Multiple Remote Selectors and Sensors**

By driving first one current and then another, I can get both the slope and the Y-intercept. Then, by leaving all the sense circuits but one floating, I can switch between the different zones—I use open-collector devices such as 7406s to supply the ground side of the wire to the remote and tie all the outgoing wires together. Presto, electronic switching for six zones in one $0.50 package! By adding a "zone" with a fixed-voltage Zener diode and a know resistance, it is possible to calibrate without even worrying about the precision of the driven current.

**The Controller.** This is the most microcontroller-oriented part of the decision process. I choose something from the 8051 family both because of my experience and because that is the subject of this book! Really, however, that leaves a wide range of choices. Is there a lot of programming to include? How important is space and size? What is the budget—what can be spent on the controller? How much I/O is needed? Are any special functions called for such as A-D, PWM, watchdog timer, math co-processor, or serial communications? Will the code change frequently? Does there need to be non-volatile memory? How important is power consumption or operation over a wide voltage range? Is operation over an extended temperature range required. All these are questions to ponder as a processor is chosen.

In actuality, this is the area where my design has changed most from ten years ago. *Then* I chose an 8031, adding off-chip RAM and EPROM as well as an 8255 parallel-port chip and a memory-mapped LCD display.[2] Re-

---

[2]Unfortunately it still sits beside me, constructed but not totally functional. I spent lots of time on it then but I could never get some of the functions working properly—particularly the display—and went on to other things. I would have benefited from development tools such as an ICE, but they were not available to me. Try burn-and-try debugging for a while and you too will move on to other things!

visiting it *now*, I lean toward a single-chip controller because the one-time-programmable (OTP) versions are now available and are much more reasonable in price.

The next step is to count pins for ports. The LCD, if not memory mapped, needs either 4 or 8 bits for data and two or three lines for control (the third one if you want to read back as well as write). I am proposing four user pushbuttons (the next block) which can take three or four lines. The sensors take seven multiplexed A-D lines or else three select lines and a V-F input. The motor drives take six lines or else three select lines, possible with some feedback sensors. Finally, the outputs to the relays that interact with the thermostat take two output lines and two inputs (to tell if heat or cool is requested). That could be thirty-four I/O lines total, 7 of which must be A-D inputs. I have just exceeded the capability of all but the largest members (a 8xc552 would work, for example).

Here is where clever hardware design comes in. The motor drives take six lines plus a possible six more for feedback if done *directly*. Likewise, the sensing takes seven lines. But none of this needs to happen simultaneously. Room temperature does not change quickly so there is no need to read more than one sensor at a time or to run more than one motor at a time. There is no need for more than one A-D reading at a time. As will be shown later, judicious use of decoders can save many pins. In addition, use of open-collector drivers allows very inexpensive selection without analog multiplexing. In the end, I reduced the requirement to twenty I/O lines. That *almost* rules out memory expansion unless a port chip is added, since the thirty-two lines drop to twenty-four with only EPROM and twenty-two with RAM or other added devices.

Now to code space. A quick scan of Appendix 5 in *Volume I* shows that the 87C51 itself has 4-K bytes of on-chip code, while the 87C52 has 8-K. The 80C51FB and FC have 16 K and 32 K, respectively. The only ones under 4 K are the very small 87C750 (1 K) and the '748,'749,'751, and '752 (all 2 K). If all the code were developed first, the smaller devices might suffice, but it appears that the traditional 40-pin devices leave several EPROM options that are all pin-compatible. Since I expect to have some user displays available on the central controller (perhaps high and low temperature each day, the percentage of on-time for the furnace each day, and other statistics), I do not want to limit myself to 2 K in advance.[3] Alphanumeric message strings can eat up code space rapidly.

---

[3]As I re-write this the code has exceeded 2K (barely), but the RAM has pushed beyond the 128 bytes of the 80C51. Issues of the detail of setback times and temperatures as well as statistics could reduce this, but it is not worth the fight for this relatively high-value product. The situation is quite different for the light meter discussed at the end of this chapter.

As to RAM, I think 128 bytes is enough. Even with some statistics, I think there can be enough. The math will avoid floating-point (as discussed early in Chapter 6) and I do not expect heavy stack use. I cannot decide on non-volatile memory yet. I suspect I do not need it since there are no user settings that are stored internally (the beauty of pots for temperature setting!), and the loss of statistics is not critical.

As far as extra features, a watchdog timer would be nice to somehow avoid tying the furnace on if the processor crashed. I weighed individual A-D converters, but none of the controllers have more than eight of them, and I would need at least thirteen. If I have to add an analog multiplexer, there seem to be better options.

As far as size, I think I am stuck with at least a 40-pin DIP since the code space in the 28-pin devices seems to limit at 2 K. If I can avoid off-chip ROM and RAM, that can save board space. In reality, there is likely to be all the room that is desired for the controller module. Even if it mounts on a central wall, there should not be more than 20 $in^2$ of circuit board required. Tiny is not important.

Environmental concerns are minimal. The controller will be inside a house. Unless the roof leaks or there is a major flood (then who cares?!), moisture should not be a problem. The ambient should not get above 120°F (unless it is mounted directly in the sun) or below 32°F in any normal operation. I suppose a northern-Canada winter retreat cabin could be started up with an ambient of –20°F, but at that point the main problem would not be the operation of the LCD and presumably such a cabin would not have zone heat anyway! I will arbitrarily limit my design to 0° to 50°C operation. I will ignore ruggedness because the controller will mount on a wall or in a basement—it will not be carried about and should not be dropped.

I do not think power consumption matters much. There was no way to derive enough power directly from the 24-V lines to the furnace and air conditioner relays without activating those coils[4] and it is too much trouble and expense to recharge a battery slowly from those sources. If the 24 can be fed up in an extra wire then it could work nicely, but again, there is no demand for very low power consumption. Since I will supply regulated power, there is no need for a wide voltage range in the microcontroller—that would only really apply to a direct battery-powered system.

So I will plan a forty-pin single-chip standard (8051/52) device, with the final choice depending on the amount of code and on-chip RAM.

---

[4]I suspect the coils might activate if 20 or 30 mA were drawn and there might be a super-sensitive model—it would be a poor system that drew enough current to continually activate the furnace!

**User Inputs.** As hinted, I want the ability to report statistics such as the high and low outside temperature each day, the percentage of time the furnace is on and perhaps the temperatures in each zone—set and actual. The sun might never set on this programming activity, but all I have to provide now is a way to walk through a menu. People wanting such information would be technically oriented, so I will not ponder long over the user-friendliness of the controls—quite unlike the individual room temperature settings where a pot was deemed most suitable for general users. I will not waste the money on a keypad or keyboard since these inputs to access statistics will not be used often.

With six buttons (*up, down, left, right, setup,* and *setdown*—I added two later), there is little gain in multiplexing. As discussed in Chapter 4, while 16 is 4 × 4 (eight lines), 6 is 3 × 2 (five lines). I can just as easily use the same 1-in-8 decoder that scans the sensors to select the individual buttons while reading them into a separate port pin. I just talked myself into saving 4 port pins. If I can get below 2-K of code space then I might be able to drop to the 87C752. The general multiplexing scheme is shown below.

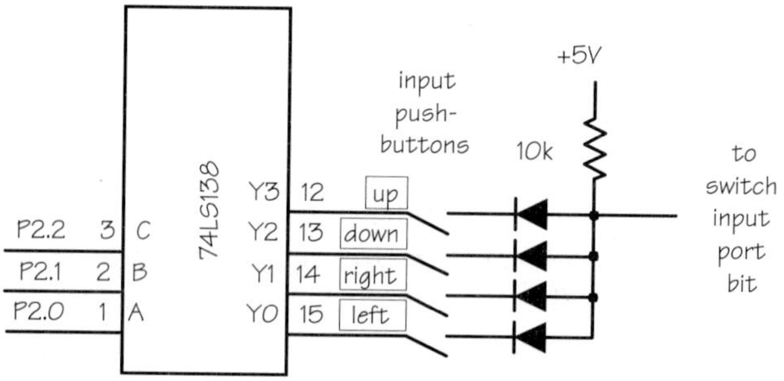


**Pushbutton Multiplexing to Save Three Port Pins**

It should be mentioned that a multiplexed scheme like this leaves no room for an interrupt to indicate a button has been pushed and the scanning of the six lines also activates the scan of six sensors. Since that involves settling time and sufficient time to get a reading out of the V-F converter, when *real* measurements are being taken of the sensors, the scanning of the pushbuttons must be slowed down to a rate where the frequency can be measured—more on that later.

**Display.** An alphanumeric LCD is about the only economical choice. I did not know initially what I wanted to show, but with a very limited set of buttons, there will have to be a significant amount of user prompting. This display cannot readily be numeric-only, but it does not need to be full graphic nor does it need to be highly visible as with vacuum fluorescent versions. It can be small and need not be a custom format. While it could be done with a 1 × 8 or 1 × 16 display, later planning brought me to a 4 × 20 which is not significantly more expensive.[5] In many ways, this is the least critical decision because all the alphanumeric LCD modules from 1 × 8 through 2 × 40 have virtually the same interface.

One decision remains—do I use the 4-bit or 8-bit mode in addressing the LCD? The latter is slightly simpler in software because you do not have to break the commands and data bytes into two nibbles to send them and it takes less time and steps, but the former saves four port pins—not an insignificant item if the port usage is crowding the use of the simplest microcontrollers. For now I stay with 8-bit mode because there are enough port pins with the 40-pin ICs, but I will keep it in mind if the final result indicates that a 2-K code limit can be achieved.[6]

**Motor Drivers.** I have begged the question by choosing DC gearmotors to run the dampers. In the early phases of the planning a decade ago there was a lot of consideration of other approaches. For example, solenoids could activate stop pins at different points in the travel of a freely-hinged flap in the air duct, but the solenoids would require continuous activation whenever the air was flowing and there would have to be several of them at each zone (one-quarter, one-half, or three-quarters open, say). Likewise geared stepper motors could be used, but they require more drive lines and could slip or lose reference position, still making some encoder function a necessity. I suppose deflecting dampers could be installed near the main trunk to distribute the air selectively near the main blower (that would make installation much more accessible, but also very custom for each house). I will not get into the mechanical design considerations of such things as worm drives—I went with a center-pivoted damper much like the tuning dampers commonly installed in ducts for permanent balancing. By driving such dampers directly off a DC gearmotor, the duct can go around endlessly, so I need not have bi-directional motor drivers.

One possibility is to install the dampers directly into floor register grill covers. Then there would be no duct cutting or access problems. The wire

---

[5]For student projects where cost is critical and where future volume production is irrelevant, I recommend obtaining display modules from surplus and salvage suppliers such as *Marlin P. Jones*. In the software section I revisit the user interface and display. Based on the number of parameters to show, a switch to 2 × 40 is proposed as well as 2 additional buttons.

[6]The lightmeter example illustrates the use of the 4-bit mode.

might even be fed through the duct to the central furnace. Unfortunately, most zones involve more than one outlet, so the preferred method would be to insert a single damper nearer to the trunk. On the other hand, if ease of installation in an after-market situation is important, the system could be revised to run multiple dampers in each zone. This trade off is discussed more later.

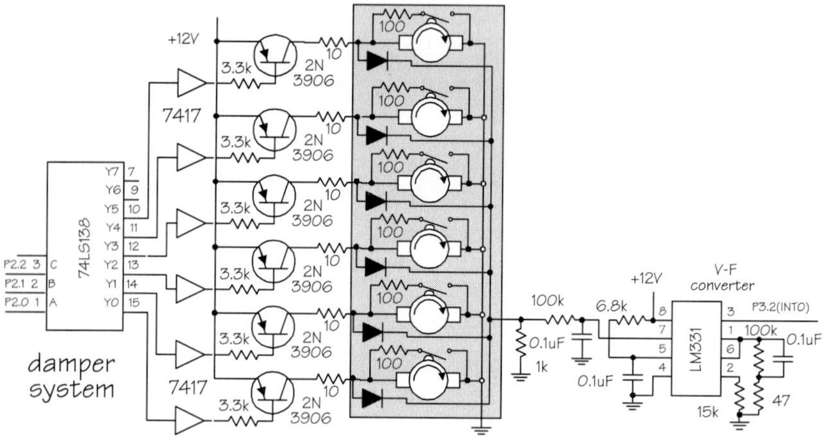


**Motor Drivers, Position Feedback, and V-F Converter**

With unidirectional drives, I decided to include current sensing to determine if the motor is connected and moving (as opposed to stalled). This comes into play with the position sensing described next.

**Position Sensing.** Since the flaps or dampers driven by DC motors have no position feedback, I needed some way to tell, at least, when the damper is fully open or fully closed. I could have put full-shaft encoders on each damper (directly on the motor is no use since it makes many turns into the gear reduction assembly). It would be nice to have such detailed information, but it is sufficient knowing only when the damper is, say, fully closed. With a rotating damper, the airflow is blocked twice in a 360° rotation. From the schematic on the previous page, you can see that I have added mechanical switches (driven by screw heads on the damper shaft) to switch in a parallel low resistance that increases the total current draw. This current rise is the indication of the shaft encoder "mark."

I am tempted to explore dampers that could install directly in register covers for ease of installation. What could be easier that installing new duct covers and running the drive wires directly in the ducts to the furnace (assuming high-temperature wire is used)? Such a mechanical configuration might allow only partial turns rather than full 360-degree travel. If so, the drivers would have to be bi-directional, but the end of travel could be sensed by detecting stall current in the motors. I really need a good mechanical designer, but like most projects, the electrical designer has to work hand-in-hand with the mechanical person.

The second drawback with register-cover-mounted dampers might be where several floor registers supply one zone. Then the controller might have to drive several register covers at the same time. This might increase the cost of the system. How can I weigh ease of installation against cost of components? Considering the possibility of having up to four registers per zone or twenty-four total outputs, I could easily go to 74154 decoders (16 outputs) at the cost of one more addressing port bit. With thirty-two total selects I could drive the 8 sensor inputs and still have drivers for twenty-four damper outputs—up to four each for each of the six zones. The added cost for each output is then the motor, the switch, a few diodes, a transistor, and one-sixth of a 7417. The configuration could use only as many of the four as needed in each zone and the microcontroller can easily sense any outputs not connected. On the circuit board would be only the transistors, resistors, and diodes. The motors and switches could be part of the dampers, which would be purchased as needed for each installation.

**Power Supply.** I am proposing only a 110-V 60 Hz source, transformed to 12 V and 5 V. The system *could* be powered off the 24 V AC commonly used for heating systems. Since the proposed controller only drives one thing at a time, I think no more than 500 mA of 12 V is needed and perhaps 300 mA of 5 V. Unless the contrast adjust needs negative voltage, no other supply voltages are needed.

**Thermostat/furnace Interface.** Rather than struggle with polarities and ground references, I propose to use relays. One relay will default back to the normal thermostat's operation whenever the power is cut to the controller or when the microcontroller crashes. Two single-pole relays parallel the normal thermostat to request heating or cooling and two additional ones report on the setting of the heat/cool switch in the regular thermostat. In general, the new system disturbs the existing thermostat as little as possible.

## Zone Heat Software

**Timed Interrupt.** I do not think a real-time operating system (kernel) as in Chapter 3 is necessary. I will just use a regular, timed interrupt to make a scheduler for the scanning and drives. Current software has the interrupt come every 10 msec. Since the time of day and day of week are maintained for setback purposes, the interrupts are counted to maintain hour, minute, second, and day of week in variables. Button scanning could be done every 50 msec to avoid double counting on button bounce. The scanning of the sensors need not come nearly that often. I suspect checking temperature every 60 sec is sufficient. The adjusting of the dampers should not happen more than every 15 min to minimize the noise around the house.

**Interrupt (Scheduler) Software**

```
void timinit(void){
  TMOD=0x51; /*t1 16-bit counter, t0 16-bit timer*/
  TH0=-10000/256;TL0=-10000%256; /*10 mSec
                interrupt*/
  TH1=TL1=0; /*allow max count*/
  ES=0;  /*no serial interrupt*/
  ET1=0;  /*no timer1 interrupts--just grab count
                every 10 mSec*/
  ET0=1;  /*timer0 is regular 10 mSec tick*/
  EX1=EX0=0;  /*no ext interrupts*/
  IP=0;  /*all(one)interrupts low priority*/
  TCON=0x50;  /*start timers*/
  EA=1; /*allow interrupts to be recognized*/
}

void clockset(void)using 1{/*every second*/
  sec=++sec%60;
  if(sec==0){
    min=++min%60;
    if(min==0){
      hrs=++hrs%24;
      if(hrs==0){
        dwk=++dwk%7;
      }
    }
  }
}

/*timer0 overflow interrupt routine--10 mSec realtime
      clock*/
void timer(void) interrupt 1 using 1{
  static uchar btncnt,seccnt,mincnt,ftncnt;
  if(goodrdg>0)--goodrdg;
```

```
    TH0=-10000/256;TL0=-10000%256; /*restart 10 mSec
         interrupt*/
    oldfcnt=newfcnt;newfcnt=(((uint)TH1)<<8)|
         TL1;fcnt=newfcnt-oldfcnt;
    ++tmscnt;
    if(--btncnt==0){btncnt=5;btnscan=1;} /*flag every
        50mSec*/
    if(--seccnt==0){ /*each second*/
      seccnt=100;clockset();
      if(--mincnt==0){
        mincnt=60;furnchk=1;/*flag check furnace every
                  minute*/
        if(--ftncnt==0){
          ftncnt=15;dmprset=1;/*flag adjust damper every
                  15 min*/
        }
      }
    }
}
```

**Frequency (Voltage) Measurement.** The V-F converter will be configured to run between 10 kHz and 100 kHz over a range of 1-V to 10-V input. I do not have the exact input voltages in mind yet. One measurement technique (see chapter 4 on measuring time intervals), since an external interrupt line is free, is to measure period. The V-F edges, or the negative-going edge) could trigger an interrupt that would run handler software to capture the instantaneous value in a free-running timer. At 10 kHz the period would be 100 µsec (count change of 100), while at 100 kHz it would be 10 µsec (a count of only 10).

That is not good enough resolution, so for the accurate measurements I could read the free-running timer every $10^{th}$ interrupt. That gives timer differences of between 100 (10 V = 100 kHz) and 1,000 (1 V = 10 kHz).

Really though, that is backwards because the resolution above is inversely related to the magnitude of the voltage into the V-F converter. I could, instead use the 10 msec interrupt (already in place for everything else) to read a count of the number of V-F pulsed during the interval. I can feed the V-F signal directly into the *external counter* input of the counter/timer that is not tied up generating the real-time interrupt. By reading the accumulated count (of V-F cycles) every 10 msec I will get a count between 100 (1 V = 10 kHz) and 1,000 (10 V = 100 kHz). That value is *directly* proportional to voltage (and thus current), so the math is simpler and the resolution is consistent across the voltage range. It is a resolution of 10 mV and is obtained in 10 msec. Also, the acquisition of the frequency reading is synchronized with the system clock. It is so simple I repeat the software here that you probably missed in the interrupt routine above!

**Software to Read Input Frequency/voltage**

```
oldfcnt=newfcnt;
newfcnt=(((uint)TH1)<<8)| TL1;
fcnt=newfcnt-oldfcnt;

/*waits for next 10 mSec event to return latest
reading*/
uint read_f(void){
  uchar start;
  for(start=tmscnt;start==tmscnt;);
  return fcnt;
}
```

**Sensor Value Extraction.** As illustrated earlier on page 380, the key of the input process is determining the Y-intercept and the slope of the voltage/current curve to determine the remote pot setting and the remote temperature sensor. The equation was described there, but now I need to add a calibration step where one node with a fixed (temperature insensitive) Zener diode and a fixed (moderately precise) resistor are measured. Once that is done, everything else can be calibrated to those two references and the precision of the current drivers is unimportant. The process cancels out typical resistance of the open-collector drivers and the impact of any slow temperature variations as well.

**Setting and Actual Temperature Software**

```
void getrdg(uchar i){
  uint flo,fhi,temp1;
  sel_meas(currentlow);sel_ch(i);flo=read_f();
  sel_meas(currenthigh);sel_ch(i);fhi=read_f();
  tset[i]=rref*(temp1=fhi-flo)/difref;
  /*100-=closed,200ohm=50degF,400=60,600=70,800=80,
       900+=open*/
  tact[i]=(vref*(temp1)-(floref*fhi)+(fhiref*flo))
       /difref;
}
```

```
void cal(void){
  sel_meas(currenthigh);sel_ch(7);fhiref=read_f();
  sel_meas(currentlow);floref=read_f();difref=fhiref-
     floref;
}

void getrdgs(void){
  uchar i;
  cal();
  for(i=0;i<6;i++)getrdg(i);
}
```

---

**Motor Drive and Feedback.** Assuming I stay with a damper system that allows full turns in one direction, calibration consists of timing a full or half turn. A better calibration might be to run a series of short on on-times to determine how many angle increment steps fit in a full turn, reflecting the inertia and coasting of the motors. I propose to never move a motor more than one increment every 15 min. For a first try, I would try to get 20 bursts to a half revolution of the damper (then I can express the setting as a single digit—0 for closed to 9 for full open).

A second factor I considered was the air-resistance effect of the damper as a function of angle. I suppose the damper rotation from full open (parallel to the flow) to around to 45 degrees has little retarding effect, while the last few degrees have a much greater impact.[7] A bit more research is needed here, but I could break the full revolution into more short motions and use several for each step near full open and only single ones near closed—a lookup table is probably called for. Interestingly, going in only one direction and wanting to avoid any unnecessary motion, if the reference mark is at full closed, the first bursts would open the damper while the ones just before the next mark would be reducing the flow. The software shown later maintains a running estimate of position, so initial calibration is very important. Re-calibration is possible without any extra motion each time the damper gets around to the mark, but the software should always allow each small motion to settle before doing the next one—in the absence of any empirical data I have to assume a continuous run might have a significantly different effect.

---

[7]Fluid dynamics are only a vague memory of my engineering past, but you already know I resist extensive calculations whenever possible.

Frankly, at this point I have no idea what to do with the motor drive current measurements. The value during the active motion should be stored for each motor so the increase (of about 100 mA) during the mark can be recognized. A study of the dynamic current during motor start is needed, since I have no data on the startup current compared to the steady-state value. The mark can only be sensed *during* the motor's motion, but I expect the inductance of the motor will result in a low current at the start.[8]

If there are to be twenty bursts of drive, I have a vague recollection that the specific gearmotors turned in about a second. Given a half turn in 500 msec, and wanting twenty bursts to a half-turn, my first try would be a drive of 30 msec (three ticks of the real-time clock). That is so convenient I would be inclined to use that duration or a larger multiple of ticks even if that fails to give precisely the twenty pulses desired. I start by leaving the motor off for at least 500 msec after a pulse to be sure it had stopped. After all, the damper adjustments only have to come every 15 min so there is no hurry.

If a user-setting pot changes significantly, there ought to be an override on the 15-min interval, since drastic re-balancing is called for and motor noise will be reassuring rather than annoying. In fact, although temperature need not be measured more often than once a minute, to give the users immediate feedback by hearing motor noise, the readings should be done every second. Since there are 100 time intervals in a second, even allowing a 10 msec interval for the V-F to settle to a new value, it is easy to get in 50 measurements, and I only have 6 zones!

The software includes a function that is passed the desired position and the specified motor. Within the function, the software first checks the current position data of that motor to determine if the damper is before or after full-open position. From that it decides whether the motor needs to go forward only one step or to go around to the opposite slope. Then the software issues the steps (with proper timing handled in the real-time interrupt), update the damper position, and returns. The main program need only deal with the desired positions for the zones and the timing of the adjustments.[9]

---

[8]Beginning electronics teaches you that the current through an inductor cannot change instantly (the same with the voltage across a capacitor). How rapidly the current can change is a function of inductance and the applied voltage.

[9]Remember this software is untested at this time! It is the *concepts* I want to show here.

**Damper Positioning Software**

```
/*run motor until switch increases current*/
void home(uchar zone){
  sel_meas(motordrive);
  sel_ch(zone);/*start motor moving*/
  ref=read_f();
  while(read_f()>=ref);/*wait for home rise in
      current*/
  sel_meas(motorstop);
}

void homeall(void){
  uchar i;
  for(i=0;i<6;i++)home(i);
}

/*calibrate time to complete 1/2 revolution*/
void caldmpr(void){
  uchar start,i;
  homeall();
  for(i=0;i<6;i++){
    start=tmscnt;
    sel_ch(i);/*start motor moving*/
    ref=read_f();
    while(read_f()>=ref);/*wait for home rise in
      current*/
    dcal[i]=(tmscnt-start)/20;
  }
}

/*move steps # of time increments*/
void dmprmove(uchar zone,uchar steps){
  uchar end;
  sel_meas(motordrive);
  end=tmscnt+steps;
  sel_ch(zone);/*starts motor*/
  while(tmscnt!=end);
}

/*close given damper one step*/
void close(uchar zone){
  uchar dist,set;
  set=dset[zone];
```

```
  if(set==0)dist=0;/*at most closed*/
  else if(set<10){/*near steps would open dmpr*/
    dist=(2*(10-set)+1);
  }
  else if(set==19){
    dist=0;
    home(zone);/*recalibrate to mark*/
  }
  else dist=1;
  dmprmove(zone,dist);
  dset[zone]=(set+dist)%20;
}

/*open given damper one step*/
void open(uchar zone){
  uchar dist,set;
  set=dset[zone];
  if(set==10)dist=0;/*already at most open*/
  else if(set>10){/*near steps would close dmpr*/
    home(zone);/*go to closed mark, then open*/
    dist=set=21-set;
  }
  else dist=1;
  dmprmove(zone,dist);
  dset[zone]=(set+dist)%20;
}
/*adjust any dampers that need change*/
void setdampers(void){
  uchar i,act,set;
  for(i=0;i<6;i++){
    act=tact[i]; set=tset[i];
    if(act>set+band){
      if(air)open(i);
      else close(i);
    }
    else if(act<set-band){
      if(air)close(i);
      else open(i);
    }
  }
}
```

**Furnace On/Off.** A temperature hysteresis or dead band is desirable for efficiency. This is usually adjustable, but I suppose 2°F is good—it could come on at 69°F and go off when 71°F is reached. The idea is that the furnace of air conditioner needs some time to get going if it is to operate efficiently. The algorithm could involve either a temperature hysteresis or a minimum time.

A second problem with a zone system is the decision to turn on or off—with six zones, which one carries the most weight? I propose a voting system where each zone gets a vote on both on and off. If the temperature is 2°F below the set point, that is one negative vote. If it is 3°F low that is two votes, and so on. Likewise, if it is 1°F above set point that is one positive vote, 2°F is two positive votes, and so on. The only exception might be where a pot is set all the way counter-clockwise—keep the zone shut off—then it has no vote. I might establish a weight so a net weight of, say, +4°F turns the furnace off (or the air conditioner on) and a –4°F turns it on. In-between the state does not change. The voting can happen perhaps every minute. Changing the a user setting pot could disrupt the vote and the normal cycling—a minimum cycle time needs to be established or a significantly changed pot setting should disqualify that input until the damper has had time to adjust (at 15 min intervals).

---

**Furnace On/Off Software**

```
/*zone error cumulative check*/
char vote(void){
  uchar i;char cum;
  for(i=0,cum=0;i<6;i++){
    cum=cum+((char)dat[22+i]-dat[28+i]);
  }
  return cum; /*set-actual*/
}

/*furnace/ac control based on vote of 6 zones
  presently 6 each way gives 2 deg hysteresis*/
void furnaceswitch(void){
  static bit ht_on,ac_on;
  if(reqheat==0){
    if(ht_on&&(vote()<-6)){
      ht_on=0;furn=1;/*shut heat off*/
    }
    if(ht_on==0&&(vote()>6)){
      ht_on=1;furn=0;/*turn heat on*/
    }
  }
  if(reqcool==0){
```

```
        if(ac_on&&(vote()>6)){
          ac_on=0;ac=1;/*shut ac off*/
        }
        if(ac_on==0&&(vote()<-6)){
          ac_on=1;ac=0;/*turn ac on*/
        }
      }
    }
```

**Time Intervals.** Since the motor adjustments do not come more often than every 15 min to reduce noise, and since you might as well be hung for a sheep as for a lamb, I propose to adjust *all* the motors in sequence at the 15-min mark. A *flag* will be set in the scheduler every 15 min (90,000 10-msec interrupts). Then the adjustments can begin based on the average temperature in each zone relative to the setting. A flag drives the entire process so it does not interfere with the timed interrupt—the main program is the *background* task. If things get too busy, the button scanning may stop until the dampers are set, for example, but no one will care since use of the buttons is part of initialization or the occasional viewing of statistics.

The turning on and off of the furnace (or air conditioning) needs to happen more often—I think every minute would be sufficient.

**Statistics.** As indicated earlier, this can be a never-ending programming job. The high/low outside could be recorded for each day, but this would get into a clock/calendar chip or that function in the software. Located in a basement there might be no light/dark indication to synchronize days so time precision would be important. For now I will omit that. It would be quite easy to indicate the current temperature in each room as well as the setting. I can imagine the user wanting to know the percentage of time the furnace ran in, say, the last 24 hr. Perhaps an accumulated hours figure could be kept to get an indication of fuel consumption. These are all frosting for the system, but probably are not difficult to provide.

**Display.** This has been the most difficult part of the software because there are so many options. Initially I considered a scrolling display covering one long line of text—the message length can then vary and it is possible to have far more than 40 or 80 characters in the total message.

Next I thought I would show only a few pieces of data.

Next I was going to run a $4 \times 40$ display, only to discover that the only choices for common alphanumeric displays are $2 \times 40$ and $4 \times 20$.[10]

---

[10] I should have realized that since the cursor address is 7 bits or 128 max—too few for $4 \times 40$.

Once the basic display format was chosen, the next step was to decide *what* to display. The tables that come next show present plans. There are four different displays (reached by pushing the left and right arrows at the same time). The default screen shows the time and day of the week (no year 2,000 problem here!) and both actual and desired temperatures in each room.

**Cursor address**

| | | | | | | | | | | | | | | | | | | | |
|---|---|---|---|---|---|---|---|---|---|---|---|---|---|---|---|---|---|---|---|
| 80 | 81 | 82 | 83 | 84 | 85 | 86 | 87 | 88 | 89 | 8A | 8B | 8C | 8D | 8E | 8F | 90 | 91 | 92 | 93 |
| C0 | C1 | C2 | C3 | C4 | C5 | C6 | C7 | C8 | C9 | CA | CB | CC | CD | CE | CF | D0 | D1 | D2 | D3 |
| 94 | 95 | 96 | 97 | 99 | 99 | 9A | 9B | 9C | 9D | 9E | 9F | A0 | A1 | A2 | A3 | A4 | A5 | A6 | A7 |
| D4 | D5 | D6 | D7 | D8 | D9 | DA | DB | DC | DD | DE | DF | E0 | E1 | E2 | E3 | E4 | E5 | E6 | E7 |

**Default display**

| | | | | | | | | | | | | | | | | | | | |
|---|---|---|---|---|---|---|---|---|---|---|---|---|---|---|---|---|---|---|---|
| w | e | d | | 1 | 0 | : | 2 | 7 | | | | | | o | u | t | | 3 | 1 |
| | | d | e | n | b | r | 1 | b | r | 2 | | l | r | | d | r | k | i | t |
| S | E | T | 7 | 3 | | 6 | 5 | | 6 | 2 | | 7 | 1 | | 7 | 3 | | 7 | 0 |
| A | C | T | 7 | 4 | | 7 | 1 | | 6 | 0 | | 6 | 9 | | 7 | 4 | | 7 | 7 |

The choice of setback (presumably for zone 0, with all others changing proportionally) was much more difficult. Currently it calls for four times to change the set temperature each day with both the time and the temperature user adjustable. That makes 56 values to set. That may be a bit much. The least expensive thermostats of that type have only two time and temperature groups—weekday, and weekend.

**Setback screen 1**

| | | | | | | | | | | | | | | | | | | | |
|---|---|---|---|---|---|---|---|---|---|---|---|---|---|---|---|---|---|---|---|
| S | | 7 | 7 | 2 | | | 9 | 6 | 5 | | | 4 | 7 | 1 | | 1 | 0 | 5 | 8 |
| M | | 6 | 7 | 2 | | 1 | 0 | 6 | 5 | | 1 | 2 | 7 | 1 | | 1 | 0 | 5 | 8 |
| T | | 6 | 7 | 2 | | | 9 | 6 | 5 | | | 4 | 7 | 1 | | 1 | 0 | 5 | 8 |
| W | | 6 | 7 | 2 | | | 9 | 6 | 5 | | | 3 | 7 | 1 | | 1 | 0 | 5 | 8 |

**Setback screen 2**

| | | | | | | | | | | | | | | | | | | | |
|---|---|---|---|---|---|---|---|---|---|---|---|---|---|---|---|---|---|---|---|
| T | | 6 | 7 | 2 | | | 9 | 6 | 5 | | | 4 | 7 | 1 | | 1 | 0 | 5 | 8 |
| F | | 6 | 7 | 2 | | 1 | 0 | 6 | 5 | | 1 | 2 | 7 | 1 | | 1 | 0 | 5 | 8 |
| S | | 8 | 7 | 2 | | | 9 | 7 | 2 | | | 4 | 7 | 2 | | 1 | 2 | 5 | 8 |
| | | | | | | | | | | | | | | | | | | | |

Finally there is a screen for statistics such as high/low for the day and cumulative degree days (the difference from the outside temperature and, say 70° F for each day, summed for successive days).

**Statistics screen**

| | | | | | | | | | | | | | | | | | | | |
|---|---|---|---|---|---|---|---|---|---|---|---|---|---|---|---|---|---|---|---|
| o | p | e | n | | | 5 | | 1 | | 4 | | 2 | | 5 | | 7 | | | |
| % | o | n | | 2 | 5 | | | | | | | | | | | | | | |
| d | e | g | | d | a | y | s | | 1 | 0 | 9 | 5 | | | | | | | |
| h | i | | 3 | 7 | | l | o | | 2 | 5 | | | | | | | | | |

The current approach to setting user-adjustable values is to move the cursor around with the four arrow buttons and use a separate up/down switch to change the value if the cursor address matches an adjustable value.

The screens are described by structures in code (ROM) that hold the address and the type of data for each display segment. Specifically the second byte of the structure is the pointer to a string in a (*code*) message array or the pointer to the value in one large (RAM) array holding the present values of all the displayable variables. This big variable array works against named variables so I include a lot of *#defines* to give useful names to the places within the array. This is not very programmer-friendly, but it is space efficient.

**Display Messages and Variables**

```
uchar dat[34];
#define dwk dat[0]
#define hrs dat[1]
#define min dat[2]
#define sec dat[3]
#define percenton dat[4]
#define dailyhi dat[5]
#define dailylo dat[6]
#define degdaysones dat[7]
#define degdayshunds dat[8]
#define outsidetemp dat[9]
/*z0=base+0,z1=base+1...*/
#define anglestepbase 10
#define zrmcodebase 16
#define zsetbase 22
#define zactbase 28
uchar stbkdat[56];
/*su=base+(0 thru 3),mo=base+(4 thru 7)...*/
```

```
#define setbacktimebase 0
#define setbacktempbase 28

uchar code msgdw[]=
/*0   1   2   3   4   5   6   7   */
{"SUN""MON""TUE""WED""THU""FRI""SAT"};
uchar code msgrm[]=
/*0   1   2   3   4   5   6   */
{"BR""LR""KIT""MBR""BR2""BR3""DEN"
/*7    8   9   10     */
 "FMR""DR""BSM""OUTSIDE"};
uchar code msgother[]=
/*0    1    2          3  4  4  5  6  7  8*/
{"SET""ACT""SETBACKS""S""M""T""W""T""F""S"
/*9      10   11            12          13        */
 "ANGLE""%ON""DEGREE DAYS""TODAY HIGH""LOW"};
struct dsp {uchar loc;uchar what;};
/*loc is address but 0 in msb means user alterable
      data;what--bits 6,7
      00=data,01=other(fixed)text;10=day of
      week;11=room;--bits 0-5:(00)index in data
      array(or setback for scr1,2);(01)index in text
      string;(10)index in data array to pick from
      dayofwk string;(11)index in data array to pick
      from room string*/
/*default--screen 0 uses dat array for data*/
struct dsp code scr0[]={
  {0x80,0x80},{0x84,0x01},{0x87,0x02},
  {0xc2,0xd0},{0xc5,0xd1},{0xc8,0xd2}, {0xcb,0xd3},
       {0xce,0xd4},{0xd1,0xd5},{0x94,0x40},
  {0x97,0x16},{0x9a,0x17},{0x9d,0x18},{0xa0,0x19},
       {0xa3,0x1a},{0xa6,0x1b},
  {0xd4,0x41},{0xd7,0x1c},{0xda,0x1d},{0xdd,0x1e},
       {0xe0,0x1f},{0xe0,0x20},{0xe3,0x21},
       {0xe6,0x22},{0x00,0x00}};
/*setbacks--screens 1 and 2 use setback array for
       data*/
struct dsp code scr1[]={
  {0x80,0x43},{0x81,0},{0x83,28},{0x86,1},{0x88,29},
    {0x8b,2},{0x8d,30},{0x90,3},{0x92,31},
  {0xc0,0x44},{0xc1,4},{0xc3,32},{0xc6,5},{0xc8,33},
    {0xcb,6},{0xcd,34},{0xd0,7},{0xd2,35},
  {0x94,0x45},{0x95,8},{0x97,36},{0x9a,9},{0x9c,37},
    {0x9f,10},{0xa1,38},{0xa4,11},{0xa6,39},
```

```
    {0xd4,0x46},{0xd5,12},{0xd7,40},{0xda,13},{0xdc,41},
      {0xdf,14},{0xe1,42},{0xe4,15},{0xe6,43}};
  struct dsp code scr2[]={
    {0x80,0x47},{0x81,16},{0x83,44},{0x86,17},{0x88,45},
      {0x8b,18},{0x8d,46},{0x90,19},{0x92,47},
    {0xc0,0x48},{0xc1,20},{0xc3,48},{0xc6,21},{0xc8,49},
      {0xcb,22},{0xcd,50},{0xd0,23},{0xd2,51},
    {0x94,0x49},{0x95,24},{0x97,52},{0x9a,25},{0x9c,53},
      {0x9f,26},{0xa1,54},{0xa4,27},{0xa6,55}};
      /*statistics--screen 3 uses dat array*/
  struct dsp scr3[]={
    {0x80,0x49},{0x86,10},{0x88,11},{0x8a,12},{0x8c,13},
      {0x8e,14},{0x90,45},
    {0xc0,0x4a},{0xc4,4},
    {0x94,0x4b},{0xa0,8},{0xa2,7},
    {0xd4,0X4c},{0xdf,5},{0xe2,0x4d},{0xe6,6}};
```

## ZONE HEAT OVERALL SOFTWARE

**Zhmain.c**

```
#include <reg51.h>
#define uchar unsigned char
#define uint unsigned int
#define uparrow 0x01
#define dnarrow 0x02
#define lftarrow 0x04
#define rtarrow 0x08
#define incr 0x10
#define decr 0x20
extern uint count;
void getrdgs(void);
void setdampers(void);
void homeall(void);
void lcdinit(void);
void timinit(void);
void displayval(uchar loc,uint val);
void display(uchar loc,msgnum);
void displayclear(void);
void furnaceswitch(void);
uchar getbtns(void);

uchar dat[34];
#define dwk dat[0]
```

```
#define hrs dat[1]
#define min dat[2]
#define sec dat[3]
#define percenton dat[4]
#define dailyhi dat[5]
#define dailylo dat[6]
#define degdaysones dat[7]
#define degdayshunds dat[8]
#define outsidetemp dat[9]
/*z0=base+0,z1=base+1...*/
#define anglestepbase 10
#define zrmcodebase 16
#define zsetbase 22
#define zactbase 28

uchar stbkdat[56];
/*su=base+(0 thru 3),mo=base+(4 thru 7)...*/
#define setbacktimebase 0
#define setbacktempbase 28

uchar code msgdofw[]=
/*0    1    2    3    4    5     6     7    */
{"SUN""MON""TUE""WED""THU""FRI""SAT"};
uchar code msgzone[]=
/*0    1    2     3     4     5     6    */
{"BR""LR""KIT""MBR""BR2""BR3""DEN"
/*7     8    9     10      */
 "FMR""DR""BSM""OUTSIDE"};
uchar code msgother[]=
/*0     1     2           3   4   4   5   6   7   8*/
{"SET""ACT""SETBACKS""S""M""T""W""T""F""S"
/*9       10    11              12             13        */
 "ANGLE""%ON""DEGREE DAYS""TODAY HIGH""LOW"};
struct dsp {uchar loc;uchar what;};
/*loc is address but 0 in msb means user alterable
      data; what--bits 6,7 00=data,
      01=other(fixed)text; 10=day of week;11=room
      bits 0-5: (00)index in data array(or setback
      for scr1,2); (01)index in text string;(10)index
      in data array to pick from dayofwk
      string;(11)index in data array to pick from
      room string*/
/*default--screen 0 uses dat array for data*/
struct dsp code scr0[]={
```

```
  {0x80,0x80},{0x84,0x01},{0x87,0x02},
  {0xc2,0xd0},{0xc5,0xd1},{0xc8,0xd2},{0xcb,0xd3},
         {0xce,0xd4},{0xd1,0xd5},
  {0x94,0x40},{0x97,0x16},{0x9a,0x17},{0x9d,0x18},
         {0xa0,0x19},{0xa3,0x1a},{0xa6,0x1b},
  {0xd4,0x41},{0xd7,0x1c},{0xda,0x1d},{0xdd,0x1e},
         {0xe0,0x1f},{0xe0,0x20},{0xe3,0x21},
         {0xe6,0x22},{0x00,0x00}};
/*setbacks--screens 1 and 2 use setback array for
       data*/
struct dsp code scr1[]={
  {0x80,0x43},{0x81,0},{0x83,28},{0x86,1},{0x88,29},
    {0x8b,2},{0x8d,30},{0x90,3},{0x92,31},
  {0xc0,0x44},{0xc1,4},{0xc3,32},{0xc6,5},{0xc8,33},
    {0xcb,6},{0xcd,34},{0xd0,7},{0xd2,35},
  {0x94,0x45},{0x95,8},{0x97,36},{0x9a,9},{0x9c,37},
    {0x9f,10},{0xa1,38},{0xa4,11},{0xa6,39},
  {0xd4,0x46},{0xd5,12},{0xd7,40},{0xda,13},{0xdc,41},
    {0xdf,14},{0xe1,42},{0xe4,15},{0xe6,43}};
struct dsp code scr2[]={
  {0x80,0x47},{0x81,16},{0x83,44},{0x86,17},{0x88,45},
    {0x8b,18},{0x8d,46},{0x90,19},{0x92,47},
  {0xc0,0x48},{0xc1,20},{0xc3,48},{0xc6,21},{0xc8,49},
    {0xcb,22},{0xcd,50},{0xd0,23},{0xd2,51},
  {0x94,0x49},{0x95,24},{0x97,52},{0x9a,25},{0x9c,53},
   {0x9f,26},{0xa1,54},{0xa4,27},{0xa6,55}};
    /*statistics--screen 3 uses dat array*/
struct dsp code scr3[]={
  {0x80,0x49},{0x86,10},{0x88,11},{0x8a,12},{0x8c,13},
    {0x8e,14},{0x90,45},
  {0xc0,0x4a},{0xc4,4},
  {0x94,0x4b},{0xa0,8},{0xa2,7},
  {0xd4,0X4c},{0xdf,5},{0xe2,0x4d},{0xe6,6}};

bit btnscan,dmprset,furnchk;
uchar screen,curloc;

/*move cursor to left in 2×40 lcd*/
void left(void){
  curloc--;
  if(curloc<0x80)curloc=0x80;
  if(curloc==0x93)curloc++;
  if(curloc==0xbf)curloc++;
```

```
  if(curloc==0xd3)curloc++;
}

/*move cursor to right in 2×40 lcd*/
void right(void){
  curloc++;
  if(curloc>0xe7)curloc=0xe7;
  if(curloc==0xd4)curloc--;
  if(curloc==0xa8)curloc--;
  if(curloc==0x94)curloc--;
}

/*move cursor up one row in 4x20 lcd*/
void uprow(void){
  if(curloc<0x94)curloc+=0x54;
  else if(curloc<0xa8)curloc+=0x2b;
  else curloc-=0x40;
}
/*move cursor down one row in 4x20 lcd*/
void dnrow(void){
  if(curloc<0xa8)curloc+=0x40;
  else if(curloc>0xd3)curloc-=0x54;
  else curloc-=0x2b;
}

uchar search(void); /*look for changeable val at
      cursor*/

void valchg(uchar dir){
  uchar varindex;
  switch(screen){
    case 0:if((varindex=search())!=0xff){
      if(dir==0)dat[varindex]++;
      else if(dat[varindex]>0)dat[varindex]--;
    }
    case 1: case 2: if((varindex=search())!=0xff){
      if(dir==0)stbkdat[varindex]++;
      else if(stbkdat[varindex]>0)stbkdat[varindex]--;
    }
  }
}
```

```
/*puts up an entire screen from structure in code
      space*/
void dspscreen(struct dsp code *scr){
  uchar i,x;
  displayclear();
  while((x=scr[i].what)!=0xff){
    switch(x & 0xc0){
      case 0:/*variable*/
        if((screen==0)||(screen==3)){
          displayval(scr[i].loc,dat[x&0x3f]);break;
        }
        else displayval(scr[i].loc,stbkdat[x&0x3f]);
          break;
      case 0x40:/*text*/
        display(scr[i].loc,msgother[x&0x3f]);break;
      case 0x80:/*day of week*/
        display(scr[i].loc,msgdofw[x&0x3f]);break;
      case 0xc0:/*zone name*/
        display(scr[i].loc,msgzone[x&0x3f]);break;
    }
  }
}

/*requested every 50mSec;pushes read in as 0; 1s for
      new pushes*/
void processbuttons(void){
  static uchar oldbtns;uchar newpushes,newbtns;
  static bit diffscreen;
  btnscan=0;newbtns=getbtns();
  if(newbtns&(uparrow|dnarrow)){/*2 buttons moves to
      next screen*/
    if(diffscreen==0){
      screen=++screen&0x03;diffscreen=1;
      switch(screen){
        case 0:dspscreen(scr0);break;
        case 1:dspscreen(scr1);break;
        case 2:dspscreen(scr2);break;
        case 3:dspscreen(scr3);break;
      }
    }
  }
  else diffscreen=0;
  if(diffscreen==0){
    newpushes=(newbtns^oldbtns)|oldbtns;
```

```
        oldbtns=newbtns;
        if(newpushes&lftarrow)left();
        if(newpushes&rtarrow)right();
        if(newpushes&uparrow)uprow();
        if(newpushes&dnarrow)dnrow();
        if(newpushes&incr)valchg(0);
        if(newpushes&decr)valchg(1);
      }
    }


    main(){
      timinit();
      lcdinit();
      homeall();
      dwk=sec=min=hrs=0;/*midnight before sunday*/
      screen=0;/*default display*/
      for(;;){
        if(btnscan)processbuttons();
        if(furnchk)furnaceswitch();/*on/off if needed*/
        if(dmprset){
          dmprset=0;
          getrdgs();/*get actual and setting
            temperatures*/
          setdampers();/*make adjustments if needed*/
        }
      }
    }
```

**zhtime.c**

```
#include <reg51.h>
#define uchar unsigned char
#define uint unsigned int
#define lcddat P1
#define dwk dat[0]
#define hrs dat[1]
#define min dat[2]
#define sec dat[3]
extern uchar goodrdg;
extern uchar msglen;
extern uchar cursorloc;
extern uchar dat[];
extern bit lcdbusy,btnscan,dmprset,furnchk;
```

```
sbit ena=0xa5; /*P2.5*/
sbit rd=0xa6; /*P2.6*/
sbit rs=0xa7; /*P2.7*/
uchar tmscnt;
static uint oldfcnt,newfcnt;
uint fcnt;

void timinit(void){
  TMOD=0x51; /*t1 16-bit counter, t0 16-bit timer*/
  TH0=-10000/256;TL0=-10000%256; /*10 mSec
interrupt*/
  TH1=TL1=0; /*allow max count*/
  ES=0;  /*no serial interrupt*/
  ET1=0;  /*no timer1 interrupts--just grab count
      every 10 mSec*/
  ET0=1;  /*timer0 is regular 10 mSec tick*/
  EX1=EX0=0;  /*no ext interrupts*/
  IP=0;  /*all(one)interrupts low priority*/
  TCON=0x50;  /*start timers*/
  EA=1; /*allow interrupts to be recognized*/
}

void clockset(void)using 1{/*every second*/
  sec=++sec%60;
  if(sec==0){
    min=++min%60;
    if(min==0){
      hrs=++hrs%24;
      if(hrs==0){
        dwk=++dwk%7;
      }
    }
  }
}

/*timer0 overflow interrupt routine--10 mSec realtime
       clock*/
void timer(void) interrupt 1 using 1{
  static uchar btncnt,seccnt,mincnt,ftncnt;
  if(goodrdg>0)--goodrdg;
  TH0=-10000/256;TL0=-10000%256; /*restart 10 mSec
       interrupt*/
  oldfcnt=newfcnt;newfcnt=(((uint)TH1)<<8)|
       TL1;fcnt=newfcnt-oldfcnt;
```

```
  ++tmscnt;
  if(--btncnt==0){btncnt=5;btnscan=1;} /*flag every
      50mSec*/
  if(--seccnt==0){ /*each second*/
    seccnt=100;clockset();
    if(--mincnt==0){
      mincnt=60;furnchk=1;/*flag check furnace every
       minute*/
      if(--ftncnt==0){
        ftncnt=15;dmprset=1;/*flag adjust damper every
       15 min*/
      }
    }
  }
}
```

**zhtemp.c**

```
/*read settings & temperatures*/
#include <reg51.h>
#define uchar unsigned char
#define uint unsigned int
#define currentlow 0
#define currenthigh 1
#define motordrive 2
#define buttons 3
#define rref 500
#define vref 276
#define band 3
#define motordrive 2
#define motorstop 0
#define rref 500
#define vref 276
extern uint tmscnt;
void sel_meas(uchar which);
void sel_ch(uchar channel);
void sel_meas(uchar which);
void sel_ch(uchar channel);
uchar ref; /*normal mtr current*/
bit air;
uchar tact[7];
uchar tset[7];
uchar dset[6];/*0-19--0=20=closed,10=full open*/
uchar dcal[6];/*1/20 time for 1/2 rev*/
uint floref,fhiref,difref,fcnt;
```

```
/*waits for next 10 mSec to return latest reading*/
uint read_f(void){
  uchar start;
  for(start=tmscnt;start==tmscnt;);
  return fcnt;
}

void getrdg(uchar i){
  uint flo,fhi,temp1;
  sel_meas(currentlow);sel_ch(i);flo=read_f();
  sel_meas(currenthigh);sel_ch(i);fhi=read_f();
  tset[i]=rref*(temp1=fhi-flo)/difref;
  /*100-=closed,200ohm=50degF,400=60,600=70,800=80,
       900+=open*/
  tact[i]=(vref*(temp1)-(floref*fhi)+(fhiref*flo))
       /difref;
}

void cal(void){
  sel_meas(currenthigh);sel_ch(7);fhiref=read_f();
  sel_meas(currentlow);floref=read_f();
  difref=fhiref-floref;
}

void getrdgs(void){
  uchar i;
  cal();
  for(i=0;i<6;i++)getrdg(i);
}

/*these are damper setting routines*/
/*run motor until switch increases current*/
void home(uchar zone){
  sel_meas(motordrive);
  sel_ch(zone);/*start motor moving*/
  ref=read_f();
  while(read_f()>=ref);/*wait for home rise in
       current*/
  sel_meas(motorstop);
}

void homeall(void){
  uchar i;
  for(i=0;i<6;i++)home(i);
}
```

```
/*calibrate time to complete 1/2 revolution*/
void caldmpr(void){
  uchar start,i;
  homeall();
  for(i=0;i<6;i++){
    start=tmscnt;
    sel_ch(i);/*start motor moving*/
    ref=read_f();
    while(read_f()>=ref);/*wait for home rise in
       current*/
    dcal[i]=(tmscnt-start)/20;
  }
}

/*move steps # of time increments*/
void dmprmove(uchar zone,uchar steps){
  uchar end;
  sel_meas(motordrive);
  end=tmscnt+steps;
  sel_ch(zone);/*starts motor*/
  while(tmscnt!=end);
}

/*close given damper one step*/
void close(uchar zone){
  uchar dist,set;
  set=dset[zone];
  if(set==0)dist=0;/*at most closed*/
  else if(set<10){/*near steps would open dmpr*/
    dist=(2*(10-set)+1);
  }
  else if(set==19){
    dist=0;
    home(zone);/*recalibrate to mark*/
  }
  else dist=1;
  dmprmove(zone,dist);
  dset[zone]=(set+dist)%20;
}
/*open given damper one step*/
void open(uchar zone){
  uchar dist,set;
  set=dset[zone];
  if(set==10)dist=0;/*already at most open*/
  else if(set>10){/*near steps would close dmpr*/
```

```
    home(zone);/*go to closed mark, then open*/
    dist=set=21-set;
  }
  else dist=1;
  dmprmove(zone,dist);
  dset[zone]=(set+dist)%20;
}
/*adjust any dampers that need change*/
void setdampers(void){
  uchar i,act,set;
  for(i=0;i<6;i++){
    act=tact[i]; set=tset[i];
    if(act>set+band){
      if(air)open(i);
      else close(i);
    }
    else if(act<set-band){
      if(air)close(i);
      else open(i);
    }
  }
}
```

**zhlcd.c**

```
/*these are all the display-related procedures*/
#include <reg51.h>
#define uchar unsigned char
#define uint unsigned int
#define cg_addr_set 0x40
#define dd_addr_set 0x80
#define no_entry_move 0x06
#define busy_flag 0x80
#define left_arrow 0x7f
#define right_arrow 0x7e
#define up_arrow 1
#define dn_arrow 2
void wrcmd(uchar cmd);
void wrdat(uchar dat);
void tmsec(uchar delay);

uchar msglen,cursorloc;

/*erases and resets display*/
void displayclear(void){
```

```
  wrcmd(0x01);
}

/*sends decimal of binary to LCD*/
void dsplayval(uchar loc,uint num){
  uint m=10000;uchar i=5;uchar ldgzero=0;
  if(loc>0)wrcmd(loc);
  do{
    if((num/m>0)||(m==1))ldgzero=1;
    if(ldgzero)wrdat(num/m+'0');
    num%=m;i--;
    if(m==10 && num==0){
      wrdat(0x07);/*custom 00*/
      m=0;
    }
  }while((m/=10)>0);
}
```

**zhdrvrs.c**

```
#include <reg51.h>
#define uchar unsigned char
#define uint unsigned int
#define currentlow 0
#define currenthigh 1
#define motordrive 2
#define buttons 3
#define lcddat P1
#define no_entry_move 0x06
#define busy_flag 0x80
#define dwk dat[0]
#define hrs dat[1]
#define min dat[2]
#define sec dat[3]
extern uchar tmscnt;
extern uchar dat[];
extern uint newfcnt,oldfcnt;/*used in rt interrupt*/

sbit ena=0xa5; /*P2.5*/
sbit rd=0xa6; /*P2.6*/
sbit rs=0xa7; /*P2.7*/
sbit btn=0xb3; /*P3.3*/
sbit furn=0xb4; /*P3.4*/
sbit ac=0xb5; /*P3.5*/
sbit reqheat=0xb6; /*P3.6*/
```

```
sbit reqcool=0xb7; /*P3.7*/

bit lcdbusy;
uchar cursor_addr=8;
uchar P2store,P3store;/*can't trust port read-back*/
uchar goodrdg;

void tmsec(uchar delay){
  uchar end;
  end=tmscnt+delay;
  while(tmscnt!=delay);
}

/*sets 1 mA or 5 mA current drive; negative logic*/
void sel_meas(uchar which){
  static uchar oldwhich;
  if(which!=oldwhich){/*change in current--allow
       settling*/
    goodrdg=2;oldwhich=which;
  }
  P2store|=0x18;
  switch (which){
    case 3:case 0:/*curent low*/
      P2store&=0xe7;break;/*10*/
    case 1:/*high current*/
      P2store&=0xf7;break;/*00*/
    case 2:/*motors*/
      break; /*11*/
  }
  P2=P2store;
}

/*sinks current from one of the 6 zone sensors;
  ref=6,outdoor=7*/
void sel_ch(uchar zone){
  P2=P2store=(P2store&0xf8)|(zone);
}

/*readings of counter saved every 10 mSec*/
uint read_f(void){
  while(goodrdg==0);
  return newfcnt-oldfcnt;
}
```

```
/*wait for busyflag to clear*/
void lcdwait(void){
  uchar bf;
  rs=0;rd=1;
  do{
    ena=1;
    bf=lcddat;
    ena=0;
  }while(bf>0x7f);
}

/*write command to LCD*/
void wrcmd(uchar cmd){
  while(lcdbusy);lcdbusy=1;
  rs=0;rd=0;lcddat=cmd;ena=1;ena=0;
  lcdwait();
  lcdbusy=0;
}

/*write data to LCD*/
void wrdat(uchar dat){
  while(lcdbusy);lcdbusy=1;
  rs=1;rd=0;lcddat=dat;ena=1;ena=0;
  lcdwait();
  lcdbusy=0;
}

/*reads buttons one at a time selected by the 138*/
uchar getbtns(void){
  uchar btns,i;
  sel_meas(buttons);
  for(i=0,btns=0;i<6;i++){
    sel_ch(i);btns=btns<<1|btn;
  }
  return btn;
}

/*zone error cumulative check*/
char vote(void){
  uchar i;char cum;
  for(i=0,cum=0;i<6;i++){
    cum=cum+((char)dat[22+i]-dat[28+i]);
  }
```

```
  return cum; /*set-actual*/
}

/*furnace/ac control based on vote of 6 zones
  presently 6 each way gives 2 deg hysteresis*/
void furnaceswitch(void){
  static bit ht_on,ac_on;
  if(reqheat==0){
    if(ht_on&&(vote()<-6)){
      ht_on=0;furn=1;/*shut heat off*/
    }
    if(ht_on==0&&(vote()>6)){
      ht_on=1;furn=0;/*turn heat on*/
    }
  }
  if(reqcool==0){
    if(ac_on&&(vote()>6)){
      ac_on=0;ac=1;/*shut ac off*/
    }
    if(ac_on==0&&(vote()<-6)){
      ac_on=1;ac=0;/*turn ac on*/
    }
  }
}

/*ititializes LCD module;8-bit mode*/
void lcdinit(void){
    tmsec(18);
    rs=0;rd=0;lcddat=0x38;ena=1;ena=0;
    tmsec(5);/*can't use busy flag yet*/
    rs=0;rd=0;lcddat=0x38;ena=1;ena=0;
    tmsec(1);/*can't use busy flag yet*/
    rs=0;rd=0;lcddat=0x38;ena=1;ena=0;
    wrcmd(0x38); /*2 lines*/
    wrcmd(0x01);
    wrcmd(0x07); /*cursor incr, disp shift--scroll*/
    wrcmd(0x0c); /*turn on display*/
  }
```

## Zone Heat Schematic[11]

Having discussed all these features, the next page shows a schematic of my current thinking. *Remember that this has not been built or tested in exactly this form, so no warranties are to be inferred.* I may change to 74154 1-in-16 decoders to gain 16 additional outputs to motors if I go to dampers in each duct outlets instead of the single-damper-per-zone approach. They would only require one extra port pin so I would shift up P2.3 and P2.4 to P2.4 and P2.5 respectively and use P2.4 as the fourth select input to the decoders.

If the code size were small enough (2 k) then I might try to fit in an 87C751 or 752, which would reduce port pins to 19 or 21 respectively.[12] In that case I would switch to 4-bit data to the LCD, but that would still only give 21 pins—the 751 would not make it (I really think the code will not stay that small anyway).

---

[11]Due to a failure of my hard drive just before completion of this chapter, the above schematic does not show the additional two buttons discussed earlier. Also, the connections to the zone in-room sensor/inputs are changed in the software so zone 0 is selected as 0 while tha calibration and outdoor sensors are moved up to addresses 6 and 7.

[12]It is not as of this revision and there is not enough economic incentive to shrink it. The code is just a shade over 2 K as shown, but software seldom shrinks without severe pressure. I might add that the demo tools from Keil which I am using refuse to link all the pieces so I had to sum the space of each module and I have no data as to the space required by the libraries. The included tools are not very useful when you want to get the real, final code (which was Keil's intent!). It is a good learning tool, but you need to take the next step to go further.

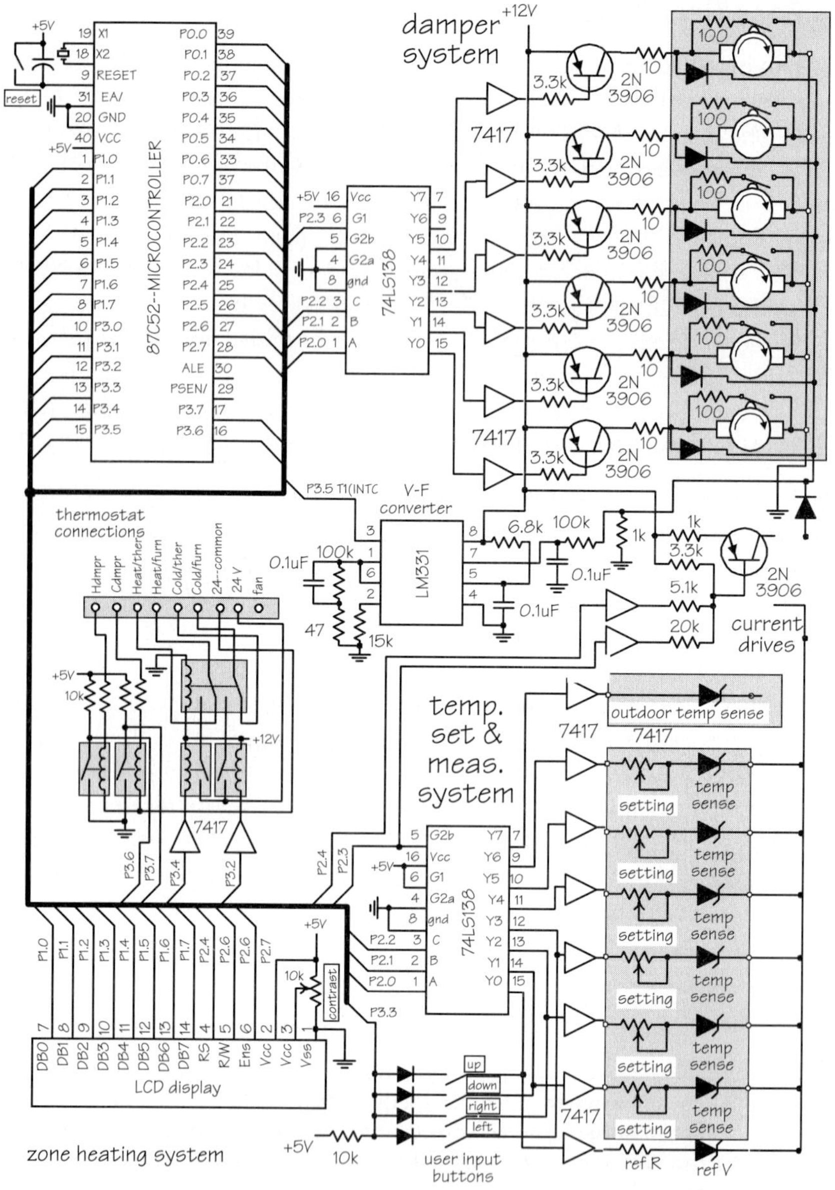


**Zone Heating System Tentative Schematic**

## LIGHT METER

This sample project has yet to be constructed, but it has given me something to plan during long boring meetings. The idea started with my interest in photography—specifically flash photography—coupled with the data sheet for a light-to-frequency chip newly released (at that time) by Texas Instruments.[13] What could make for a smaller, simpler system than a single-chip input, a microcontroller, a display, buttons, and a battery? When I realized that currently flash meters sell for about $400, it seems quite easy to develop a prototype for far less, let alone to produce a product selling for less. So, here are the planning and design thoughts for a low-cost flash meter.

### What the Customer Wants

Photographic film is identified with two speed rating systems, but the ASA number is the more common.[14] The typical speeds available range from 25 to 1600 with 100, 200, and 400 most common. It takes twice as much light exposure to get a suitable picture with 100-speed film than with 200-speed film. The higher the number the dimmer the light can be or the shorter the exposure time. Conventional exposure meters take a setting of film speed and show combinations of lens opening (aperture) and shutter speed that best suit the amount of light on the scene. Of course, there are bright spots and dark spots, so the goal is a best average.[15] For the serious photographer there are also *spot meters,* which can give the exposure for any given tiny area of the scene. Then the photographer can set exposure for the area they particularly want to be best exposed or measure the various exposures and pick a smarter average than is possible with an automatic camera.

---

[13]If you have read earlier sections on A-D, you realize I am a great proponent of bringing analog data into a microcontroller as a varying frequency square wave. Again, this is because it takes only one pin, can be evaluated coarsely quite quickly, or to higher precision over longer time, and can give a reading based on the average over the reading time, thus being more noise immune.

[14]The DIN rating can be correlated to the logarithm, going up by one when the sensitivity doubles.

[15]High priced, modern automatic cameras have different modes to average the exposure from objects near the center (for a portrait with the subject centered) or over a broad area, deemphasizing the top (for landscape scenes). Even low-cost cameras take an average more favoring the center.

When one moves to flash photography, the rules change. Since the intense light happens for a very brief time, the shutter speed is unimportant.[16] Typically a flash setting keeps the shutter open for 1/60- to 1/90-second. The photographer sets the lens opening (aperture) for the desired exposure based on the strength of the flash and the distance to the subject.[17] The conventional method of setting the aperture involved dividing the rating the flash unit (in exposure value—EV) by the distance to get the aperture setting. For example, with an EV of 160, a subject at 10 ft needs an aperture of f16, but at 20 ft a wider aperture of f8 is needed to let in more of the (weaker because of the distance) light.

That is exposure in its basic form. Unfortunately, any more complicated situation becomes a problem. Professional portrait photographers often use multiple flash sources in synchronism—one for the side, one for the hair, one for the background, and one to fill in the rest of the face! There are rules for arranging the lights in terms of distance, diffusion, and so on, but it becomes so complex that many photographers regard it as unpredictable and use instant film first to see what the combined effect is going to be. Likewise, if the flash is not a significant portion of the total exposure, there are problems. If the ambient light during the 1/60-second, for the aperture expected for the flash gives significant exposure, the entire balance can be off. Also, what can you do if the flash is needed as a "fill flash" to reduce the dark shadows? Here you have an argument for the flash meter.

So, what I would like is a meter that will read light intensity over the duration of the shutter opening time that is triggered to begin measurement when a high intensity light burst arrives (the flash). In addition to measuring the flash, though, it should include the ambient light that arrives during the entire shutter open time. For flash, the meter should allow the user to set the film speed and shutter time and report the necessary aperture. In addition, it should function as a conventional non-flash light meter. Since the subject may be in the center or include an entire area, there should be some sort of spot/area setting, and some way for the meter to be used at the subject (incident measurement) as well as at the camera (reflected measurement). It should run on only one or two batteries and last a long time—perhaps two

---

[16]The only problem is to have the shutter open when the flash intensity peaks—called synchronization. For modern electronic strobe light the flash is quite quick; the difference was with flash bulbs where the ignition of the material in the bulb took time to spread, giving a longer delay.

[17]The light intensity from a point source drops off by the square of the distance, so there is one quarter as much light on an object 20 feet away as one 10 feet.

years of incidental use. It should be quite small and lightweight with controls that are quite simple to use and learn. Finally, the parts should cost well under 100 dollars. Putting these into specifications:

Performance
- Measures conventional and flash exposure.
- Common film speeds from ASA 12 through 3200
- Shutter speeds from 1/4000 sec to 5 minutes
- Apertures from f32 to f1.0 (32,22,16,11,8,5.6,4,2.8,2.0,1.4,1.0)
- Switches to flash mode when sudden light increase detected
- Battery lasts for at least 20 hrs of continuous use
- Auto shut-off after 5 min

Physical
- Less than 1 × 3 × 4 inches
- Very few buttons
- Display 1-in. × 3-in.
- Mechanically convertible to spot meter or incident meter

Environmental
- Very rugged—survive drop to concrete floor from 3 feet
- Operates from 20 to 140° F
- Water-resistant

Cost less than 100 dollars

### Roughing Out Lightmeter

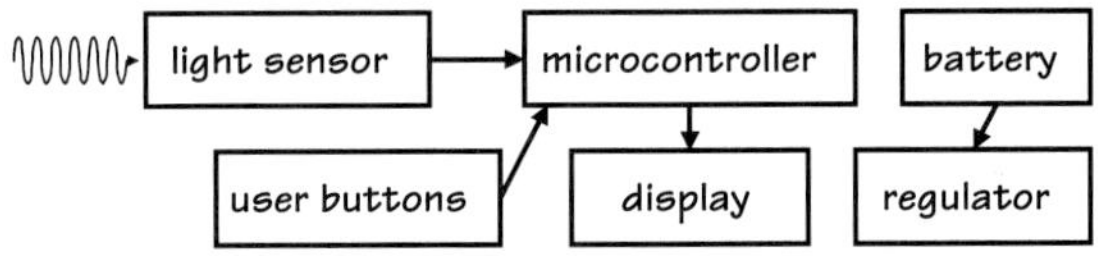


**Light Meter Hardware Block Diagram**

**Light Sensor.** The heart of the meter is to be the TSL 230 from Texas Instruments which combines a photo-diode array with a V-F converter and gives a linear frequency vs. light output over many decades of light intensity. The chip requires no external components and has programmable sensitivity and frequency range. If it were not used, then a separate photodiode, amplifier, and some sort of A-D would be needed, which could never match the two to three dollars for the TI chip.

One question is the spectral sensitivity—silicon-based sensors are inherently more sensitive in the red and infrared than in the blue. I believe filters that reduce the red and block the IR are quite expensive, but they might be a considered if further testing reveals this to be a problem. Since there seem to be no economical alternatives, I will have to warn users that the unit is calibrated for broad-spectrum light and will misread if the light is unusually warm (red) or cold (blue). Perhaps a correction factor could be supplied for early-morning or late-evening light, which tends to be warmer. Since most flash is designed to match daylight, the color sensitivity should be no problem for the flash measurements. I do not how fluorescent light will register due to the narrow lines of light output, but commonly the resulting pictures have a greenish cast if a corrective filter is not used on the camera.

There are mechanical and optical questions as well. When measuring *incident* light there should be a diffuser so the sensor registers the light arriving from a wide angle. When measuring *reflected* light (what the camera sees), the sensor should approximate the viewing angle of the camera or possibly an even smaller region. Rather than trying to compensate for the exposure with wide-angle through telephoto lenses, I think a relatively narrow viewing angle is called for, with the user moving it around if there is significant brightness variation across the scene. I assume the more conventional light metering built into modern cameras is sufficient for wide area measurements—why compete? In essence, the proposed light meter will combine a spot meter and an incident light meter for both ambient light and flash.

**User Buttons.** Less is better since size is important. Since I propose to use a small alphanumeric LCD display (custom ones are too expensive for a single prototype), I think three buttons might suffice: *up, down*, and *next menu item.* The current thinking is to have a series of menu choices such as film speed, aperture, and shutter speed with some functions that are reached by double-clicking the menu button like with a computer mouse.

A lot of attention will have to be paid to ease of use and having the meter default to the most likely settings. For example, just after a flash is detected, the place in the menu sequence should shift so the next *menu* push clears the reading. That way the meter can be used repeatedly with a flash setting while requiring only one button push.[18] Likewise, if the aperture is set, the shutter speed setting is omitted or vice versa.

---

[18]Perhaps the meter can automatically reset when it senses the next flash burst, but there is the risk that other photographers might fire off their flashes in between, as at a wedding where all the point-and-shoot cameras are fired indiscriminately.

**Microcontroller.** Here is the real reason for discussing this project here—nothing before has focussed on *really tiny* as a goal. The 87C750 (or 751) is the choice within the 8051 family because of the size and cost constraints. A processor in a 24-pin "skinny DIP" is about as small as you can get.[19] It has only one counter/timer, but it does have two external interrupts so the frequency can be easily read even from two sensors. It only fits 1-k bytes of program (non-expandable, on-chip—the 751 has 2 k), so code size is critical.

**Display.** Keeping power in mind, the display must be an LCD. With all the options envisioned from menu input, it should have alphanumeric capability. Depending on how much information must be shown at one time, either a $1 \times 8$ or a $2 \times 16$ display could suffice. The display should be as small as possible consistent with the overall meter size. I plan to use the custom-character set that can be loaded into the bottom addresses giving eight arrangements of any combination of the $5 \times 7$ dot elements.

To my knowledge, no other form of display can satisfy the power constraints. If viewing in low ambient is needed, a back-lit function can be temporarily turned on by an additional button.

**Battery.** The goal I set is to run the meter off either one or two common AA batteries. They are the most readily available and, given enough life, are preferable to rechargeable ones with the corresponding need to find a charger or build one in. The camera industry is quite comfortable with disposable batteries.

**Regulator.** Rather than trying to get the 5 V needed from the batteries directly (very inefficient with a 9 V one and too bulky with 4 1.5 V ones), a boost (switching) regulator is planned. There are several that can work from 1.5 V[20] as well as at least one charge pump regulator that can work from 3V without an inductor.[21]

[19]Of course, there is surface-mount to consider and I believe the PIC and some of the 4-bit processors are smaller, but this is a book on the 8051! Since this is to be prototyped with stock parts and since the display is to be non-custom, I allow myself enough space to fit all the electronics behind the display.

[20]I looked at the MAXIM 856 and 858 which can supply 75 and 30 mA respectively from 1.5 V and 270 and 30 mA from 3 V.

[21]The MAX 619 can provide 5 V at 50 mA; it does require 2 10 μF and two 0.22 μF capacitors.

The option of designing the entire system to operate on lower voltage is worth considering. The 750/1 can only operate on 5 V ±10 percent, but there are some members of the 8051 family that can operate from 2.7 V to 3.6 V or 6 V. The TSL 230 can operate down to 2.7 V. Unfortunately, to my knowledge, standard alphanumeric LCDs only operate on 5 V, but all my instincts say there *must* be new ones that take advantage of the stampede to low voltage driven by the cellular phone market. Some serious searching is called for since going to two batteries *without* a regulator could be a significant cost savings—the regulator costs about $1.60 in quantity and the capacitors probably bring it up to $2.50 in addition to the extra space required for the circuitry.

**Power Management.** This becomes one of the most interesting aspects of the design. A quick survey of power requirements if the system runs off 5V:

- The light sensor draws 3 mA max when active or 10 μA in power-down mode.
- The 87C750 draws a max of 18 mA (@ 12 MHz—4 mA @ 3.5 MHz), about 2 mA in idle mode (keeping peripherals, registers and RAM alive), or 50 μA in power-down mode (only keeping RAM alive).
- The LCD draws about 2 mA.
- The MAX 619 regulator draws 100 mA active (besides the current to the load) or 1 mA in shutdown mode.

The total is less than 25 mA worst case—half what the regulator can supply.

Addressing the power management further, I have to decide whether to hold the user's settings in memory when powered down. If I have the unit power down after only 5 minutes of inactivity, it would be frustrating to lose the film speed choice or the aperture/shutter speed choice. I can keep the microcontroller RAM alive with 50 μA or I could keep the setup data in the LCD controller (the non-displayed locations can be used as storage if you do not scroll the display). The latter is undesirable because the display would be lit, falsely implying activity. Since the power-down works for any voltage above 2.0 V, I can shut down the regulator and supply the microcontroller power at that time directly from the batteries through a germanium (0.3 V forward drop) diode. A second diode will keep the power-down voltage from back-feeding to the rest of the circuitry. Since the 87C750 is rated 5 V ±10 percent, the 4.7 V after passing through the diode is enough above the 4.5 V minimum. This arrangement will stop virtually all the current to the LCD, the buttons, and the sensor without any special shutdown steps for them.

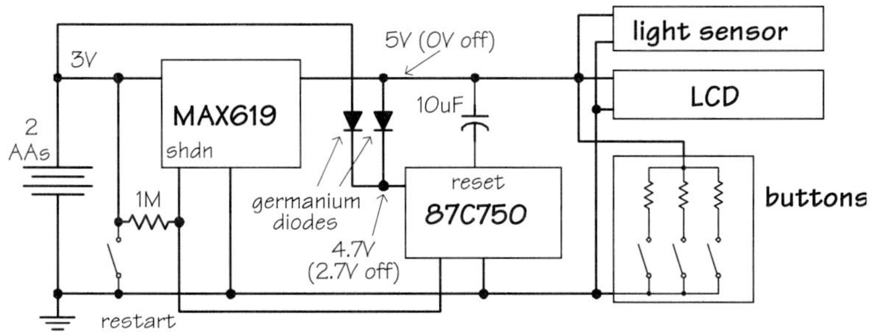


**Light Meter Power-down Scheme**

The shutdown sequence is to first write the port bit connected to the *shdn* high (a logic zero keeps the regulator running) and immediately write a one to the PD bit (bit 1) of the *PCON* (*sfr* address $87_{16}$) register. A capacitor at the regulator output will easily keep the microcontroller powered long enough to complete the shutdown instruction.

Having shut the system down, restarting involves pulling *shdn* to ground with a switch and allowing the resulting power-up to reset the microcontroller—the only way to exit the power down. I have to ensure that the power-up code from the C compiler does not automatically zero all RAM (as an ANSI standard C program would have to do), since that would destroy the very settings that were saved! A variable named *warm* holds a code that is unlikely to come up by accident on first power-up, but will be present at any other restart if the power-save mode has retained RAM contents.

```
        /*Check for Recovery from Power-down*/
if (warm!=0xaa){
  warm=0xaa;
  stpASA=6;stpf=2;stpspd=0;
  for(i=0;i<16;ltcnts[i++]=0);
}
```

Battery life in shut down mode should be quite long—pulling about 50 µA out of two-AA cells should give a battery life of about 6 years with no use.[22] Assuming a 75 percent efficiency when regulating, given a load of 25 mA @ 5 V or a battery draw of about 50 mA or a continuously active life

[22]My information is that a AA cell drives 100 Ω above 0.9 V for at least 170 hr—about 2500 mAH. That sounds like a lot for a single cell.

of about 45 hours. Since the unit will auto shut down in 5 min, that translates into about 540 on cycles. If the start button simply has the microcontroller restore the display based on the RAM contents the 5 min can be shortened—perhaps to 2 min, giving 1,350 uses per set of batteries. This will have to be explored further to determine what the user will accept—display residual time traded against frequency of battery replacement.

## Lightmeter Software

**Frequency-to-Light Conversion.** Given only one timer (the 750 has only one!) and two external interrupts, one option for measuring the TSL 230 frequency is to capture the timer value each time (or every tenth time) the frequency signal edge arrives at an external interrupt pin. Alternately with more capability a timer could run a real-time interrupt and a separate counter could accumulate the pulses from the sensor.[23] Then the interval would have to be quite short because a flash has duration much shorter than 1/100 sec—perhaps a 1-msec interval would be enough. The counts difference (duration of one or several cycles from the TSL 230) or count (number of cycles in a fixed time) can be converted to light intensity or the log of intensity, which can lead to the exposure settings.

To catch flash events, the periods of the signal from the TSL 230 are saved in a cyclic buffer. Then the arrival of a flash is detected by a significant change in the period of the new cycle compared to one perhaps 8 cycles before.

**Interrrupt Software (First Option)**

```
void area(void)interrupt 0 using 1{
  ltcnt++;
}

void spot(void)interrupt 2 using 1{
  ltcnt++;
}

/*regular 1mSec interrupt*/
void clock(void) interrupt 1 using 1{
  static uchar samplesindex;uchar old;
  waitcount--;
```

[23]This is impossible with the 87C750 which has only one counter, but could be done with the 87C751 which has a regular counter and a second fixed timer that can be used if the $I^2C$ bus is not involved.

```
    if(((flash>0)&&(clear>0))||flash==0){
      samplesindex=++samplesindex&0x07;
      old=ltcnts[samplesindex];
      ltcnts[samplesindex]=ltcnt;
    }
    if(clear>0)clear--;/*mark when buffer is filled*/
    if(var[3]!=1){/*flash or combination*/
      if((ltcnt>>3)>old){/*detect a flash by larger
count*/
        flash=1;clear=8;
      }
    }
}
```

**Light Sensor Scaling.** Since high irradiance can result in a frequency of up to 1 MHz, the built-in pre-scalar in the TSL 230 will be used for high intensities to keep the max frequency to 10 kHz—much easier to measure with a 1 MHz clock to the counter. It gives a resolution of 100 at the highest intensity and the least sensitivity. From there the divide can be eliminated and the sensitivity increased to get (with a 16-bit counter) a relative period of $6.5 \times 10^6$ or a relative intensity of 0.000000015—an intensity range of about eight decades. I do not know what the actual illumanance may be, nor can I, in advance, correlate it to the film exposure without some experimental work, since the light–gathering of the optics are an unknown to me. A look-up table handles the steps in sensitivity of the TSL 232 because the 4 bits involved do not operate in a logical sequence.

```
      /*See if Sensitivity Needs Changing*/
   uchar code sens[]={0xd0,0x90,0xa0,0xb0,0x30};
   if(flash==0){
   if(ltcnt>0xc0){
     sensstep--;
   }
   else if(ltcnt<0x05){
     sensstep++;
   }
   else return;
   sensstep%=4;
   P3=sens[sensstep];
   clear=8;/*refill buffer after sens change*/
```

**Calculating Exposure.** Once the light is determined, the appropriate combination of aperture and shutter speed, the exposure, can be determined. The relation between aperture and shutter speed is constant for a given film—f2@1/1000 Sec is the same to the film as f2.8@1/500; f4@1/250; f5.6@1/125; f8@1/60; f11@1/30; f16@1/15; f22@1/8; or f32@1/4.

Actually, a logarithmic unit is established called **exposure value** (EV) where EV1 indicates f1.4@1 Sec. with EV2 indicating f2@1 Sec and so on. Likewise, the **film speed** steps double the sensitivity so if f4@1/250 works for ASA 100 then f5.6@1/250 is good for ASA 200; f8@1/250 for ASA 400; F11@1/250 for ASA 800; and f16@1/250 for ASA 1600. In other words, looking at the steps as indices of arrays, the *sum* of the aperture and speed arrays must be *constant* for a given film speed and the *value* of the sum goes up by one for each step up in film speed.

Since the goal is very small code, I have a function below which avoids both floating point variables and log functions. Since multiplying numbers is the same as *adding* their logarithms, I keep the sensitivity as a factor to add. By building up the EV including tenths, the individual components can be fit together more precisely. Since the EV goes up by 1 (or 10 tenths) for each doubling of light, I work directly from the period sum. The bigger the period the less the light. Since binary numbers nicely represent factors of two, I just shift the total light period left, adding 10 to the EV, until the msb of the period reaches the top bit position. The more shifts, the smaller the period and the more intense the light. Once there, the remaining bits of the period reduce the value back down from the last 1 (or 10 tenths). I use only the last three bits below the top one. Having gotten 10*log 2 of the light, the remaining step both adjusts for the sensitivity setting (from a lookup table) and throws in an overall correction factor for the entire system. The latter that must be empirically determined based on how much light, as indicated by the TSL 230 frequency, relates to what exposure related to film sensitivity.

```
              /*Compute Exposure Value*/
void update(void){
  uchar i;uint light;
  for(i=light=0;i<8;i++)light+=ltcnts[i];
  if(light>0)for(ev=0;light<0x8000;ev+=10)light<<=1;
  ev=ev-(10*((light&0x7000)>>12))/8;
  ev=ev+evcal+evsens[sensstep];
}
```

In the interrupt routine can be a comparison to recognize the arrival of a flash burst and to begin accumulating the exposure over the shutter open time (1/60 or 1/90 second). Failing in detecting a flash, the software would accumulate the intensity values for a fixed time and convert that to either shutter speed or aperture setting, given that the user has selected the other.

**Light to Settings.** A very simple relationship exists between the shutter speed steps and the aperture steps. For a given film speed and light intensity the product remains constant. I favor tables as shown below, but I have just run over the 1K of space so I may have to go to multiplication reflecting the very regular sequence of the data. The tables eat up about 80 bytes of code space! The intensity will be a logarithmic value, but rather than importing a floating-point logarithm function, I would have a logarithmic table correlated to the frequency and use linear interpolation between values.

```
              Film, Aperture & Speed Arrays*/
uchar stpASA,stpf,stpspd;
uint code
         ASA[]={12,25,50,100,200,400,800,1600,3200};
uint code
         f[]={320,220,160,110,80,56,40,28,20,14,10};
uint code spd[]={4000,2000,1000,500,250,125,
  60,30,15,8,4,2,1,2,4,8,15,30,60};
```

**Display Functions.** There is no shortage of port pins so this will be 8-bit interfaced. If the 1 × 8 configuration is used then custom characters will have to be written to the LCD to show [.8] or [.6] in one digit. Close attention to message size is important to limit code size. The tables below show the sort of displays anticipated. As you can see, 8 characters is not a lot, but it should be possible to toggle through.

The mode of operation is determined by the mode when the entry is stopped. After ASA and mode are set, the meter will compute either the f stop (if that is left last showing) or the shutter speed (if that is last up). While the computed value will be shown, the last entered value will be stored so the user can adjust the setting of either of the fixed

| | | |
|---|---|---|
| Af2.8 | A 2sec | C f16 |
| F f1.4 | Ambient | A 1/4 |
| A1/1000 | Flash | F f8 |
| ASA 400 | Combnatn | A 1/30 |

I had planned to use custom characters as shown below, but I am hurting for code space if I fit in 1 K. The custom characters take up 64 bytes of space plus the small bit of code to load the characters. Since I have 8 (or even 16) characters in the display, it may be better to eliminate the custom characters and replace the custom ones with two ordinary ones.

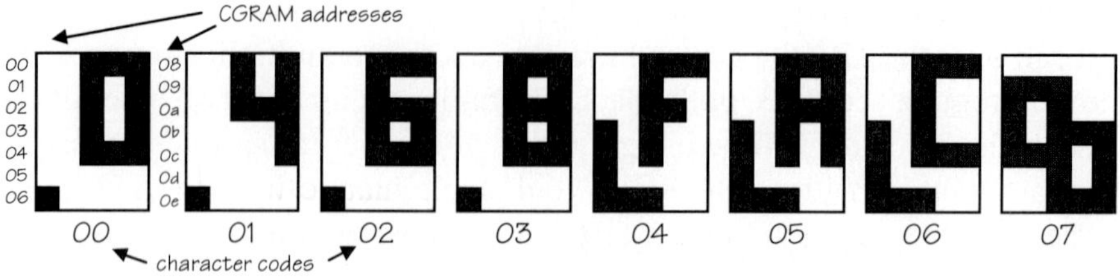


**Custom Characters for LCD**

```
              Custom Character Array
uchar code chars[]={ /*custom characters*/
0x07,0x05,0x05,0x05,0x07,0x00,0x10,0x00,/*.0*/
0x05,0x05,0x07,0x01,0x01,0x00,0x10,0x00,/*.4*/
0x07,0x04,0x07,0x05,0x07,0x00,0x10,0x00,/*.6*/
0x07,0x05,0x07,0x05,0x07,0x00,0x10,0x00,/*.8*/
0x07,0x04,0x06,0x14,0x14,0x10,0x16,0x00,/*FL*/
0x07,0x05,0x07,0x15,0x15,0x10,0x16,0x00,/*AL*/
0x07,0x04,0x04,0x14,0x17,0x10,0x16,0x00,/*CL*/
0x00,0x1C,0x14,0x17,0x1c,0x05,0x07,0x00};/*00*/
/*loading custom characters*/
for(i=0;i<64;i++)LCDdatw(chars[i]);
```

**Button/menu Flowchart**

**Button Reading.** Were the only timer tied up reading the frequency input, the button readings could be spaced out by a *software* delay loop (horrible thought!). Actually, that is not so bad since there is little else to do except respond to buttons. In the chosen solution, however, the scan can run off the real-time interrupt. The only requirement is to read the buttons no more often than every 40 msec to achieve debouncing.

```
              Button Processing
   new=btn;
   newpushes=(new^old)|old;
   switch(newpushes){
```

```
        case 1:/*right*/
          ent=++ent>steps[ent]?0:steps[ent];break;
        case 2:/*left*/
          ent=--ent<0xff?ent:steps[ent];break;
        case 4:/*down*/
          var[ent]--;break;
        case 8: /*up*/
          var[ent]++;break;
        case 0x10:; /*clear*/
      }
```

The two program pieces below show alternative ways of adjusting the value of a menu item up or down by one. Despite the more expanded look in C, it turns out that the first version is about 60 percent of the length of the second (I haven't checked run time because it is irrelevant here). You may say, "So what?" but in the case where the code space must be very limited, this can matter very much.

**Alternative Menu Update Techniques**

```
if(btnup)ASAstep++; else ASAstep--;
if (ASAstep>18)ASAstep=17;
ASAstep%=17;
break; /*28 bytes*/
```

```
if(btnup)ASAstep=(++ASAstep>17)?17:ASAstep;
else ASAstep=(--ASAstep>17)?0:ASAstep;
break; /*39 bytes/
```

## OVERALL SOFTWARE FOR LIGHTMETER

**ltmtr.c**

```
#include <reg750.h>
#define uint unsigned
#define uchar unsigned char
#define btn P1&0x1f
#define modes 7
#define LCDena P02
#define LCDrs P00
#define LCDrw P01
#dcfinc LCDport P3 /*lower 4*/
#define sensport P3 /*upper 4*/
#define shdn P17
```

```
uchar ltcnt,warm,waitcount,ent,flash,sensstep,clear,
       ev;
uint reading;
uchar stpASA,stpf,stpspd;
uchar ltcnts[8];
uchar var[4];/*ASAstep, fstep, shutterstep, ltmode*/
uchar code sens[]={0xd0,0x90,0xa0,0xb0,0x30};
uchar code evsens[]={0,33,66,100,133};
uchar code evcal=3;/*tbd for overall illumination*/
uchar code steps[4]={8,10,18,2};
uint code
         ASA[]={12,25,50,100,200,400,800,1600,3200};
uint code
         f[]={320,220,160,110,80,56,40,28,20,14,10};
uint code spd[]={4000,2000,1000,500,250,125,
             60,30,15,8,4,2,1,2,4,8,15,30,60};
uchar code msg[]={"ASA ""f "" sec""Ambient""Flash"
       "Combnatn"};

void area(void)interrupt 0 using 1{
  ltcnt++;
}

void spot(void)interrupt 2 using 1{
  ltcnt++;
}

/*regular 1mSec interrupt*/
void clock(void) interrupt 1 using 1{
  static uchar samplesindex;uchar old;
  waitcount--;
  if(((flash>0)&&(clear>0))||flash==0){
    samplesindex=++samplesindex&0x07;
    old=ltcnts[samplesindex];
    ltcnts[samplesindex]=ltcnt;
  }
  if(clear>0)clear--;/*mark when buffer is filled*/
  if(var[3]!=1){/*flash or combination*/
    if((ltcnt>>3)>old){/*detect a flash by larger
      count*/
      flash=1;clear=8;
    }
  }
  /*see if sensitivity needs changing*/
```

```
    if(flash==0){
      if(ltcnt>0xc0){
        sensstep--;
      }
      else if(ltcnt<0x05){
        sensstep++;
      }
      else return;
      sensstep%=4;
      P3=sens[sensstep];
      clear=8;/*refill buffer after sens change*/
    }
  }

void wait(uchar msec){
  waitcount=msec;while(waitcount>0);
}
uchar LCDdatr(void){
  uchar temp;
  LCDrw=1;LCDena=1;
  temp=LCDport<<4;
  LCDena=0;LCDena=1;
  temp=temp|(LCDport&0x0f);
  LCDena=0;return temp;
}

uchar LCDaddrr(void){
  LCDrs=0;return LCDdatr();
}

void LCDwait(void){
  LCDrs=0;while((LCDaddrr()&0x80)>0);
}

/*uchar LCDdatr(void){
  LCDwait();LCDrs=1;
  return LCDr();
}*/

void LCDw(uchar cmd){
  LCDport=sens[sensstep]|(cmd>>4);
  LCDrw=0;LCDena=1;LCDena=0;
  LCDport=sens[sensstep]|(cmd&0x0f);
  LCDena=1;LCDena=0;
```

```
  LCDwait();
}

void LCDcmdw(uchar cmd){
  LCDrs=0;LCDw(cmd);
}

void LCDdatw(uchar dat){
  LCDrs=1;LCDw(dat);
}

void display(uchar loc,uchar *string){
  if(loc!=0xff)LCDcmdw(loc|0x80);/*set cursor addr*/
  for(string=0;*string!=0;string++)LCDdatw(*string);
}

void dspval(uint num){
  uint m=10000;uchar i=5;uchar ldgzero=0;
  do{
    if((num/m>0)||(m==1))ldgzero=1;
    if(ldgzero)LCDdatw(num/m+'0');
    num%=m;i--;
    if(m==10 && num==0){
      LCDdatw(0x07);/*custom 00*/
      m=0;
    }
    if(m==1 &&  ent==1){/*custom .0 .2 .4 or .8*/
      LCDdatw(num/2);
      m=0;
    }
  }while((m/=10)>0);
}

/*compute exposure*/
void update(void){
  uchar i;uint light;
  for(i=light=0;i<8;i++)light+=ltcnts[i];
  if(light>0)for(ev=0;light<0x8000;ev+=10)light<<=1;
  ev=ev-(10*((light&0x7000)>>12))/8;
  ev=ev+evcal+evsens[sensstep];
}

void initialize(void){
  uchar i;
  /*check for recovery from power-down*/
```

```
  if (warm!=0xaa){
    warm=0xaa;
    stpASA=6;stpf=2;stpspd=0;
    for(i=0;i<16;ltcnts[i++]=0);
  }
  /*set up timer & interrupts*/
  RTL=0x00;RTH=0x04;/*divide by ~1000*/
  TL=0x00;TH=0x04;
  TCON=0x15;/*enable timer, extints edge trig*/
  IE=0x87;/*enable all interrupts*/
  /*initialize LCD in 4-bit mode*/
  wait(15);
  LCDrs=0;LCDrw=0;LCDena=1;
  LCDport=3;LCDena=0; /*send 3*/
  wait(5);
  LCDena=1;LCDena=0; /*send 3 again*/
  LCDcmdw(0x32);
  LCDcmdw(0x20); /*4-bit, 1 line, 5x7*/
  LCDcmdw(0x04); /*display off*/
  LCDcmdw(0x01); /*reset*/
  LCDcmdw(0x06); /*incr cursor*/
  LCDcmdw(0x40); /*write to char gen RAM*/
  clear=8; /*capture 8 readings before deciding on
      light intensity*/
}

void main(void){
  static uchar new,old;
  uchar newpushes;
  initialize();
  for(;;){
    new=btn;
    newpushes=(new^old)|old;
    switch(newpushes){
      case 1:/*right*/
        ent=++ent>steps[ent]?0:steps[ent];break;
      case 2:/*left*/
        ent=--ent<0xff?ent:steps[ent];break;
      case 4:/*down*/
        var[ent]--;break;
      case 8: /*up*/
        var[ent]++;break;
      case 0x10:; /*clear*/
    }
    LCDcmdw(0x80);/*home cursor*/
```

```
        display(2,msg[ent]);
        update();
        dspval(var[ent]);
    }
}
```

## Lightmeter Schematic

A *tentative* schematic follows. Notice that it involves only 3 8-pin DIPs, a "skinny" 24-pin DIP, the display, and a few small discrete components (3 electrolytic capacitors, 2 ceramic capacitors, an on-board pot, 5 resistors, and a crystal). The design below uses *two* sensors so one can be located at the focus of a lens for spot metering and the second can be used for incident and area metering. I suspect a final design might use only one and mechanically shift in a lens or a diffuser.

**Mechanical Design.** The issues of packaging and circuit layout are not addressed here. If *very* tiny size is important I expect that the microcontroller and regulator are available in surface mount packages. The sensors are limited to 8-pin DIPs, probably because the optics takes up some space. Buttons are a mechanical design issue, but they all can be simple push-to-make contacts that could allow cost reduction in an integrated design. I would suppose that a highly integrated, high-volume design would absorb the display into a single package with the rest of the electronics, at which point the display could be customized and the entire design would need changes.

I maintain that the design should include a spot meter function, which would suit a round package with a telescope appearance. By sliding a translucent spherical cover over the nonfocused sensors, the meter could read incident light. Optical design is not really my area, so I stop here!

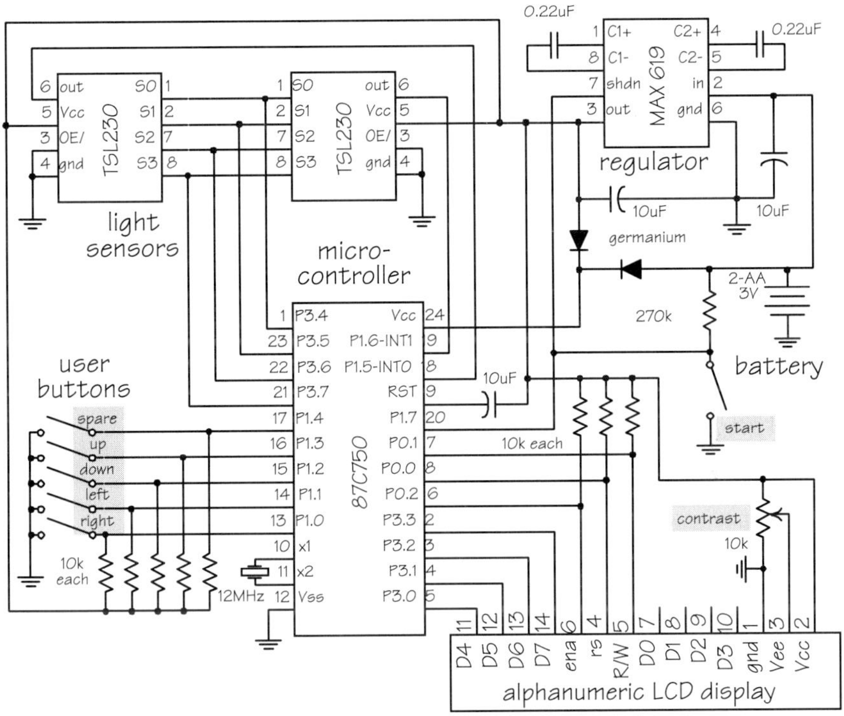


**Tentative Light Meter Schematic**

## CONCLUSIONS

Having walked through two designs, you can see that there are many trade-offs. First you have the hardware decisions to make, and then you still have software choices to make and then revise. Perhaps you can see why I seem to be long on ideas and short on finished, working products. It is a sort of disease—I can do it better, faster, more simply, less expensively, and lastly, more efficiently. That is why this book is subtitled, "*Building Efficient Applications*." You have to have a passion for efficiency or you will throw expensive, complicated solutions at the problem and never stop to really think. You have to consider what is *really* needed rather than just using the approaches of the past.

In fact, I am probably obsolete with my emphasis on efficiency because the new trend is to put more and more into custom silicon. Personal computers now have far fewer chips on the motherboard that the original PC. It matters not that software is less efficient, the computers are faster and larger to more than make up for it. Of course, if your product volume is measured in millions of units, the entire design game has changed. Perhaps the new technologies will make the 8051 family obsolete—why use 8 bits when you can have 32 for the same price? Why be efficient if you can be faster and bigger?

Just remember that the exploding performance and cost gains are only for silicon—other hardware doesn't change that way, and new humans seem to keep being born about the same way. I am particularly conscious of that, seeing students for about 20 years now. They seem to be younger every year and painfully ignorant, until I remember I have had an extra 35 years to learn!

Finally, if you are developing a product, take care of your priorities and be sure to leave time for integration and debugging, as discussed in Chapter 1. If I were getting paid to develop the two products here, I would apply myself full time (and have to quit teaching). Even then, I know there would be frequent changes in direction and, as the product was being shipped out the door, I would be thinking of all the things I would like to change! I hope you never lose your interest in new designs.

# A1 Addresses, Phones, and Products

It would be foolish to think that this list will remain complete for 6 months, let alone for the (hopefully long) life of this book. However, from experience it would seem that some leads are better than none, and one good lead can open up to a host of others. The following information is what I have available in late '98.

## MAGAZINES

| | |
|---|---|
| Circuit Cellar Ink<br>The Computer Applications Journal<br>PO Box 698<br>Holmes,PA 19043-9613<br>(800)269-6301<br>(860)875-2751 fax(860)871-0411<br>http://www.circellar.com/ | EDN<br>8773 South Ridgeline Blvd<br>Highlands Ranch, CO 80126-2329<br>(303) 470-4445 fax (303) 470-4280<br>cahners.subs@denver.cahners.com |
| Embedded Systems Programming<br>Miller Freeman Publications<br>600 Harrison Street<br>San Francisco, CA 94107<br>(888)847-6177 | Midnight Engineering<br>1700 Washington Ave.<br>Rocky Ford, CO 81067<br>(719)254-4558 |

# COMPILERS & SOFTWARE TOOLS

| | |
|---|---|
| Archimedes Software<br>303 Parkplace Ctr. Suite G125<br>Kirkland,WA 98033<br>(425)822-6300 fax(425)822-8632<br>sales@archimedesinc.com<br>http://www.archimedesinc.com<br>*C compilers,development environment* | Avocet Systems Inc. (also 2500AD)<br>120 Union St (POB 490)<br>Rockport, ME 04856<br>(800)448-8500; (800)843-8144<br>(207)236-9055; fax (207)236-6713<br>sales@avocetsystems.com |
| Franklin Software, Inc.<br>Saratoga Ave. #2<br>San Jose CA 95129<br>(800)880-8051<br>(408)296-8051 fax(408)296-8061<br>info@fsinc.com<br>http://www.fsinc.com<br>*C compilers,development environment* | IAR Systems Software, Inc.<br>One Maritime Plaza Suite 1770<br>San Francisco,CA 94111<br>(800)427-8868<br>(415)765-5500; fax(415)765-5503<br>info@iar.com<br>http://www.iar.com<br>*C compilers,development environment* |
| Keil Software, Inc.<br>16990 Dallas Pkwy. Suite 120<br>Dallas,TX 75248<br>(800)348-8051<br>(214)735-8052 fax(214)735-8055<br>sales@keil.com<br>http://www.keil.com<br>*C compilers,development environment* | Systronics<br>555S. 300 East #21<br>Salt Lake City,UT 84111<br>(801)534-1017 fax(801)534-1019<br>sales@systronics.com<br>http://www.systronics.com<br>*BASIC compiler* |
| Tasking<br>333Elm St..<br>Dedham,MA 02026<br>(800)458-8276<br>(781)320-9400 fax(781)320-9212<br>http://www.tasking.com<br>*C compilers,development environment* | Intel Corporation<br>3065 Bowers Ave.<br>Santa Clara, CA 95051<br>(800)628-8686<br>(408)986-8086 fax (408)727-2620<br>http://www.intel.com<br>*may still offer PLM-51 compiler* |
| ChipTools Inc<br>1232 Stavebank Rd.<br>Mississauga, Ont.,Canada L5G 2V2(905)274-<br>6244 fax(905)891-2715<br>Borland-like debugger | Production Languages Corp.<br>PO Box 109<br>Weatherford,TX 76086<br>(800)525-6289<br>(817)599-8363 fax(817)599-5098<br>*development environment, debugger* |
| Richard Barnett, PE,PhD<br>Purdue University<br>EET-KNOY<br>West Lafayette, IN 47907<br>rbarnett@purdue.edu<br>*EET Monitor Software* | |

## REALTIME OPERATING SYSTEMS

| | |
|---|---|
| Avocet Systems Inc. (also 2500AD)<br>120 Union St (POB 490)<br>Rockport, ME 04856<br>(800)448-8500; (800)843-8144<br>(207)236-9055; fax (207)236-6713<br>sales@avocetsystems.com | CMX Company<br>19 Indian Head Heights<br>Framingham, MA 01701<br>(508)872-7675 fax(508)620-6828<br>cmx@cmx.com<br>http://www.cmx.com |
| Byte-BOS Integrated Systems<br>12032 Broadway Terrace<br>Oakland, CA 94611<br>(800)788-7288<br>(510)658-7797; fax (510)658-9334<br>info@bytebos.com<br>http://www.bytebos.com | IAR Systems Software, Inc.<br>One Maritime Plaza Suite 1770<br>San Francisco,CA 94111<br>(800)427-8868<br>(415)765-5500; fax(415)765-5503<br>info@iar.com<br>http://www.iar.com |
| Franklin Software, Inc.<br>Saratoga Ave. #2<br>San Jose CA 95129<br>(800)880-8051<br>(408)296-8051 fax(408)296-8061<br>info@fsinc.com<br>http://www.fsinc.com | Keil Software, Inc.<br>16990 Dallas Pkwy. Suite 120<br>Dallas,TX 75248<br>(800)348-8051<br>(214)735-8052 fax(214)735-8055<br>sales@keil.com<br>http://www.keil.com |
| US Software Corp.<br>14215 N.W. Science Park Drive<br>Portland, OR 97229<br>(800)356-7097<br>(503)641-8446 fax(503)644-2413<br>sales@ussw.com<br>http://www.ussw.com | |

## CHIPS

| | |
|---|---|
| Atmel<br>2325 Orchard Pkwy.<br>San Jose,CA<br>(800)292-8635<br>(408)436-4295<br>sales@atmel.com<br>http://www.atmel.com | Dallas Semiconductor<br>4401 S. Beltwood Pkwy.<br>Dallas, TX 75244<br>http://www.dalsemi.com<br>8051 family-faster, integrated RAM |
| Intel Corporation<br>3065 Bowers Ave.<br>Santa Clara, CA 95051<br>(800)628-8686<br>(408)986-8086 fax (408)727-2620<br>http://www.intel.com<br>*8051 family-originator* | Matra MHS (Temic)<br>2201 Laurelwood Rd.<br>Santa Clara,CA 94115<br>(800)554-5565x3927<br>(408)567-8220 fax (408)748-0439<br>http://www.temic.de<br>*8051 family-especially high speed* |
| Oki Semiconductor Inc.<br>785 N. Mary Ave.<br>Sunnyvale, Ca 94086<br>(408) 737-6590<br>http://www.okisem.com | Philips Company<br>811 E. Arques Avenue<br>PO Box 3409<br>Sunnyvale, CA 94088-3409<br>(800)447-1500x1000<br>(408)991-3518<br>http://www.semiconductors.phillips.com<br>*largest variety of 8051 family* |
| Siemens Components, Inc<br>Integrated Circuit Division<br>10950 N. Tantau Ave.<br>Cupertino, CA 95014<br>(800)777-4363<br>(408)777-4910x166 fax(708)296-4805<br>http://www.sci.siemens.com | Silicon Systems<br>Tustin, CA<br>(714)573-6659<br>http://www.ssll.com |
| WSI, Inc.<br>47280 Kato Rd.<br>Fremont,CA 94538<br>(800)832-6974<br>(510)656-5400 fax (510)657-8495<br>http://www.wsipsd.com<br>*programmable peripherals (not micros)* | |

## COMMERCIAL 8051-FAMILY BOARDS

| | |
|---|---|
| Rostek Inc.<br>PO Box 6719<br>South Bend, IN 46660<br>(219) 272-3239<br>jrosheck@rostek.com<br>http://www.rostek.com | Keil Software, Inc.<br>16990 Dallas Pkwy. Suite 120<br>Dallas,TX 75248<br>(800)348-8051<br>(214)735-8052 fax(214)735-8055 sales@keil.com<br>http://www.keil.com |
| Allen Systems<br>2346 Brandon Rd.<br>Columbus, OH 43221<br>(614) 488-7122<br>*very low cost 8031 board* | Binary Technology<br>PO Box 67<br>Meriden, NH 03770<br>(603) 469-3232 |
| Blue Earth Research<br>1416 N. Riverfront Dr.<br>Mankato, MN 56001<br>(507) 387-4001<br>http://www.blueearthresearch.com<br>*very small boards* | Ceibo USA<br>7 Edgestone Ct.<br>Florissant,MO 63033<br>(800)833-4084<br>(314)830-4084 fax(314)830-4083<br>usa@ceibo.com<br>http://www.ceibo.com<br>*750/1 board* |
| Cottage Resources<br>1405 Stevenson Dr.<br>Suite 3-672<br>Springfield, IL 62703<br>(217) 529-7679 | DPR Microtech<br>POB 1272<br>Monrovia, CA 91017<br>(888)495-7655<br>dpryan0@gte.net<br>*development boards* |
| HiTech Equipment Corp.<br>9400 Activity Rd.<br>San Diego,CA 92126<br>(619)566-1892 fax(619)530-1458<br>info@hte.com<br>http://www.the.com | Intel Corporation<br>2200 Mission College Blvd.<br>Santa Clara, CA 95052-8119<br>(800) 468-8118 fax(408) 727-2620<br>*evaluation boards* |
| Micromint<br>4 Park Street<br>Vernon, CT 06066<br>(800)635-3355<br>(860)871-6170 fax(860)872-2204<br>sales@circuitcellar.com<br>http:www.micromint.com<br>*large line of boards* | New Micros, Inc.<br>1601 Chalk Hill Rd.<br>Dallas,TX 75212<br>(214)339-2204 fax(214)339-1585<br>general@newmicros.com<br>http://www.newmicros.com |

| | |
|---|---|
| Software Science<br>3750 Roundbottom Rd.<br>Cincinnati, OH 45244<br>(888)561-2060<br>info@softwarescience.com<br>http://www.softwarescience.com | SYLVA Control Systems<br>519 Richard ST.<br>Thunder Bay Ontario, CAN P7A 1R2<br>sylva@baynet.net<br>http://www.sylvacontrol.baynet.net |
| Systronics<br>555S. 300 East #21<br>Salt Lake City,UT 84111<br>(801)534-1017 fax(801)534-1019<br>info@systronics.com<br>http://www.systronics.com<br>*Dallas device development board* | |

## 8051-FAMILY DEVELOPMENT HARDWARE

| | |
|---|---|
| Cactus Logic<br>14120 E. Live Oak Ave. Suite A2<br>Baldwin Park,CA 91706<br>(800)847-1998<br>(818)337-4547 fax(818)337-0689<br>info@softaid.net<br>http://www.softaid.net<br>*emulator, turboC-like environment* | Ceibo USA<br>7 Edgestone Ct.<br>Florissant,MO 63033<br>(800)833-4084<br>(314)830-4084 fax(314)830-4083<br>usa@ceibo.com<br>http://www.ceibo.com<br>*in-circuit emulators, debuggers* |
| Eastern Systems Inc. (Ashling)<br>CA<br>(408)747-0440<br>*in circuit emulator* | Emulation Technology, Inc.<br>2344 Walsh Ave. Bldg. F<br>Santa Clara,CA 95051<br>(800)995-4381<br>(408)982-0660 fax(408)982-0664<br>in circuit emulators |
| HiTech Equipment Corp.<br>9400 Activity Rd.<br>San Diego,CA 92126<br>(619)566-1892 fax(619)530-1458<br>info@hte.com<br>http://www.the.com<br>*in circuit emulators* | Intel Corporation<br>3065 Bowers Ave.<br>Santa Clara, CA 95051<br>(408)986-8086 fax(408)727-2620<br>*in circuit emulators* |
| Kontron Elektronik GmbH<br>Oskar-von-Miller Str. 1<br>85386 Eching, Germany<br>+8165 77-444 fax +8165 77-128<br>*in circuit emulator* | Lauterbach Datentechnik GmbH<br>Fichtenstr. 27<br>D-85649 Hofolding, Germany<br>(508)620-4521 fax(508)620-4522 (USA contacts)<br>*in circuit emulators* |

| | |
|---|---|
| MetaLink Corp.<br>PO Box 1328<br>Chandler,AZ 85254-1329<br>(800)638-2423<br>(602)926-0797 fax(602)926-1198<br>info@metaice.com<br>http://www.metaice.com<br>*in circuit emulators* | Microtek International<br>3300 NW 211th Terrace<br>Hillsboro,OR 97124-9953<br>(800)886-7333<br>(503)886-7333 fax(503)629-8460<br>info@mikrotekintl.com<br>http://www.mikrotekintl.com<br>*in circuit emulator* |
| Noahu (say "no how")<br>51E. Campbell Ave.<br>Campbell, CA 95008-2053<br>(888)886-6428<br>(408)866-1820 fax(408)378-7869<br>info@noahu.com<br>http://www.noahu.com<br>*in circuit emulators* | Signum Systems<br>11992 Challenger Ct.<br>Moorpark, CA 93021<br>(800)838-8012<br>(805)523-9774<br>sales@signum.com<br>http://www.signum.com<br>*in circuit emulators* |
| Advin<br>1050-L East Duane Ave.<br>Sunnyvale, CA 94086<br>(800)627-2456<br>(408)243-7000 fax(408)736-2503<br>sales@advin.com<br>heep://www.advin.com<br>*device programmer* | BP Microsystems Inc<br>1000N. Post Oak Rd. Suite 225<br>Houston,TX 77055-7237<br>(800)225-2102<br>(713)688-4600 fax(713)688-0920<br>info@bpmicro.com<br>http://www.bpmicro.com<br>*device programmers* |
| Intronics, Inc<br>Box 13723<br>Edwardsville,KS 66113<br>(800)339-5152<br>(913)422-2094 fax(913)441-1623<br>ge@netweb.net<br>http://www.intronics.com<br>*device programmer* | Logical Devices, Inc.<br>1221 S Clarkson, Suite 200<br>Denver, CO 80210<br>(800)331-7766<br>(303)279-6868 fax(303)279-6869<br>http://www.logicaldevices.com<br>*device programmers* |
| Needham Electronics Inc<br>4630 Beloit Dr. #20.<br>Sacramento,CA 95838<br>(916)924-8037 fax (916)924-8065<br>sales@needhams.com<br>heep://www.needhams.com<br>*device programmers* | Philips Semiconductor<br>(800)774-4547<br>www.-ussemiconductors.philips.com<br>*device programmers* |
| Tribal Microsystems<br>44388 S. Grimmer Blvd.<br>Fremont,CA 94538<br>(510)623-8859 fax(510)623-9925<br>sales@tribalmicro.com<br>http://www.tribalmicro.com<br>*device programmers, emulators* | EmuTec Inc.<br>2707 Col;by Ave. Suite 901<br>Everett,WA 98201<br>(206)337-3283<br>emutec@emutec.com<br>http://www.emutec.com<br>*EPROM emulator (not 8051 specific)* |

| | |
|---|---|
| Incredible Technologies Inc.<br>1600 Hicks Rd.<br>Rolling Meadows, IL 60008<br>(800)262-0232<br>(847)870-7027 fax (847)870-0120<br>http://www.itsgames.com<br>*EPROM emulators(not 8051 specific)* | Scanlon Design Inc.<br>5224 Blowers St.<br>Halifax,NS,CAN BJ3 1J7<br>(800)352-9770<br>(902)425-3938 fax(425)-4098<br>info@scanlondesign.com<br>http://www.scanlondesign.com<br>*EPROM emulators(not 8051 specific)* |
| Tech tools<br>POB 462101<br>Garland, TX 75046<br>(972)272-9392 fax(972)494-5814<br>http://www.tech-tools.com<br>*EPROM emulators(not 8051 specific)* | Logical Systems Corp.<br>PO Box 6184<br>Syracuse,NY 13217-6184<br>(315)478-0722 fax(315)479-6753<br>sales@logicalsys.com<br>http://www.logicalsys.com<br>*socket adapters* |
| EDI Corp<br>PO Box 366<br>Patterson,CA 95363<br>(209)892-3270 fax(209)892-3610<br>*socket adapters* | |

## OTHER ELECTRONIC AND MECHANICAL HARDWARE

| | |
|---|---|
| Digi-Key<br>701 Brooks Ave. South<br>Thief River Falls;, MN 56701-0677<br>(800) 344-4539<br>(218) 681-3380 fax (218)681-3380<br>http://www.digikey.com<br>*new electronic components* | Jameco Electronic Products<br>1355 Shoreway Rd.<br>Belmont, CA 94002-4100<br>(800) 831-4242 fax (800)237-6948<br>(650) 592-8097 fax (650)592-2503<br>http://www.jameco.com<br>*new electronic components* |
| Marlin P. Jones & Assoc. Inc.<br>PO Box 12685<br>Lake Park, FL 33403-0685<br>(800) 652-6733 fax (800) 432-9937<br>(561) 848-8236 fax (561) 844-8764<br>*used/new electronic & mechanical components* | ABRA Electronics Inc.<br>1320 Route 9<br>Champlain, NY 12919<br>(800) 717-2272 fax (800) 898-2272<br>http://www.abra-electronics.com<br>sales@abra-electronics.com<br>*new electronic components* |
| Herbach & Rademan<br>16 Roland Ave.<br>Mt. Laurel, NJ 08054-1012<br>(800) 848-8001<br>(609) 802-0422 fax (609) 802-0465<br>http://www.herbach.com<br>sales@herbach.com<br>*used electronic & mechanical components* | Newark Electronics<br>4801 N. Ravenswood<br>Chicago, IL 60640<br>(800) 463-9275<br>http://www.newark.com<br>*electronic distributor* |

# Index